The Problems of Sulphur

IEA Coal Research

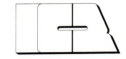

IEA Coal Research was established in 1975 under the auspices of the International Energy Agency (IEA) and is currently supported by fourteen countries.

IEA Coal research provides information and analysis of all aspects of coal production and use, including:

- supply, mining and geosciences.
- transport and markets.
- coal science.
- coal utilisation technology.
- environmental effects.
- by-products and waste utilisation.

IEA Coal Research produces:

- periodicals including *Coal abstracts*, a monthly current awareness journal giving details of the most recent and relevant items from the world's literature on coal, and *Coal calendar*, a comprehensive descriptive calendar of recently held and forthcoming meetings of interest to the coal industry.
- technical assessments and economic reports on specific topics throughout the coal chain.
- bibliographic databases on coal technology, coal research projects and forthcoming events, and numerical databanks on reserves and resources, coal ports and coal fired-power stations.

Specific enquiries about this report should be address to: Frank Derbyshire, Research and Development Director, Sutcliffe Speakman Carbons Ltd, Guest Street, Leigh, Lancashire WN7 2HE

General enquiries about IEA Coal Research should be addressed to:

Dr David Merrick,
Head of Service,
IEA Coal Research,
14–15 Lower Grosvenor Place,
London, SW1W 0EX,
United Kingdom
Telephone: 01–828 4661
Telex: 917624 NICEBA G
Fax: 01–828 9508

ML

Reviews in Coal Science

The Problems of Sulphur

IEA Coal Research, London

Published by **Butterworths**
London Boston Singapore Sydney Toronto Wellington
on behalf of
IEA Coal Research

PART OF REED INTERNATIONAL P.L.C.

First published 1989

British Library Cataloguing in Publication Data

The Problems of Sulphur.
1. Coal. Chemical properties
I. IEA Coal Research
662.6' 22

ISBN 0–408–04041–6

Library of Congress Cataloging-in-Publication Data

The Problems of sulphur.
p. cm.-- (Reviews and coal sciences)
Includes biliographies and indexes.
ISBN 0–408–04041–6 :
1. Coal--Desulphurization. I. IEA Coal Research. II. Series.
TP325.P888 1989
662.6'23--dc19 88–8140
CIP

Typeset by Scribe Design, Gillingham, Kent.
Printed in Great Britain at the University Press, Cambridge

Preface

This book is the first in a series of reviews on coal science. It brings together in one volume work on all aspects of the problems associated with sulphur in coal. Part 1 addresses the forms of sulphur in coal and evaluates processes directed at the chemical removal of sulphur. Part 2 expands on this to look at alternative means of removing sulphur both physically and biologically, sulphur removal during the combustion of coal and flue gas desulphurisation processes. Part 3 looks at the role of sulphates in the atmosphere from the points of view of their formation, transport and deposition and of their effects on health, materials and the atmosphere. The work was undertaken by IEA Coal Research, an independent and internationally respected authority on coal science and technology.

Related titles

Coke

Monograph on Victorian Brown Coals

Introduction to Carbon Science

Contents

Part 1 Chemical desulphurisation of coal

Part 2 Control of sulphur oxides from coal combustion

Part 1

Chemical desulphurisation of coal

Summary

The recent literature relating to the chemical desulphurisation of coal is reviewed. Following an assessment of the forms of organic and pyritic sulphur in coal and the potential chemical reactions these may undergo, fifteen processes for the chemical cleaning of coal which are currently under development are described and evaluated. These include the PETC process; the Ames wet oxidation process; the Ledgemont process; the ARCO promoted oxidative process; the TRW Meyers desulphurisation process; the TRW Gravichem process; the JPL chlorinolysis process; the KVB process; the Battelle hydrothermal process; the TRW Gravimelt process; the General Electric microwave treatment process; the Aqua-refined coal process; the Magnex process; the IGT flash desulphurisation process; and a chemical comminution process. It is concluded that with the present state of knowledge the additional sulphur removed by chemical cleaning processes compared with that removed by physical cleaning processes is insufficient to justify the additional complexity and hence cost of preparation. Much more fundamental research is required to elucidate the modes of occurrence and chemistry of sulphur in coal, particularly of organic sulphur, before the efficiency of the chemical cleaning of coal can be improved.

Chapter 1

Introduction

The combustion of coal for the generation of electric power and process heat necessitates the control of pollutants which are potentially harmful to man or the environment. Although the control of emissions of particulates, nitrogen oxides and trace elements is receiving attention, most effort has been directed towards the control of emissions of sulphur in the form of sulphur dioxide.

Emissions of sulphur dioxide from coal combustion may be controlled by one or more of the following:

1. The use of low sulphur content coal.
2. The pretreatment of coal to remove sulphur.
3. The retention of sulphur during combustion.
4. The post-combustion treatment of flue gases.
5. The conversion of coal into a liquid or gaseous form.

Reserves of coals which contain sufficiently low concentrations of sulphur to enable sulphur dioxide emission standards to be met when the coals are burned are limited and restricted to specific geographical locations. The costs of coal conversion processes preclude, at present, the use of coal derived liquids or gases for combustion. Processes which retain sulphur during combustion are currently being developed. The sulphur retention property of limestone is being employed in fluidised bed combustion and in the production of limestone/coal pellets for use in stoker fired boilers (Gaimmar *et al.*, 1980). The most widely adopted method of controlling emissions of sulphur dioxide is by the use of flue gas desulphurisation (FGD). However, the reliability of flue gas scrubbers is often such that two or three systems in series are required to ensure sufficient control of sulphur dioxide. The cost of FGD represents a significant proportion of the equipment cost of a coal fired power station. This proportion increases as the size of the plant decreases making FGD prohibitively expensive for the small industrial coal fired boiler (Prior, 1977).

The remaining option for controlling sulphur emissions is that of pretreating coal to remove sulphur prior to combustion. The physical cleaning of coal to remove mineral matter by washing and gravity separation techniques is already widely practised in the coal industry. These processes also remove between 30 and 90% of the pyritic sulphur associated with the mineral matter in coal (McCandless and Cantos, 1979). However, because coal is inevitably rejected with the higher density material of high sulphur content the yield of coal may be reduced by as much as 40%. In addition, as much as 50% of the sulphur in coal may be associated with the organic portion of coal and is therefore not removed. In order to decrease the sulphur content of such coals to acceptable levels it is necessary to treat the coal chemically to remove both organic and pyritic sulphur. This is chemical coal cleaning (Ruether, 1979). No such process is, as yet, available commercially although several are under active development.

The majority of chemical coal cleaning processes are designed to produce coals with sufficiently low sulphur contents to permit their combustion for power generation without the use of FGD. However, several of the processes also remove mineral matter, trace elements and other compounds as well as modifying the basic combustion properties of the coal. Chemically cleaned coal may therefore find other applications such as in the production of coal/oil mixtures, coal/water mixtures or for use in magnetohydrodynamic (MHD) power generation (Meyers, 1980).

This part presents the requirements for a chemically cleaned coal, descriptions of processes under development and an evaluation of these processes. Chapter 7 discusses subsequent research into the chemistry of desulphurisation reactions and its implications for coal cleaning. Firstly however, it is necessary to understand the nature of sulphur in coal.

Chapter 2

Sulphur in coal

Sulphur is recognised as one of the major impurities in coal. Its concentration in world coal resources varies between 0.19% and 10% (Yurovskii, 1960) although the range for the world's economically recoverable coal reserves is narrower, at 0.38 to 5.32% (Meyers, 1977).

Sulphur occurs in coal as:

1. pyritic sulphur;
2. organic sulphur;
3. sulphate salts;
4. elemental sulphur.

Pyritic and organic sulphur together account for the large majority of sulphur in coal. Sulphate sulphur occurs chiefly as gypsum and iron sulphate, the latter normally resulting from the oxidation of iron pyrite during the storage of coal. The concentration of sulphate sulphur in coal is generally less than 0.1%. This, together with its solubility in water, means that this form of sulphur is of little concern during coal cleaning. The concentration of elemental sulphur in coal is also small, generally less than 0.2% (Greer, 1979).

Pyritic sulphur

Iron sulphide, FeS_2, the primary inorganic source of sulphur in coal, occurs in two crystalline habits: pyrite (cubic) and marcasite (orthorhombic). Pyrite is the more common so that pyrite and marcasite are often referred to collectively as pyrite or iron pyrites (Greer, 1978a).

Pyrite occurs in coal as:

1. Narrow seams or veins up to 150 mm thick and up to several hundred millimetres long.
2. Nodules. These consist of framboids (after *framboise*, the French for raspberry) which range from a few to several hundred micrometres in diameter, with most of them approximately 10 to 20 μm in diameter. The framboids are assemblies of octahedral crystals.
3. Discrete crystals. These individual pyrite euhedra may be widely dispersed throughout the coal. They occur predominantly in a size range of about 1–40 μm in diameter, with the large majority of the crystals being between 1–2 μm in

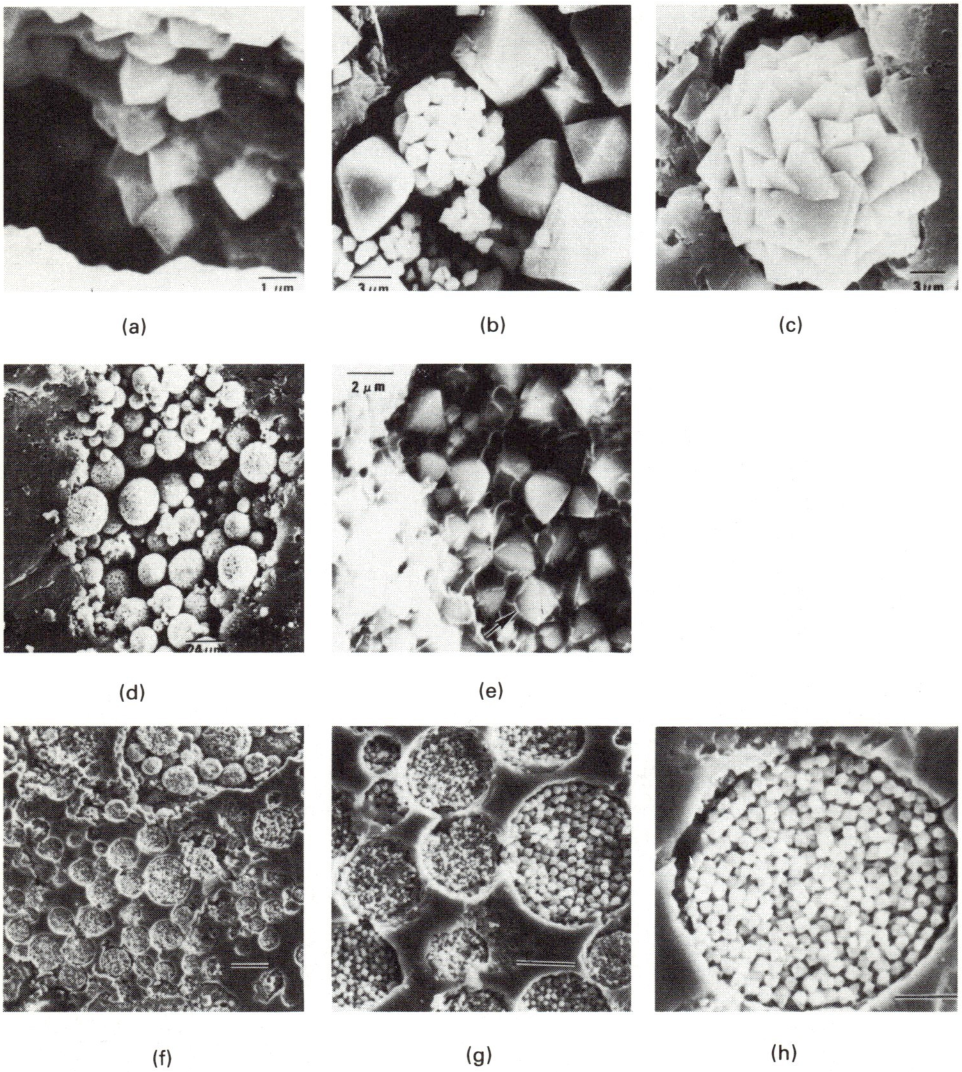

Figure 2.1 Framboidal and crystalline forms of pyrite. **a**, Pyrite crystals are seen attached to the upper surface of a cavity in coal. **b**, Pyrite crystals occurring as a cavity filling. Both individual crystals and spherical assemblies of crystals (framboids) are shown. There is infilling of material between the crystals and the framboid. **c**, Polycrystalline pyrite masses, infrequently observed. **d**, Crack filled with individual framboids of various sizes from approximately 4 to 30 μm in diameter. There is no infilling of material between framboids. **e**, A freshly fractured surface showing pyrite crystals within coalified plant cell network. The arrow indicates a broken cell wall revealing a single pyrite crystal within this particular cellular feature. **f,g,h**, Increasing magnification series for an etched surface of centimetre-sized pyrite nodule or concretion. The nodule is composed of an assembly of variously sized framboids which in this case have been joined by an infilling of additional pyrite. Within each spherical assembly of crystals (framboids) the pyrite crystal size is remarkably uniform, and yet comparison of different framboids within this field of view indicates that the pyrite crystal size varies from framboid to framboid. (Photographs courtesy of Ames Laboratory, Ames, IA, USA)

Figure 2.2 Multi-use fuel processing test plant, San Capistrano, CA, USA (photograph courtesy of TRW Systems and Energy Corp.)

diameter. This form of pyrite is not removed completely by physical coal cleaning methods as it is uneconomic to grind coal to this order of size.

The framboid and crystalline forms of pyrite are well illustrated in Figure 2.1 which shows micrographs obtained by Greer during his studies of the microstructure of coal (Greer *et al.*, 1980; Greer, 1979, 1978a,b, 1977a,b, 1976).

Organic sulphur

There exists very little data on the organic sulphur functions in coal (Davidson, 1980). However, it is generally assumed that the organic sulphur can be described by the following classification (Markuszewski *et al.*, 1980):

1. Aliphatic or aromatic thiols (mercaptans, thiophenols): R–SH, Ar–SH
2. Aliphatic, aromatic or mixed sulphides (thioethers): R–S–R, Ar–S–Ar, R–S–Ar
3. Aliphatic, aromatic or mixed disulphides (bisthioethers): R–S–S–R, Ar–S–S–Ar, R–S–S–Ar
4. Heterocyclic compounds of the thiophene type (dibenzothiophene):

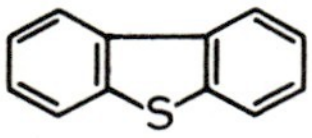

There is evidence to suggest that the content of thiols is substantially larger in lignites and high volatile bituminous coals than in low volatile coals. The proportion of thiophenic sulphur is also greater in higher ranked coals than in lower ones (Table 2.1, Attar and Dupuis, 1979).

In general, the ratio of pyritic to organic sulphur increases as the total sulphur content of the coal increases (Table 2.2, Meyers, 1977). Hence, physical coal cleaning is useful in reducing the sulphur content of high sulphur coals but is less effective when applied to low sulphur coals.

It should be noted that significant inaccuracies can occur in the determination of pyritic and organic sulphur contents by standard analytical techniques. The total sulphur content of a coal is usually determined by a wet chemical method. The pyritic sulphur is then leached from the coal by acid and the organic sulphur determined by difference. Greer (1979) compares the organic sulphur contents of various coals determined by a standard chemical method (ASTM) with those

Table 2.1 Distribution of organic sulphur groups in five coals (after Attar and Dupuis, 1979)

Coal	*Organic sulphur* w/w%	*Organic sulphur accounted* %	*Thiolic* %	*Thiophenolic* %	*Aliphatic sulphide* %	*Aryl sulphide* %	*Thiophenes** %
Illinois	3.2	44	7	14	18	2	58
Kentucky	1.43	46.5	18	6	17	4	55
Martinka	0.60	81	10	25	25	8.5	21.5
Westland	1.48	97.5	30	30	25.5	—	14.5
Texas Lignite	0.80	99.7	6.5	21	17	24	31.5

*Corrected for 'uncounted for' sulphur

Table 2.2 Sulphur forms in selected bituminous coals (after Meyers, 1977)

Country	*Location or mine*	*Sulphur content* w/w %[a]			*Ratio of pyritic to organic sulphur*
		Total	*Pyritic*	*Organic*	
Australia	Lower Newcastle	0.94	0.15	0.79	0.19
Brazil	Santa Caterina	1.32	0.80	0.53	1.5
Canada	Fernie	0.60	0.03	0.57	0.053
China (mainland)	Taitung	1.19	0.87	0.32	2.7
Germany	—	1.78	0.92	0.76	1.2
India	Tipong	3.63	1.59	2.04	0.78
Japan	Miike	2.61	0.81	1.80	0.45
Malaysia	Sarawak	5.32	3.97	1.35	2.9
Poland	—	0.81	0.30	0.51	0.59
Republic of South Africa	Transvaal	1.39	0.59	0.70	0.84
UK	Derbyshire	2.61	1.55	0.87	1.8
USA	Eagle No. 2	4.29	2.68	1.61	1.7
USSR	Shakhtersky	0.38	0.09	0.29	0.031

[a]Moisture-free basis, pyrite plus sulphate reported as pyrite

obtained by electron microprobe X-ray analysis (EPM). Values obtained by the EPM method are up to 30% lower than those indicated by the ASTM determination. It is believed that encapsulation of the discrete pyrite crystals occurs during the acid leaching step. This residual inorganic sulphur is credited to organic sulphur. In addition, a portion of the sulphur determined as organic by the ASTM method can be elemental sulphur. Although the elemental sulphur content of coals is usually small (<0.2%), Greer (1979) quotes an exceptional example of a high sulphur coal (>6% total sulphur) in which more than 50% of the alleged organic sulphur is actually elemental sulphur. It is also difficult to obtain a representative coal sample, and this may add to the errors.

Conventional methods of analysis may not be applicable to chemically cleaned coals. Sulphur may be transformed into a form which is not present in raw coal, and therefore it may behave in an unpredictable manner when the treated coal is subjected to analysis (Wheelock, 1981).

Chapter 3

Requirements for chemically cleaned coal

The production of coals which comply with environmental regulations for SO_2 emissions when burned in large utility boilers remains the main incentive for the development of chemical coal cleaning processes. In the early 1970s, when most of the development work was initiated, the US environmental regulations were less stringent than they are today. The 1971 New Source Performance Standards (NSPS) required that emissions of SO_2 to the atmosphere be limited to 0.52 kg/GJ (1.2 lb/10^6 Btu) for large scale utility coal fired boilers. An emission of 0.52 kg/GJ is approximately equivalent to burning a coal containing between 0.6 and 0.7% sulphur without emission control equipment (Meyers, 1977). These standards were met partially by the installation of flue gas desulphurisation systems (scrubbers) and partly by the use of Western US coals of low sulphur content. To prevent the rapid depletion of these reserves of low sulphur content coals and their transportation over a long distance the NSPS were revised in 1979. A requirement for a 90% reduction in potential SO_2 emissions was introduced except when emissions to the atmosphere are less than 0.26 kg/GJ (0.6 lb/10^6 Btu) when a 70% reduction in potential emissions is required.

Standards for SO_2 emissions in other countries have become progressively more stringent during the last decade. It is against these standards that processes under development for the chemical cleaning of coal must be assessed (Giberti *et al.*, 1979; Gathen, 1979).

Chapter 4

Chemistry of coal cleaning

Meyers presents a fairly comprehensive review of the literature on chemical reactions of organic and pyritic sulphur suitable for use in the extraction of sulphur from coal (Meyers, 1977). However, one important omission in this review is the work of Yurovskii (1960). Yurovskii describes work concerning the origin and form of sulphur in coal, the sulphur contents of world coals and possible chemical treatments for the removal of sulphur from coal.

Reactions of pyritic sulphur

Meyers (1977) categorises the reactions of pyrite which are theoretically capable of proceeding at temperatures below the decomposition temperature of coal (approximately 400°C) as:

1. displacement reactions;
2. acid–base neutralization;
3. oxidation;
4. reduction.

The mechanism and even the products of reacting iron pyrite with a base such as NaOH are not well defined. However, sodium sulphide and sodium thiosulphate have been identified in the products of the reaction, suggesting the overall reaction (Attar 1980):

$$8FeS_2 + 30NaOH \rightarrow 4Fe_2O_3 + 14Na_2S + Na_2S_2O_3 + 15H_2O \quad (4.1)$$

The exposure of coal to air results in the natural oxidation of pyrite to sulphate. When the sulphate is subsequently leached by water the process is known as 'acid mine drainage' (Agarwal *et al.*, 1975). Virtually all the mechanisms reported in the literature for the removal of pyrite from coal are attempts to enhance this natural oxidation process. Oxidants ranging from metal ions (Fe^{3+}, Hg^{2+}, Ag^{+}) and strong acids (HNO_3 and $HClO_4$) to O_2, Cl_2, SO_2 and H_2O_2 have sufficient oxidation potential to react with pyrite (Meyers, 1977). The use of ferric sulphate as an oxidant has been investigated by Mixon and Vermeulen (1979) and applied in the TRW Meyers process (Meyers, 1977).

Several reactions involving the reduction of pyrite are reported in the literature. However, the consumption of expensive reducing agents precludes the commercial application of this type of reaction (Mesher and Peterson, 1979).

Mesher and Peterson (1979) investigated a fifth type of reaction using sodium sulphide and sodium polysulphide to extract zero valent sulphur from pyrite to form ferrous sulphide (FeS):

$$N_2S + FeS_2 \rightarrow Na_2S_2 + FeS \quad (4.2)$$
$$Na_2S_2 + 2FeS_2 \rightarrow Na_2S_4 + 2FeS \quad (4.3)$$

The coal was subsequently washed with a weak HCl solution to decompose the ferrous sulphide to H_2S and $FeCl_2$:

$$FeS + 2HCl \rightarrow FeCl_2 + H_2S \quad (4.4)$$

However, only 50% of the pyritic sulphur was removed after several hours of treatment at 250°C.

Reactions of organic sulphur

Because organic sulphur is present in petroleum in structural forms analogous to those in coal, Meyers (1977) suggested the following reactions for the removal of organic sulphur from coal, based on the experience of the oil refining industry:

1. solvent partition;
2. thermal decomposition;
3. acid–base neutralization;
4. reduction;
5. oxidation;
6. nucleophilic displacement.

Although examples of the removal of organic sulphur from coal by solvent partition, reduction, oxidation and displacement reactions appear in the literature (Meyers, 1977), oxidation and displacement reactions are used almost exclusively in the chemical coal cleaning processes currently under development.

Oxidation

Oxidation reactions form the basis of the majority of the chemical coal cleaning processes discussed in this part. Air or oxygen is used to remove up to 30% of the organic sulphur in the PETC, Ames, Ledgemont, and ARCO processes. Greater organic sulphur removal is claimed for the JPL chlorinolysis process which uses chlorine as oxidant.

Displacement

The use of caustic to remove organic sulphur from coal was proposed in the mid-1960s. Masciantonio (1965) investigated the treatment of bituminous coal with molten caustic (sodium and potassium hydroxides or mixtures of sodium, potassium and calcium hydroxides) at temperatures of between 150 and 400°C. Organic sulphur was removed from a Pittsburgh seam coal at temperatures above 200°C. Increasing the residence time and temperature increased the percentage of organic sulphur removal. However, the severe conditions used resulted in low yields of clean coal particularly for the Eastern Interior Basin and Western US coals. Only 52% of a Wyoming coal was recovered from the molten caustic.

Reggel *et al.* (1972), working at the Pittsburgh Energy Technology Center (PETC), investigated the desulphurisation of high organic sulphur bituminous coals by treatment with 10% (2.5 M) aqueous sodium hydroxide. Their results, presented in Table 4.1, show no consistent removal of organic sulphur. Indeed, in four of the examples shown the concentrations of organic sulphur were increased by the treatment. Despite these inconclusive results the Battelle Institute developed a variation of this process using $Ca(OH)_2$. This is discussed later.

Table 4.1 Effect of aqueous sodium hydroxide on the organic sulphur content of coal (after Meyers 1977)

Mine	*Treatment*	*Acidifying agent*[a]	*Mineral matter* w/w%	*Organic sulphur* w/w%, maf
River King	Native coal		9.8	2.19
	NaOH	CO_2	12.4	2.06
	$Ca(OH)_2$	HCl	8.16	2.14
	NaOH	HCl	0.67	2.44
	H_2O	CO_2	9.46	1.96
	H_2O	HCl	8.76	1.98
	NaOH	SO_2	0.72	1.99
	NaOH	H_2SO_4	0.52	2.33
Elliot	Native coal		18.15	1.05
	NaOH	CO_2	22.84	2.44
	NaOH	HCl	5.11	1.75
	NaOH,325°C	HCl	8.43	2.46
	NaOH	H_2SO_4	7.26	1.94

[a]Treatment involves contacting with 10% aqueous NaOH for 2 h at 225°C in an autoclave followed by acidification

Chapter 5

Processes for the chemical cleaning of coal

The processes discussed in this section are grouped according to their chemistry. First oxidation reactions and displacement reactions are discussed. The section entitled 'other processes' includes assessments of the Magnex process which is based on the chemical enhancement of the magnetic susceptibility of pyrite; a chemical comminution process; and an IGT process employing hydrodesulphurisation. The major processes under development for the chemical cleaning of coal are summarised in Table 5.1. There are also many organisations and institutions undertaking research on a much smaller scale, a comprehensive list of which is given by Contos *et al.* (1978).

It will become clear from the process descriptions below and the evaluations of these processes presented in Chapter 6 that there is much development work yet to be completed on the chemical cleaning of coal. Unfortunately, articles often appear in the technical press which claim that a simple, efficient process already exists. The author has been unable to substantiate these claims which must therefore be treated with caution and scepticism. It is important to distinguish these articles from the papers presented on the *bona fide* research described in this chapter.

Processes based on the oxidation of sulphur

PETC process

The PETC process is the outcome of a research project for the chemical beneficiation of coal initiated in 1970 at the US Department of Energy's Pittsburgh Energy Technology Center (PETC) (Friedman *et al.*, 1977; Warzinski *et al.*, 1979). It incorporates the oxidation of sulphur compounds by air and the subsequent neutralisation with lime of the sulphuric acid produced. The majority of the pyritic sulphur and part of the organic sulphur are removed.

$$4FeS_2 + 15O_2 + 8H_2O \rightarrow 2Fe_2O_3 + 8H_2SO_4 \qquad (5.1)$$

The so-called 'oxydesulphurisation' step has been tested at bench scale under batch, semi-continuous and continuous operation. Tests at pilot plant scale are underway at the TRW San Capistrano site (Figure 2.2). The bench scale tests were conducted at PETC in a one litre capacity, magnetically stirred stainless steel autoclave (Warzinski *et al.*, 1979). For the batch operation the autoclave containing the coal slurry was pressurised with air at ambient temperature to the desired oxygen partial pressure and subsequently heated to reaction temperature. After the desired residence time the reaction was quenched by means of an internal cooling coil. The main disadvantage of the batch mode of operation is that the partial pressure of oxygen decreases during the reaction. This led to the development of the semi-continuous operation whereby the reactor was charged with coal slurry as before but purged with inert gas, sealed and heated to reaction temperature. Once this temperature was attained the autoclave was pressurised rapidly with air to the desired operating pressure and a continuous flow of air maintained until the completion of the reaction. In all cases the soluble sulphates were removed with distilled water and the treated coal dried before analysis.

Twenty four different coals have been oxydesulphurised at PETC. Data for sulphur removals are presented in Table 5.2. The balances of the ultimate analyses are given in Table 5.3 (Warzinski *et al.*, 1979). All coals, unless otherwise stated were treated for one hour at the temperature indicated under pressures of either 5.52 MPa in the batch mode or 6.89 MPa in the semi-continuous mode. Within Table 5.2 the coals are grouped into those containing low concentrations of organic sulphur (<1%), those containing medium concentrations of organic sulphur (1–1.5%) and those containing high concentrations of organic sulphur (>1.5%). The oxydesulphurisation step is most effective in reducing the sulphur content of

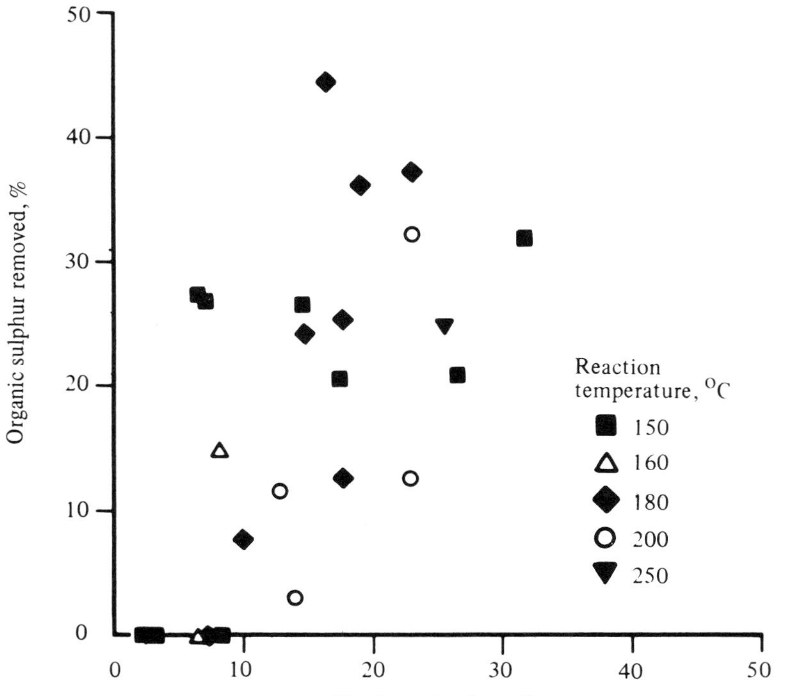

Figure 5.1 Efficiency of organic sulphur removal for various coals treated by the PETC oxydesulphurisation process (after Warzinski *et al.*, 1979)

Table 5.1 Summary of processes for the chemical cleaning of coal

Process name	*Developer*	*Funding*	*Operating principle*	*Chemicals used*	*Coal size*
PETC oxydesulphurisation process	Pittsburgh Energy Technology Center	US DOE	Oxidation	Air, lime	70–80% 74 μm
Ames wet oxidation process	Ames Laboratory, Iowa State Univ.	US DOE	Oxidation	Oxygen, sodium carbonate	70–80% 74 μm
Ledgemont oxygen leaching process	Hydrocarbon Research Inc.	HRI	Oxidation	Oxygen, lime, ammonia	70–80% 74 μm
ARCO promoted oxydesulphurisation	Atlantic Richfield Company	ARCO	Oxidation	Oxygen, lime, promoter	<595 μm
TRW Meyers process	TRW Systems and Energy	TRW	Oxidation	Oxygen, ferric sulphate, acetone, lime	<1.4 mm
JPL cholorinolysis process	California Institute of Technology	US DOE	Oxidation	Chlorine 1,1,1-trichloroethane	74–149 μm
KVB process	Research Cottrell, Inc.	Research Cottrell, Inc. US DOE	Oxidation Displacement	Nitrogen dioxide, sodium hydroxide	0.6–1.2 mm
Battelle hydrothermal coal process	Battelle Columbus Laboratories	Battelle, US EPA	Displacement	Sodium hydroxide, calcium hydroxide, lime, carbon dioxide	<297 μm
TRW gravimelt process	TRW	TRW	Displacement	Potassium hydroxide, sodium hydroxide	149 μm
General Electric microwave treatment process	General Electric Company	GE, US DOE	Microwave heating Displacement	Sodium hydroxide	149–590 μm
Magnex process	Nedlog Technology Group	Nedlog Technology Group	Chemical enhancement of magnetic susceptibility	Iron pentacarbonyl	<1.41 mm
IGT hydro-desulphurisation process	Institute of Gas Technology	US EPA	Hydro-desulphurisation	Hydrogen, iron oxide	<1.41 mm
Chemical comminution	Syracuse Research Corp.	Syracuse Research Corp.	Chemical comminution	Ammonia	0.149–380 mm

[a]JPL process using 1,1,1-trichloroethane
[b]JPL process using water
[c]Organic sulphur removal for two reaction steps
[d]Original developer of this process was Hazen Research, Inc.
[e]Single stage desulphurisation
[f]Second stage of desulphurisation applied to product of first stage
N/A Not available

Process conditions			*Sulphur removal*		*Mineral matter removal*	*Trace element removal*	*Scale*	*Source of data*
Temp. °C	*Press.* MPa	*Res. Time* s	*Pyritic*	*Organic*	%			
<150–200	5.52–6.89	3600	95	13	20	0	pilot plant test of desulphurisation step	Wheelock, 1980; Warzinski *et al.* 1979, 1980
<150	2.07	3600	80–90	<20	0	0	pilot plant test desulphurisation step	Wheelock, 1980; Markuszewski *et al.*, 1979, 1980
<200	2.07	3600	80–90	<20	0	0	bench scale tests of desulphurisation step	Wheelock, 1980; Sareen, 1975
120[e] 343[f]	2.17[e] 16.07[f]	3600[e] 3600[f]	88–98[e] 65–89[f]	0[e] 23–30[f]	12–50[e] 0–62[f]	N/A N/A	1.2 kg/h pilot plant test of desulphurisation step	Beckberger, 19
100–130	0.30–0.61	5–8 hours	84–99	0	2–4	As, Cd, Mn, Ni, Pb, Zn	8 t/d pilot plant tests of integrated process	Meyers *et al.*, 1979[a,b,c]
60–130	0.10–0.52	2700	71–95[a] 72–95[b]	46–98[a] to 24[b]	1–4	As, Hg	2 kg/h continuous integrated process	Kalvinskas *et al.*, 1980; Wheelock, 1980
100	0.10	1800–3600	60–100	30–50	N/A	N/A	50 g batch test of desulphurisation step	Wheelock, 1980; Guth, 1979
250–350	4.14–17.25	600–1800	90–95	20–70	N/A	As, Be, B, Pb, Th, V	10 kg/h continuous desulphurisation	Wheelock, 1980; Stambaugh *et al.*, 1975, 1977
370	N/A	1800	90	75	N/A	N/A	laboratory scale	Meyers *et al.*, 1980; Meyers and Hart, 1980
N/A	N/A	30–60	90	50–70[c]	N/A	N/A	10–500 g test of desulphurisation step	Wheelock, 1980; Zavitsanos *et al.*, 1978
170	0.10	1800–7200	57–92	0	7–71	N/A	91 kg/h continuous integrated process	Wheelock, 1980; Porter and Goens, 1979
800	0.10	3600	90% total sulphur		N/A	N/A	45 kg/h operated continuously	Wheeler, 1980; Flemming *et al.*, 1977
85	0.91	7200	50–70	0	50–60	N/A	23 kg batch operation	Contos *et al.*, 1978; Howard and Datta, 1977

Table 5.2 Sulphur removal from coals treated by the PETC oxydesulphurisation process (after Warzinski *et al.*, 1979)

Coal origin and treatment							*Sulphur content, w/w %*						*Heating value MJ/kg*		*Potential SO_2 emission kg/GJ*	
							Total S		*Pyritic S*		*Organic S*					
No.	*Coal seam*	*Mine*	*State*	*ASTM Rank*	*Temp (°C)*	*Recovery*	*Untreated*[a]	*Treated*[b]	*Untreated*	*Treated*	*Untreated*	*Treated*	*Untreated*	*Treated*	*Untreated*	*Treated*
1	Black Creek	Natural Bridge	AL	Mvb	180[c]	94	1.22	0.65	0.42	0.16	0.69	0.47	31.62	27.20	0.77	0.48
2	[d]	Peacock	CO	HvBb	200[c]	92	1.88	0.67	0.95	0.10	0.60	0.57	30.52	25.59	1.23	0.52
3	Iboden	Paramount Elkhorn No. 1 Strip	VA	HbAb	150[c]	100	1.19	0.95	0.26	0.04	0.78	0.79	33.20	32.51	0.72	0.58
4	Lower Freeport	Luciusboro Strip	PA	HvAb	180[c]	99	2.83	0.75	2.03	0.02	0.65	0.68	29.71	27.78	1.90	0.54
5	Lower Kittanning	[e]	PA	Lvb	150	98	0.96	0.57	0.53	0.08	0.39	0.47	33.94	31.70	0.57	0.36
6	Middle Kittanning	Congo Strip	OH	HvCb	180[c]	89	1.08	0.60	0.26	0.04	0.78	0.55	26.01	22.44	0.83	0.53
7	Mammoth	Storm King	MT	SbA	150	92	0.83	0.57	0.32	0.18	0.45	0.36	27.38	25.38	0.61	0.45
8	Pittsburgh	Bruceton	PA	HvAb	150	100	1.31	0.80	0.61	0.05	0.68	0.71	32.96	31.24	0.80	0.51
9	Upper Freeport	Baker	MD	Mvb	200[c]	99	1.58	0.54	0.82	0.02	0.56	0.50	29.41	25.86	1.07	0.42
10	Upper Freeport	Coal Junction Strip	PA	Mvb	200[c]	95	2.14	0.63	1.37	0.04	0.49	0.50	26.18	23.76	1.63	0.53
11	Brookville	Humphrey	PA	HcAb	180	96	4.20	1.17	3.06	0.13	1.11	1.01	30.82	26.39	2.73	0.89
12	Lower Freeport	West Valley Strip	PA	HcAb	180	98	4.14	1.04	3.09	0.22	1.01	0.78	30.50	26.50	2.71	0.79
13	Pittsburgh	Pitkulski Strip	PA	HvAb	160	100	1.67	0.89	0.71	0.03	0.82	0.83	27.10	25.33	1.23	0.70
14	Pittsburgh	No. 43 Strip	OH	HvAb	180	98	3.88	1.05	2.36	0.19	1.48	0.84	29.44	25.07	2.64	0.84
15	Pittsburgh	No. 43 Strip	OH	HvBb	180	98	3.01	0.98	1.93	0.16	1.05	0.80	29.88	25.09	2.02	0.78
16	Whitebreast	Lovilia No. 4	IA	HvCb	150[f]	81	5.85	1.07	3.95	0.18	0.90	0.76	25.28	21.26	4.63	1.01
17	Wyoming No.9	Reliance	WY	HvCb	150	101	1.75	0.90	0.38	0.06	1.14	0.82	28.87	26.70	1.21	0.67
18	Bevier	No. 22 Strip	KS	HvAb	150	93	5.00	1.98	2.92	0.36	2.04	1.60	28.38	28.43	3.52	1.39
19	Illinois No. 5	[g]	IL	HvCb	150	90	3.34	2.03	0.92	0.12	2.06	1.82	29.42	26.98	2.27	1.50
20	Illinois No. 6	River King	IL	HvBb	150[h]	89	3.69	2.12	1.13	0.11	2.25	2.00	28.35	23.33	2.60	1.82
21	Indiana No. 5	Enos	IN	HvBb	250[h]	91	3.27	1.84	0.70	0.20	1.98	1.64	28.70	23.48	2.28	1.56
22	[i]	Homestead	KY	HvAb	160	93	4.80	2.34	1.08	0.12	2.33	2.14	26.47	26.17	3.63	1.79
23	Minshall	Chrisney No. 1	IN	HvCb	200	85	5.65	1.43	3.01	0.10	1.53	1.22	26.33	23.80	4.29	1.20
24	Pittsburgh	Ireland	WV	HvAb	180	99	3.89	2.09	1.38	0.02	2.18	2.03	31.15	28.35	2.50	1.47

[a]Untreated, moisture free basis.
[b]Treated, moisture treated basis.
[c]Modified mode, all others batch mode.
[d]Uncorrelated coal seam.
[e]From Cambria slope preparation plant.
[f]Four repressurisations in batch mode.
[g]Information not available.
[h]10.34 MPa initial air batch mode.
[i]Blend of Kentucky seams Nos. 9, 11 and 13

Table 5.3 Ultimate analyses of coals treated by PETC oxydesulphurisation process (after Warzinski *et al.*, 1979)

Coal[a]	*Moisture free weight per cent*									
	Ash		*Carbon*		*Hydrogen*		*Nitrogen*		*Oxygen*	
	Untreated[b]	*Treated*[c]	*Untreated*	*Treated*	*Untreated*	*Treated*	*Untreated*	*Treated*	*Untreated*	*Treated*
1	3.7	2.6	76.9	71.6	5.2	3.8	1.6	1.6	11.4	19.8
2	6.5	5.2	73.7	68.0	5.3	3.6	1.6	1.9	10.9	20.5
3	3.7	3.4	80.2	79.6	5.1	4.9	1.5	1.4	8.4	9.9
4	15.1	13.3	72.5	70.2	4.5	3.9	1.3	1.3	3.7	10.6
5	7.1	7.0	83.5	79.8	4.4	4.1	1.4	1.4	2.6	7.2
6	16.9	15.3	64.6	59.6	4.4	3.4	1.2	1.1	11.8	20.0
7	9.5	7.5	68.4	66.4	4.5	4.0	1.2	1.2	15.6	20.4
8	5.5	4.6	79.4	76.5	5.3	4.8	1.5	1.5	7.0	11.7
9	16.2	14.6	72.3	67.7	4.3	3.3	1.3	1.2	4.3	12.7
10	22.4	21.4	65.0	62.0	3.6	3.0	1.1	1.2	5.9	11.8
11	9.3	6.6	73.3	68.2	5.2	3.9	1.7	1.4	6.3	18.7
12	11.7	8.8	73.5	68.2	4.9	3.7	1.3	1.2	4.5	17.0
13	21.4	20.1	65.4	63.0	4.4	3.9	1.3	1.3	5.9	10.7
14	12.7	10.4	70.7	66.1	4.8	3.8	1.6	1.4	6.4	17.2
15	11.2	9.3	72.1	66.1	4.8	3.6	1.8	1.3	7.0	18.7
16	17.4	16.4	63.0	56.3	4.2	3.3	1.2	1.1	8.3	21.9
17	3.2	2.1	72.4	69.2	4.7	4.2	1.6	1.5	16.4	22.0
18	14.8	12.0	68.6	69.3	4.8	4.7	1.1	1.2	5.7	10.8
19	8.7	6.6	70.8	68.2	5.0	4.3	1.4	1.4	10.7	17.6
20	11.6	11.0	68.4	61.0	4.6	3.4	1.2	1.1	10.6	21.2
21	9.1	11.5	69.4	63.4	4.9	3.1	1.6	1.5	11.8	18.7
22	14.1	11.4	63.8	65.4	4.6	4.2	1.3	1.4	11.3	15.3
23	15.9	11.1	62.8	63.2	4.6	3.3	1.2	1.3	9.8	19.7
24	7.8	6.8	73.4	71.0	5.1	4.5	1.3	1.2	8.5	14.4

[a]See Table 5.2
[b]Untreated
[c]Treated

those coals containing the least organic sulphur because these contain a higher proportion of pyritic sulphur which is almost totally removed.

With increasing severity of operating conditions increasing amounts of organic sulphur are removed but with an accompanying decrease in the heating values of the treated coals. This is especially true for the lower ranked coals. Figure 5.1 is a plot of organic sulphur removed against heating value loss. The large amount of scatter is due to the differing coal ranks and operating temperatures (Warzinski *et al.*, 1979).

The data in Table 5.3 show an increase in oxygen content and a decrease in the carbon and hydrogen contents for the treated coals. These effects are illustrated in Figures 5.2 and 5.3. In Figure 5.2 the number of moles of oxygen taken up by the coal per mole of carbon in the product coal increases with the loss of heating value. From Figure 5.3 it is seen that for the range of heating value losses encountered, between 0 and 30%, hydrogen is preferentially removed from the organic matrix. Therefore, it may be concluded that loss of heating value is attributable to both the consumption of hydrogen and oxygen to final oxidation products (carbon dioxide and water) and the partial replacement of hydrogen by oxygen in the coal (Warzinski *et al.*, 1979).

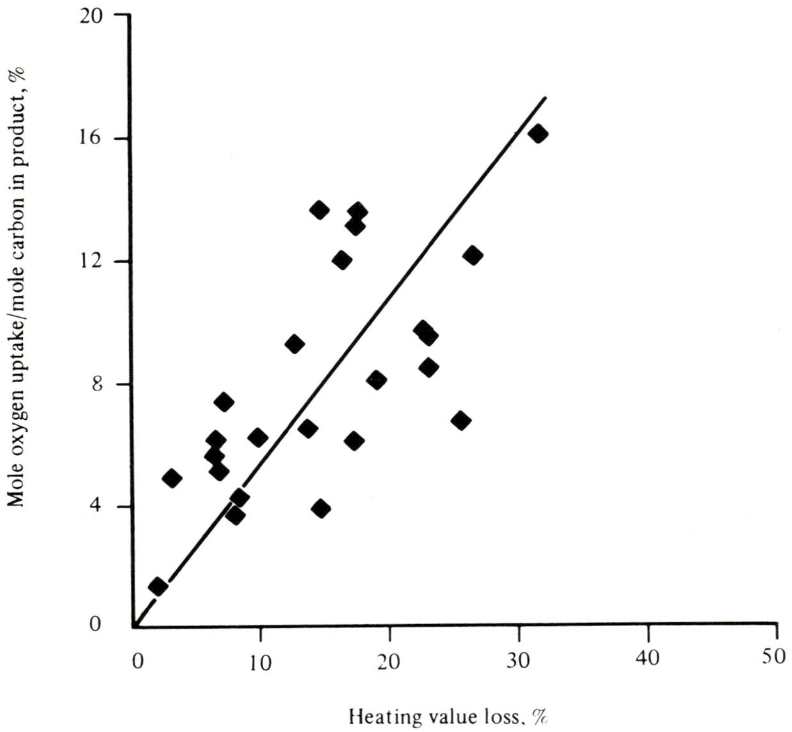

Figure 5.2 Degree of oxidation of coals treated by the PETC oxydesulphurisation process (after Warzinski *et al.*, 1979)

The washing of the treated coal to remove sulphates also results in a reduction in the mineral matter content of the coal. The average reduction shown in Table 5.3 is about 20% although a maximum reduction of 41% was recorded. The caking properties of the coal are also destroyed by the process.

As a result of recent cutbacks in the US Department of Energy (DOE) budget for chemical desulphurisation this project is to be terminated (Meyers, 1981a).

Ames wet oxidation process

This desulphurisation process has been under development at the Ames Laboratory of Iowa State University since 1975. Initial work was sponsored by the State of Iowa, but more recent work has been supported by the US DOE. The process is essentially similar to the PETC process except an alkaline leaching solution is employed and oxygen is used in place of air (Chuang *et al.*, 1980a,b; Chen, 1979; Tai *et al.*, 1977a,b).

The chemical reaction for the oxidation of pyritic sulphur is the same as that for

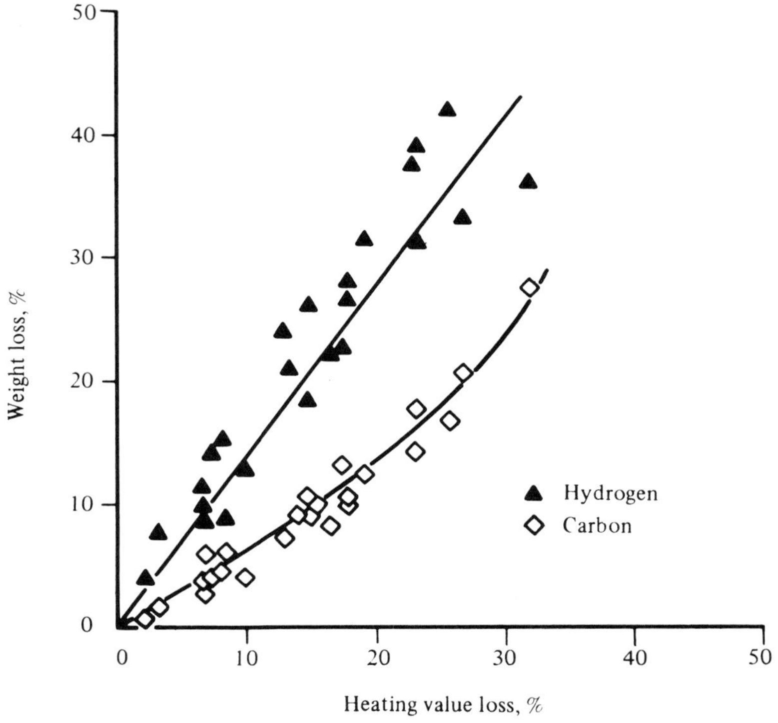

Figure 5.3 Carbon and hydrogen losses for coals treated by the PETC oxydesulphurisation process (after Warzinski *et al.*, 1979)

the PETC process, shown in equation (5.5). However, immediately following this reaction the sulphuric acid is neutralised by sodium carbonate.

$$H_2SO_4 + Na_2CO_3 \rightarrow Na_2SO_4 + CO_2 + H_2O \qquad (5.2)$$

In the conceptual, integrated process the treated coal is separated and the effluents treated with lime to produce solid wastes ($CaSO_4$ and $CaCO_3$) and a solution containing NaOH. The NaOH solution is converted to Na_2CO_3 by contacting with CO_2 and recycled.

The use of an alkaline leaching solution is claimed to improve the rate of extraction of pyritic sulphur and to assist in the removal of organic sulphur (Wheelock, 1980). Hence, somewhat milder operating conditions to those of the PETC process may be used. The alkaline solution is also less corrosive to materials of construction. However, retention of calcium in the coal may have implications for subsequent control of sulphur emissions during the combustion of the desulphurised coal (Markuszewski *et al.*, 1980).

Several coals from the Appalachian and midwestern regions of the USA and iron pyrites isolated from coal have been tested in bench scale autoclaves operating semi-continuously. The effect on the rate and extent of desulphurisation of process conditions such as reaction time, partial pressure of oxygen, concentration of Na_2CO_3 solution and temperature are shown in Figures 5.4, 5.5, 5.6 and 5.7

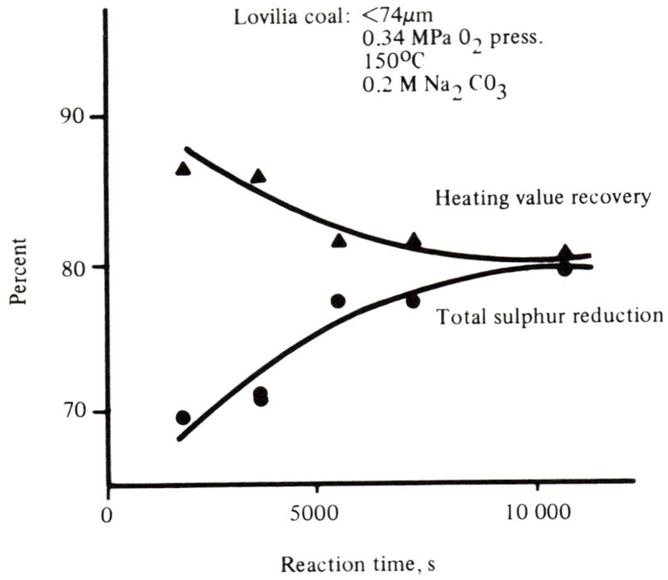

Figure 5.4 Effect of reaction time on the oxydesulphurisation of coal by the Ames process (after Markuszewski *et al.*, 1979)

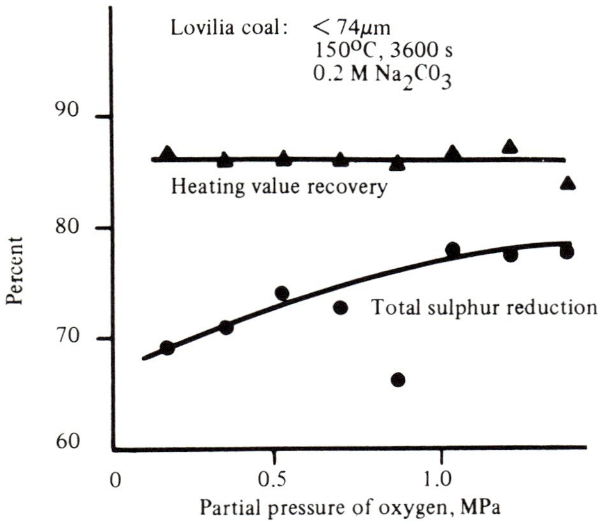

Figure 5.5 Effect of oxygen partial pressure on the oxydesulphurisation of coal by the Ames process (after Markuszewski *et al.*, 1979)

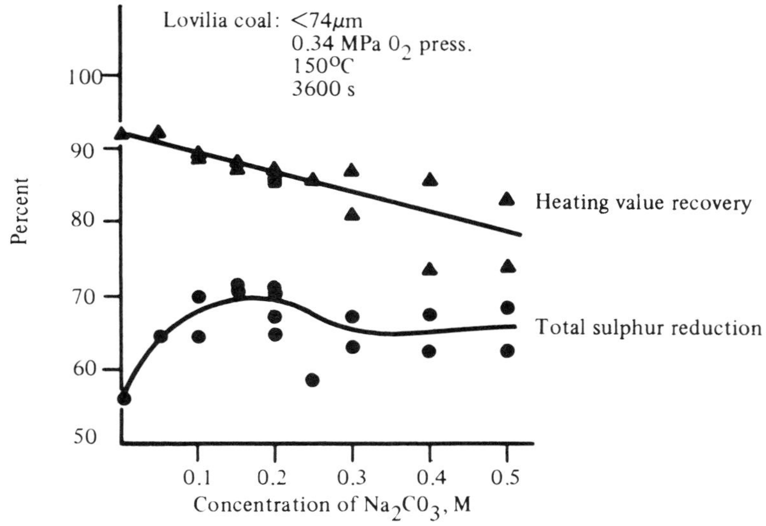

Figure 5.6 Effect of sodium carbonate concentration on the oxydesulphurisation of coal by the Ames process (after Markuszewski *et al.*, 1979)

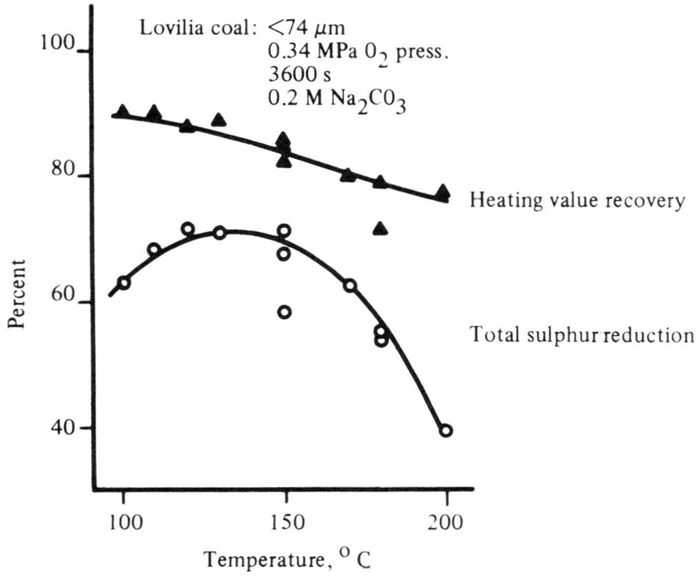

Figure 5.7 Effect of temperature on the oxydesulphurisation of coal by the Ames process (after Markuszewski *et al.*, 1979)

(Markuszewski *et al.*, 1979). Typical leaching conditions are 150°C, a partial pressure of oxygen of 1.42 MPa, a residence time of 1 h, and a 0.2 M sodium carbonate solution. Generally, 80–90% of the pyritic sulphur is extracted and 80–90% of the heating value of the coal recovered. The removal of organic sulphur is highly variable, ranging from 0–50% for different coals (Wheelock, 1980).

As for the PETC process, the greater the severity of the operating conditions the lower the heating value of the coal product (Figures 5.4, 5.6 and 5.7). These losses are attributed almost exclusively to carbon oxidation. The hydrogen and oxygen contents of the treated coal are essentially unchanged. The omission of a washing stage subsequent to desulphurisation results in the removal of little mineral matter (Markuszewski *et al.*, 1979).

A modification of this process has been investigated. This involves the addition of a second leaching step, in which the alkaline coal slurry from the first leaching step is raised to a higher temperature (>200°C) in a nitrogen atmosphere, to decompose any oxidized organic sulphur compounds (Wheelock *et al.*, 1979). Preliminary results are shown in Table 5.4. In each of the three tests the pyritic

Table 5.4 Desulphurisation of coal by the modified Ames process (after Ames Laboratory, 1980)

Treatment[a]	*Heating value* MJ/kg	*Ash* %	*Sulphur distribution,* kg/GJ				*Sulphur reduction*		*Heating value recovery* %
			Pyritic	*Sulphate*	*Organic*	*Total*	*Total*	*Organic*	
Untreated[b]	30.27	4.7	0.37	0.06	1.32	1.75	—	—	—
Two-step 200°C	28.24	11.1	0.16	0.05	1.13	1.33	2.40	14.9	82.6
Two-step 220°C	28.35	10.8	0.11	0.03	1.17	1.31	25.5	11.7	80.8
Two-step 240°C	28.30	10.9	0.17	0.02	1.06	1.16	33.8	19.8	77.4

[a]Step 1: Leached 1 h, at 150°C and 0.34 MPa O_2 with 0.2 M Na_2CO_3
Flow rate of O_2 was 1.18×10^{-5} m^3/s
Step 2: Leached 1 h with 0.2 M Na_2CO_3 in nitrogen atmosphere at indicated temperature
[b]Physically precleaned

sulphur content was reduced by between 71 and 54%. The organic sulphur content was reduced by nearly 20% at a second stage temperature of 240°C. The heating value of the product coal decreased slightly with increasing second stage temperature (Ames Laboratory, 1980).

Ledgemont oxygen leaching process

Bench scale development of this process was undertaken at the Ledgemont Laboratory of Kennecott Copper Corporation between 1971 and 1975 (Agarwal *et al.*, 1975). More recent development has been undertaken by Hydrocarbon Research Inc. which has acquired exclusive rights to the process. There are two principal versions of this process. In one version ground coal is slurried with water and contacted with oxygen under pressure. This is essentially the same as the PETC process except that oxygen is used in place of air. In the second version, coal is leached under similar conditions but with a solution of ammonia. The use of an alkaline leach solution is very similar to the Ames process. It is also the most relevant of the two Ledgemont variants because some organic sulphur is removed.

In the conceptual integrated process the feed coal is cleaned physically to remove coarse refuse particles and then ground to a top size of 150 μm, mixed with recycled solution and leached at 130°C for between 1 and 2 h. The leaching solution is saturated with oxygen under a partial pressure of between 1.01 and 2.02 MPa. Under these conditions iron pyrites is oxidised to soluble sulphates, primarily ferric sulphate and sulphuric acid.

If ammonia is present the acid is neutralised and some organic sulphur is converted to soluble sulphates. The leached coal is subsequently recovered by filtration, washed, agglomerated or pelletised and dried. Lime is used to precipitate gypsum and iron compounds from the wash water and leach liquor. The filtrate is recycled.

$$4FeS_2 + 16NH_3 + 13H_2O + 15O_2 \rightarrow 16NH_4 + 8SO_4^{2-} + 4Fe(OH)_3 \quad (5.3)$$

Determination of process conditions has been undertaken in a batch mode in high pressure autoclaves (Sareen *et al.*, 1975). The effects of retention time and ammonia concentration on sulphur removal from bituminous coal (Illinois No. 6) are shown in Figure 5.8 (Sareen, 1977). Approximately 90% of the pyritic sulphur

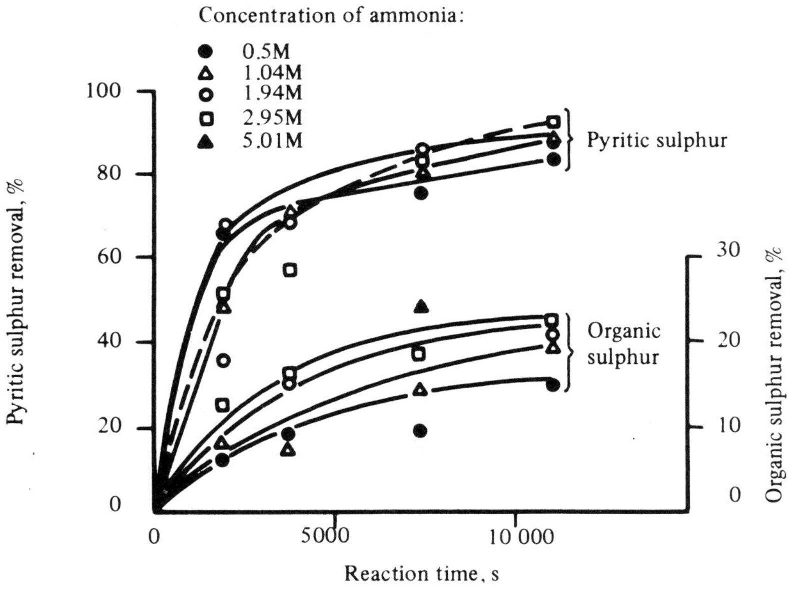

Figure 5.8 Sulphur removal as a function of ammonia concentration and time for the Ledgemont process (after Sareen, 1977)

is removed regardless of the concentration of ammonia. However, organic sulphur removal increases as the concentration of ammonia solution increases.

As in other oxydesulphurisation processes part of the carbon in the coal is oxidised to carbon monoxide (CO) carbon dioxide (CO_2) and soluble coal acids. The greater propensity for the formation of coal acids in basic systems than in acidic systems and the long residence time (2 h) of this process result in a 10% carbon loss. As well as combining with the carbon in the coal, oxygen is also taken up by the coal complex during the desulphurisation process. A breakdown of the

Table 5.5 Operating data for the Ledgement oxygen leaching processes[a] (after Sareen, 1977)

Parameter	H_2O/O_2 *process*	NH_3/O_2 *process*
Sulphur removal: pyritic	>90%	>90%
organic	0%	25%
Oxygen consumption (kg O_2/kg coal):		
oxygen for pyrite reaction	0.035	0.0375
oxygen uptake	0.054	0.0340
oxygen to form CO_2	0.032	0.0350
Heating value loss (maf)	8%	13%
Carbon loss (maf)	6.5%	10%

[a]Illinois No. 6 coals, 2% pyritic sulphur

oxygen consumption for the process is shown in Table 5.5 (Sareen, 1977). The loss of carbon and the oxygen uptake in the coal result in a loss in heating value of as much as 13% for the treated coal. Hydrocarbon Research Inc. are planning to design, construct and operate a pilot plant which will demonstrate the integrated process at a scale of 2–5 t/day.

As a result of recent cutbacks in the US DOE budget for chemical desulphurisation development of the Ledgemont process is to be terminated (Meyers, 1981a).

ARCO promoted oxydesulphurisation process

The Atlantic Richfield Company (ARCO) has been developing a proprietary process for the oxydesulphurisation of coal since 1967. The basic process is similar in concept to the Ledgemont and PETC processes except that oxidation and leaching are conducted in a solution of an iron-complexing agent, such as sodium oxalate, which serves as a reaction promoter. A proprietary promoter, which is low in cost and therefore need not be recovered, has recently been substituted for sodium oxalate (Wheelock, 1980). The use of a promoter is claimed to enhance the oxygen selectivity and enable relatively mild operating conditions to be employed. A variation of the basic process involves the addition of a second reaction step with a proprietary reagent under more severe conditions than the first leaching step. The work is sponsored jointly by the Electric Power Research Institute (EPRI) and ARCO.

Operating conditions for the basic and modified processes have been determined in bench scale autoclaves operated in both batch and continuous modes. In batch leaching tests under conditions of 120°C, 217 kPa and 1 h residence time, between 88 and 98% of pyritic sulphur and between 78 and 94% of the iron were removed from five bituminous coals. The mineral matter contents of four ROM coals were reduced by between 32 and 50% and of one pre-washed coal by 12% (Beckberger *et al.*, 1979). Applying the second reaction stage to the three coals from the first leaching step with the highest organic sulphur contents further reduced the pyritic sulphur by between 65 and 89% and the organic sulphur by between 23 and 30%. The second reaction step also further reduced the mineral matter contents of two of the coals by 57 and 62%. Comparing the product from the second step to the feed coal produces sulphur removals of between 97 and 99% pyritic sulphur, between 19 and 31% organic sulphur and between 65 and 74% on a total sulphur basis.

The relatively mild operating conditions and enhanced oxygen selectivity offered by the use of a promoter result in higher carbon and heating value recoveries than achieved with other oxydesulphurisation processes described in this review. For the single step process carbon recoveries exceed 98% while typical increases in coal oxygen content vary between 2–5%. The resultant heating value recoveries were 95% or greater. Carbon recoveries for the two-step process were similar to those for the one-step process but oxygen contents were about the same or lower than those of the feed coal. Hence, heating value recoveries were somewhat better than in the one-step process.

ARCO has also determined the effect of oxydesulphurisation on coal constituents and properties other than the sulphur content. Analyses of the feed and product coals are given in Table 5.6 (Beckberger *et al.*, 1979). In general,

Table 5.6 Composition and properties of two coals treated by the ARCO promoted oxydesulphurisation process (after Beckberger, 1979)

	Western Kentucky 9/14						*Sewickley*					
	Feed coal		*One-step product*		*Two-step product*		*Feed coal*		*One-step product*		*Two-step product*	
Ash composition, wt %	*Ash*	*dcb**	*Ash*	*dcb**	*Ash*	*dcb**	*Ash*	*dcb**	*Ash*	*dcb**	*Ash*	*dcb**
P_2O_5	0.20	0.03	0.16	0.01	0.46	0.01	0.06	0.01	0.10	0.01	0.12	0.01
SiO_2	27.61	4.79	57.65	3.63	19.80	0.54	42.11	7.75	55.65	6.97	31.86	1.53
Fe_2O_3	34.42	4.39	3.96	0.25	11.40	0.31	24.26	4.46	2.82	0.35	4.12	0.20
Al_2O_3	20.51	2.61	27.21	1.71	12.28	0.34	20.37	3.75	24.64	3.09	13.55	0.65
TiO_2	0.79	0.10	1.50	0.09	4.04	0.11	0.86	0.16	1.10	0.14	3.19	0.15
CaO	1.67	0.21	3.45	0.22	1.60	0.04	5.42	1.00	7.02	0.88	1.32	0.06
MgO	0.66	0.08	0.95	0.06	1.66	0.05	1.19	0.22	0.97	0.12	0.98	0.05
SO_3	1.58	0.20	1.11	0.07	20.10	0.55	2.20	0.40	4.02	0.50	21.64	1.04
K_2O	2.13	0.27	2.51	0.16	1.42	0.04	1.91	0.35	2.24	0.28	2.42	0.12
Na_2O	0.35	0.04	0.52	0.03	26.40	0.72	0.70	0.13	0.68	0.09	19.90	0.95
Undetermined	0.08	0.01	0.98	0.06	0.84	0.02	0.92	0.17	0.76	0.10	0.90	0.01
	100.00	12.73	100.00	6.29	100.00	2.73	100.00	18.40	100.00	12.53	100.00	4.80
Fusion temperatures of ash °C												
Reducing atmosphere:												
Initial deformation	1077		1360		1035		1149		1260		1038	
Softening	1138		1482 +		1063		1182		1332		1071	
Fluid	1282		1482 +		1116		1266		1466		1118	
Oxidising atmosphere:												
Initial deformation	1332		1432		1113		1302		1299		1116	
Softening	1371		1482 +		1143		1313		1396		1149	
Fluid	1421		1482 +		1196		1352		1482 +		1216	
Slagging fouling and viscosity indexes:												
Basic/acidic	0.67		0.13		0.18		0.53		0.17		0.59	
Slagging index (R_3)	2.45		0.24		1.43		2.20		0.27		0.69	
Fouling index (R_f)	0.23		0.07		0.65		0.37		0.12		11.7	
SiO_2/Al_2O_3	1.83		2.12		1.61		2.07		2.26		2.35	
T_{250} (°C)	1213		1593		1177		1253		1549		1229	
Silica percentage	50.57		87.35		57.46		57.70		83.73		83.23	

*dry coal basis

application of the one-step process results in beneficial changes in the coal handling and combustion properties. This is in contrast to the effects of the second desulphurisation step.

The concentrations of the major mineral matter constituents are generally reduced by the one-step process. The greatest reduction occurs for iron although substantial reductions are obtained for silica and alumina. Compared with the one-step process products the two-step products are generally lower in mineral matter. Major changes due to the second step include a reduction in calcium content and an increase in sodium content for the treated coal. Generally silica and alumina are also reduced during the second reaction step.

The slagging and fouling indexes of the product coals were determined by the method developed by Attig and Duzy (1969). The one-step process substantially reduced the slagging and fouling indexes for all the coals. Treating the coals by the second reaction step increased the slagging indexes for two coals but reduced that of a third. The increased slagging factors were due, almost entirely, to the absorption of sodium from the sodium based chemical used in the second reaction step.

The slagging index of a coal is also indicated by the temperature at which molten ash viscosity reaches 250 P, T_{250}. The T_{250} point defines approximately the upper temperature limit of the ash plastic zone. The values of T_{250} were increased by the one-step process but decreased by the second reaction step (Table 5.6, Beckberger *et al.*, 1979).

The effect of processing on the storage stability of the coals was also determined. The oxidation rate for the one-step process product was significantly lower than for the feed coal whereas the rate for the two-step process product was found to be higher. In addition, the grindability of the one-step process product increased compared with that of the feed coal, whereas that of the two-step process product decreased. Some reduction in particle size also occurred during the second reaction step for Illinois No. 6 coal.

TRW Meyers desulphurisation process

Development of the TRW Meyers process commenced in 1971, sponsored by the then US Air Pollution Control Administration (APCA), now part of the Environmental Protection Agency (EPA) (Meyers *et al.*, 1979a; Meyers, 1975). The basis of the process is the oxidation of the pyritic sulphur in coal by a hot solution of ferric sulphate. Organic sulphur is not removed. Coal is crushed to a top size of about 1.4 mm and mixed with ferric sulphate solution [$Fe_2(SO_4)_3$]. The resulting slurry passes to a reactor where it is held at 100 to 130°C for several hours. Iron pyrites is oxidized to ferrous sulphate, sulphuric acid, and elemental sulphur.

$$5FeS_2 + 23Fe_2(SO_4)_3 + 24H_2O \rightarrow 51FeSO_4 + 24H_2SO_4 + 4S \qquad (5.4)$$

Oxygen is introduced to reoxidise the ferrous sulphate back to ferric sulphate. Whereas the sulphates are soluble in the aqueous solution, the elemental sulphur is not.

$$4FeSO_4 + 2H_2SO_4 + O_2 \rightarrow 2Fe_2(SO_4)_3 + 2H_2O \qquad (5.5)$$

After recovering the coal from the leachant by filtration and washing the coal to remove any remaining solution, the dewatered coal is extracted with acetone to remove the elemental sulphur. The original leach solution, now containing ferrous sulphate and sulphuric acid, is neutralised by the addition of lime to remove the excess of sulphates and restore the concentration to its original value:

$$Fe_2(SO_4)_3 + CaO \rightarrow 3CaSO_4 + Fe_2O_3 \quad (5.6)$$

$$FeSO_4 + CaO \rightarrow CaSO_4 + FeO \quad (5.7)$$

The calcium sulphate and iron oxides so produced are inert and insoluble. The leach solution is recycled.

Initial development of this process was conducted at bench scale (Meyers *et al.*, 1979c). A survey was undertaken for the US EPA in which 32 US coals containing an average 2.02% pyritic sulphur and 3.05% total sulphur were subjected to the TRW Meyers process. Between 83 and 99% of the pyritic sulphur and between 25 and 80% of the total sulphur were removed (Meyers, 1977; Hamersma *et al.*, 1977; Contos *et al.*, 1978).

In 1977 TRW completed the construction of a pilot plant, known as the Reactor Test Unit, at their Capistrano test site in California, USA This test unit, capable of processing between 100 and 300 kg/h of coal continuously, was used to demonstrate the mixing of coal and reagent, the primary pyrite reaction and reagent regeneration, the secondary pyrite reaction, and slurry filtration (Meyers *et al.*, 1979b; Santy and Van Nice, 1979; Van Nice *et al.*, 1977). The final process step, solvent extraction of the elemental sulphur, was not included. Over 22 t of bituminous coal from the Martinka mine in West Virginia were treated during 250 h of operation. Leaching/regeneration was conducted at 110–132°C and 0.30 to 0.61 MPa pressure with a residence time of between 5 and 8 h. The pyritic sulphur content was reduced from 1% in the feed coal to 0.16% in the product without measurable loss of coal. Operation of the integrated process was demonstrated despite several mechanical problems and corrosion of the primary reactor.

The use of a ferric sulphate leach solution has been shown to reduce the mineral matter and trace element contents of several coals (Meyers, 1977; Hamersma *et al.*, 1977). The coals studied had initial mineral matter contents of between 7.55 and 49.28% and final mineral matter contents of between 3.37 and 43.46%. In each case there was a small but significant (<5%) removal of mineral matter in excess of that which could be accounted for by removal of pyrite. Hamersma *et al.* (1977) studied the removal of trace elements during pyrite leaching at 100°C for 20 US coals by analysis of treated and untreated coal. Analyses were performed for antimony, arsenic, beryllium, boron, cadmium, chromium, copper, fluorine, lead, lithium, manganese, mercury, nickel, selenium, silver, tin, vanadium and zinc. They also measured the trace element removal which could be obtained by physical cleaning by analysing coal samples from gravity separation equipment. The following general conclusions were drawn:

1. As, Cd, Mn, Ni, Pb and Zn were removed to a significantly greater extent by the Meyers process.
2. F and Li were partitioned to a greater extent by physical separation procedures.
3. Ag and Cu were removed with a slight preference for gravity separation.
4. Cr and V were removed equally by both processes.

As little or no reaction of the reagents with the coal matrix occurs during the TRW Meyers process it is unlikely that the combustion properties of the treated coals will be modified extensively (Contos *et al.*, 1978).

A modification of the TRW Meyers process has been investigated for the removal of pyrite from metallurgical coal (Betancourt and Hancock, 1978;

Hancock and Betancourt, 1978). This application has the advantage that the elemental sulphur, produced by the oxidation of pyrite with ferric sulphate, is removed during the coking process. Leaching with acetone is eliminated. However, passivation of the pyrite occurred reducing the rate of oxidation. This had a detrimental effect on the final efficiency of the process.

Despite the relatively advanced stage of development of the TRW Meyers process no further work is planned (Meyers, 1980; Meyers, 1981b). This is due mainly to the inability of the process to remove organic sulphur.

TRW Gravichem process

This process is essentially the TRW Meyers process with the addition of a gravity separation step. The ferric sulphate leach solution has a specific gravity of between 1.3 and 1.5. The float coal, which is low in pyritic sulphur and ash, is removed, leaving only the sink fraction of the feed coal to undergo the TRW Meyers process (Hart *et al.*, 1978a,b; Meyers, 1979).

A modification of the Gravichem process was later suggested by George *et al.* (1979) of the US DOE, Morgantown Energy Technology Center (METC). This involved the substitution of water-only cyclones to effect a gravity separation in a water medium rather than in a chemical solution. In effect this is merely cleaning the coal by physical means prior to the TRW Meyers process.

JPL chlorinolysis process

The development of a chlorinolysis process for desulphurising coal commenced in 1976 at the Jet Propulsion Laboratory (JPL) of the California Institute of Technology (Kalvinskas and Hsu, 1979). There are two variants of the process although both are based on the oxidation of sulphur by chlorine. Preliminary work on the process, which is capable of removing both organic and pyritic sulphur, was internally funded by the National Aeronautics and Space Administration (NASA). More recent work has been funded by the US DOE.

In one version of the process coal pulverised to between 94 and 19 μm is mixed with 1,1,1-trichloroethane in a weight ratio of 2 : 1. The slurry is then treated with chlorine gas under conditions of between 60 and 130°C and between 0.10 and 0.52 MPa for 45 min.

The chemistry of the process is somewhat complex. The following reactions have been proposed (Contos *et al.*, 1978; Hsu *et al.*, 1977)

$$R\text{–}S\text{–}R' + Cl^{+}\text{–}Cl^{-} \overset{H^{+}}{\rightleftharpoons} RSCl + R'Cl \qquad (5.8)$$

$$RS\text{–}SR' + Cl^{+}\text{–}Cl^{-} \overset{H^{+}}{\rightleftharpoons} RSCl + R'SCl \qquad (5.9)$$

Sulphenyl chloride is oxidised to sulphonate or sulphate according to the following reactions:

$$RSCl \xrightarrow{Cl_2,\ H_2O} RSO_2Cl \xrightarrow{H_2O} RSO_3H + HCl \qquad (5.10)$$

$$RSCl + 2Cl_2 + 3H_2O \rightarrow RSO_3H + 5HCl \quad (5.11)$$

$$RSCl \xrightarrow{Cl_2,\ H_2O} RSO_2Cl \xrightarrow[Cl_2,\ H_2O]{\Delta} SO_4^{2-} + RCl \quad (5.12)$$

$$RSCl + 3Cl_2 + 4H_2O \rightarrow RCl + H_2SO_4 + 6HCl \quad (5.13)$$

The pyritic sulphur reactions may be summarised as follows:

$$FeS_2 + 2Cl_2 \rightarrow FeCl_2 + S_2Cl_2 \quad (5.14)$$

$$2FeS + 7Cl_2 \rightarrow 2FeCl_3 + 4SCl_2 \quad (5.15)$$

$$2FeS_2 + 10SCl_2 \rightarrow 2FeCl_3 + 7S_2Cl_2 \quad (5.16)$$

$$S_2Cl_2 + 8H_2O + 5Cl_2 \rightarrow 2H_2SO_4 + 12HCl \quad (5.17)$$

$$RH + S_2Cl_2 \rightarrow RS_2Cl + HCl \quad (5.18)$$

$$2FeS_2 + 15Cl_2 + 16H_2O \rightarrow 2FeCl_3 + 4H_2SO_4 + 24HCl \quad (5.19)$$

At the temperatures used in this process the S_2Cl_2 formed by the chlorination of FeS_2 is readily converted to HCl and H_2SO_4. At room temperature, without the presence of adequate moisture, this reaction is slow and S_2Cl_2 might react with organic compounds to form organo-sulphur compounds. On the other hand, in an organic solvent, at a slightly elevated temperature the rate of chlorination of coal is slower than in aqueous media at room temperature.

The 1,1,1-trichloroethane is subsequently recovered by steam distillation and recycled.

The coal, now suspended in water, is filtered, washed and heated to 400°C in a stream of inert gas such as nitrogen. Careful consideration will have to be given to the design of the coal handling equipment for this step in a commercial process.

The coal is dechlorinated by the high temperature treatment with the production of hydrogen chloride.

$$RH + R'Cl \rightarrow RR' + HCl \quad (5.20)$$

It is proposed to treat the hydrogen chloride by a modified Deacon process to generate chlorine for reuse. It is also proposed to distill the filtrate from the coal recovery step to recover additional hydrogen chloride for reprocessing and to treat the remaining liquor with lime to precipitate gypsum, an easily disposable waste product. The second version of the process uses water in place of 1,1,1-trichloroethane thus eliminating the use of steam distillation for solvent recovery.

Initial development work was carried out in laboratory scale equipment using 100 g batches of coal suspended in 1,1,1-trichloroethane. More than twelve different coals including bituminous, subbituminous and lignite were processed. The reduction in total sulphur content was between 34 and 76%, in pyritic sulphur content between 0 and 92% and in organic sulphur content between 0 and 72%. More recent tests have been conducted in a bench scale reactor of 2 kg coal capacity with water or 1,1,1-trichloroethane as the suspending medium. Five bituminous coals have been subjected to chlorination, solvent recovery, filtration and washing, and dechlorination steps during forty-four test runs. Total sulphur reductions were between 52 and 63% with 1,1,1-trichloroethane and between 45 and 66% with water. Organic sulphur reductions were between 46 and 89% with 1,1,1-trichloroethane and between 0 and 24% with water for four of the five coals. No organic sulphur removal was detected for the fifth coal with either 1,1,1-trichloroethane or water. The maximum pyritic sulphur reductions were between 71 and 95% with 1,1,1-trichloroethane and between 72 and 95% with water (Kalvinskas *et al.*, 1980).

An overall material balance on the combined chlorination, solvent distillation, coal filtration and water wash for forty-one test runs provided the following material recoveries:

Raw coal: 100 ± 3.5%
Organic coal fraction: 99.3 ± 3.5%
Ash: 114%
Sulphur: 94%
Chlorine: 86%
1,1,1-trichloroethane (17 runs): 83 ± 7%

The high figure for ash is due to exposed metal and test coupon corrosion products. The low solvent recovery may be explained partly by the loss of solvent during sampling.

Proximate analyses of the five processed coals showed a decrease in volatile matter of between 8 and 15%, a reduction in mineral matter of between 1 and 4%, an increase in fixed carbon of between 9 and 20% and heating value losses in the range of 0 to 3.02 MJ/kg.

Analyses for the trace elements lead, arsenic, selenium, mercury and vanadium were performed on three of the coals tested (Table 5.7). Variations in trace

Table 5.7 Trace element removal for three coals treated by the JPL chlorinolysis process (Kalvinskas *et al.*, 1980)

Element ppm	*Trace element removal*[a], %		
	Ohio No. 8	*Illinois No. 6*	*Island Creek*
Pb	−107 ± 200	31 ± 39	—
As	50 ± 10	60 ± 15	−33
Se	−8 ± 30	−25 ± 25	−33
Hg	−30 ± 85	80 ± 18	95
V	—	—	—

Detection limts: As, 2 ppm; Hg, 0.1 ppm; Pb, 50 ppm; Se, 2 ppm; V, 50 ppm.
[a]with a standard deviation expressed in %

element content between raw coal samples as well as a lack of precision in the analyses resulted in questionable trace element reductions. Only arsenic and mercury were reduced significantly (by between 50–60% and 80%, respectively). The concentration of vanadium was below the detection limit of 50 ppm of the equipment. Elemental analyses for the five processed coals showed a reduction in the concentration of iron of between 43 and 76% and of calcium of between 35 and 86%.

One of the main problems of the JPL chlorinolysis processes is the retention of chlorine in the coal. After thirty-six separate dechlorination runs the chlorine contents of the product coals were between 0.23 and 5.1%. The raw coals contained chloride concentrations in the range of 0.1 to 0.5%. A chlorine concentration in coal of 0.1% or less is generally considered necessary to obviate fireside corrosion problems during the combustion of the coal. In addition, product coal recoveries for the dechlorination step averaged only 88 ± 5%. However, provisions for recovery of coal fines from the nitrogen purge gas to the scrubber and improvements in containing coal during handling, charging and removal from the dechlorinator are expected to eliminate the majority of the loss (Kalvinskas *et al.*, 1980).

Another potentially serious problem associated with the JPL processes is that of corrosion. Chlorine, hydrogen chloride, hydrochloric acid and sulphuric acid are all utilised or produced by the JPL process. All are highly acidic and corrosive in nature and will require special, and therefore expensive, materials of construction.

As a result of recent cutbacks in the US DOE budget for chemical desulphurisation development work on the JPL processes is to be terminated (Meyers, 1981a).

KVB process

The KVB process, also known as the Guth process after Eugene Guth in whose name the original patent is filed, is in the initial stages of development by Research Cottrell, Inc, CA, USA (Guth, 1979). The process incorporates the oxidation of sulphur compounds using NO_2 as a carrier for oxygen and the use of a caustic wash to remove up to 40% of the organic sulphur. Initial work on this process was funded internally. The proposed funding of an expansion of the project by the US DOE is unlikely to be forthcoming (Meyers, 1981a).

The KVB process differs from other oxidation-based processes described in this review in that the oxidation is conducted under dry conditions. Coal, pulverised to between 0.6 and 1.2 mm, is treated with a gas stream containing NO_2 and O_2 for between 30 and 60 min at 100°C and atmospheric pressure. The NO_2 selectively oxidises part of the pyritic and organic sulphur.

$$FeS_2 + 6NO_2 \rightarrow FeSO_4 + SO_2 + 6NO \qquad (5.21)$$

The SO_2 is swept away in the gas stream whereas the iron sulphate remains with the coal. Although various methods of removing the SO_2 from the recycle gas have been proposed these have yet to be demonstrated (Wheelock, 1980). The NO is rapidly reoxidised to NO_2 in the presence of O_2. The coal can then be washed with hot water to remove soluble iron sulphur compounds, some trace elements and reduce the concentration of mineral matter.

The final step in the process involves treating the washed coal with a caustic solution for 1 h at 93°C to remove organic sulphur. The precleaning of the coal by oxidation and washing minimises the formation of iron hydroxide during treatment with caustic. The formation of iron hydroxide not only consumes caustic solution but can also block the pores and voids in the coal structure impeding the contact of caustic with the oxidised organic sulphur compounds.

The oxidation step has been demonstrated in a 25 mm diameter tubular reactor with a 300 mm high static coal bed. Using a gas containing between 5 and 10% NO_2 and a temperature of 93°C, between 30 and 90% of the pyritic sulphur was removed from four bituminous coals. When the treated coals were subsequently washed with hot water, the overall removal of sulphur in both steps corresponded to a reduction in pyritic sulphur of between 60 and 100%. Further washing of the treated coals with hot caustic solution reduced the organic sulphur content by between 30 and 50%.

Considerably more laboratory work is required to determine the optimum conditions for a larger number of coals. Also, the addition of nitrogen to the coal may increase NO_x emissions when the product is burned. Nitrogen additions are uncertain and require future study and possible design modifications (Bechtel National Inc., 1980).

Processes based on the displacement of sulphur

Battelle hydrothermal process

Research on the Battelle hydrothermal process, which commenced in 1969, is funded by both Battelle and the US EPA (Stambaugh, 1977). Raw coal is ground to a size of less than 297 μm and mixed with a caustic solution containing 10% w/w sodium hydroxide and 2 to 3% w/w calcium hydroxide. The slurry is heated to between 250 and 350°C at a pressure of between 4.14 and 17.25 MPa for between 10 and 30 min. After cooling, the coal is removed by filtration, washed and dried. The rate of filtration is increased by the addition of a dispersant such as sodium lauryl sulphate to the slurry. The product is washed with a saturated solution of lime water to facilitate the removal of sodium. Even with these modifications to the filtration and washing steps the product recovery circuit requires a large number of filtration and washing stages (Wheelock, 1980; Stambaugh *et al.*, 1979).

The spent leachant containing the extracted sulphur as sodium sulphide (Na_2S) can be regenerated for recycling in several ways. One procedure involves treatment of the spent leachant with carbon dioxide to remove the sulphur as H_2S. This is subsequently converted to elemental sulphur by the Claus or Stretford process. The resulting liquor containing sodium carbonates is treated with lime to convert the carbonates back to sodium hydroxide which is recycled to the desulphurisation section. The calcium carbonate is decomposed thermally to lime and carbon dioxide for recycling to the leachant regeneration section.

The coal leaching step has been demonstrated in both batch and continuous flow, bench scale reactors. The continuous flow reactor is capable of treating 10 kg/h of coal. Between 90 and 99% of pyritic sulphur and between 20 and 70% of organic sulphur have been removed from a wide variety of bituminous coals from the midwestern and eastern United States. The heating values recovered in the products were in the range of 90 to 95% of that of the feed coals, depending on the coal and processing conditions. The remainder of the heating value, recovered during leachant regeneration, may be used to generate process heat. The product filtration and washing steps and the leachant regeneration step have been tested in the laboratory.

In addition to the removal of sulphur the Battelle hydrothermal process also removes a number of trace metals as shown in Table 5.8 (Stambaugh *et al.*, 1975).

Hydrothermally treated coal may contain up to 0.5% sodium and between 2 and 3% calcium which are claimed to act as sulphur scavengers during subsequent combustion of the product. During the desulphurisation operation the coal structure is opened up to give the product a sponge-like morphology. The porous structure allows the alkali to penetrate the coal particles and subsequently to react with the functional groups in the coal such as the carboxylic acid groups. The alkali is also deposited physically within and on the coal particles. Small scale combustion tests have shown that the alkalis can capture between 57 and 100% of any sulphur remaining in the coal after the hydrothermal process (Stambaugh *et al.*, 1976).

Table 5.8 Trace element removal by the Battelle hydrothermal process (after Stambaugh *et al.*, 1975)

Element	*Concentration,* ppm		*Removal* %
	Raw coal	*Leached product*	
Lithium	15	3	80.0
Beryllium	10	3	70.0
Boron	75	4	94.7
Phosphorus	400	80	80.0
Chlorine	20	2	90.0
Potassium	5000	200	96.0
Vanadium	40	2	95.0
Arsenic	25	2	92.0
Molybdenum	20	5	75.0
Barium	25	4	84.0
Lead	20	5	75.0
Thorium	3	0.5	83.3

However, excess sodium and calcium retention in the coal can lead to slagging and fouling by ash when the coal is burned in utility boilers. A value of 0.5%, or less, residual sodium has not been achieved for some coals (Contos *et al.*, 1978).

Severe corrosion problems may occur in the desulphurisation reactor because alkalis at high temperatures (>250°C) in the presence of water are notorious for initiating stress corrosion failures of materials. Battelle has found only one material, Inconel 671 alloy, which might have the capability of withstanding the desulphurisation reaction conditions without undergoing rapid failure. However, this material, which is extremely expensive, has not been evaluated in any long term tests (Contos *et al.*, 1978).

Work on the hydrothermal process has been suspended. The developers see a need to demonstrate both the scavenging action of the residual alkali in larger scale combustion tests and the integrated process at bench scale before proceeding to the pilot plant stage of development (Wheelock, 1980).

TRW Gravimelt process

Development of this process, which is based on the extraction of sulphur from coal by hot molten caustic, has only recently commenced. The actual sequence of processing has not yet been defined. However, the results of laboratory scale tests, which are presented in Table 5.9, will be discussed (Meyers *et al.*, 1980; Meyers and Hart, 1980).

The Gravimelt process involves the treatment of coal with a mixture of 50% potassium hydroxide and 50% sodium hydroxide at 370°C for 1800 s followed by washing with water. The chemistry is postulated to proceed according to the following equations:

$$8FeS_2 + 30MOH \rightarrow 4Fe_2O_3 + 14M_2S + M_2S_2O_3 + 15H_2O \quad (5.22)$$

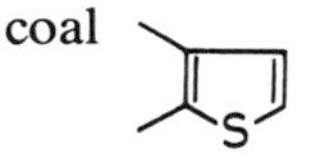

$+ 2MOH \rightarrow M_2S +$

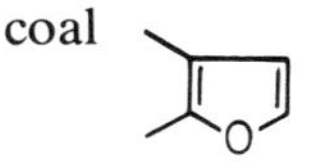

$+ H_2O$ (5.23)

As these reactions are competitive and the reaction between pyrite and caustic proceeds much more readily than the reaction between thiophene and caustic it is

Table 5.9 Organic and pyritic sulphur removal by the TRW Gravimelt process (after Meyers and Hart, 1980)

Example	*Coal*	*KOH/NaOH*[a] treatment	*Sulphur content* kg/GJ				*Removal*[e] %		
			Total	*Pyritic*	*Sulphate*	*Organic*	*Organic*	*Pyritic*	*Total*
1	Run of mine	None	1.42	0.64	0.09	0.69	—	—	—
2	Run of mine	Yes	0.47	0.09	0.00	0.43	23	89	67
3	Float[b]	None	0.69	0.04	0.09	0.56	—	—	—
4	Float[b]	Yes	0.17	0.04	0.00	0.13	77	0	75
5	Float[b]	Yes	0.26	0.04	0.00	0.21	62	0	63
6	Float[b]	Yes	0.17	0.04	0.00	0.13	77	0	75
7	Sink[c]	None	2.11	0.99	0.34	0.77	—	—	—
8	Sink[c]	Yes	1.12	0.56	0.00	0.56	28	44	47
9	Sink[c] $Fe_2(SO_4)_3$ leach[d]	None	0.86	0.13	0.09	0.64	(17)	87	59
10	Sink[c] $Fe_2(SO_4)_3$ leach[d]	Yes	0.34	0.04	0.00	0.30	53	67	60

[a]Reaction of coal with 4:1 w/w of a 50:50 mixture of KOH and NaOH at 370°C for 1800 s, followed by washing with water.
[b]The coal that floats in a liquid of 1.33 specific gravity.
[c]The coal that sinks in a liquid of 1.33 specific gravity.
[d]Reaction of coal with aqueous $Fe_2(SO_4)_3$ at 100°C followed by washing with water.
[e]Calculated on the basis of differential sulphur content/unit heat content between treated coal and substrate coal. Substrate coal for examples 4, 5 and 6, is 3; substrate coal for example 2, is 1; substrate coal for example 9, is 8; substrate coal for example 10, is 9.

necessary to remove part of the pyritic sulphur from the coal to maximise the removal of organic sulphur.

Treatment of an as-received, uncleaned coal with caustic resulted in the removal of 89% of the pyritic sulphur and 23% of the organic sulphur. However, treatment of the 'float' product from a gravity separation step, which contained little pyritic sulphur, resulted in the removal of 72% of the organic sulphur. As expected, treatment of the 'sink' product, rich in pyritic sulphur, resulted in relatively small reductions in both the pyritic and organic sulphur contents. However, pretreating the sink coal with a ferric sulphate leach to extract pyritic sulphur (essentially the TRW Meyers process) improved the removal of pyritic and organic sulphur during subsequent treatment with caustic (see Table 5.9).

More recent work using pure molten sodium hydroxide as well as mixtures of sodium hydroxide and potassium hydroxide has confirmed total sulphur removals of between 80 and 90% for four coals. Concentrations of residual Na and K are claimed to be equal to or less than the starting alkali metal content of the coal. Additionally, it is claimed that the TRW Gravimelt process breaks down the mineral matter in the coal to forms which are insoluble in water but soluble in a proprietary liquid. Removal of between 95 and 96% of the mineral matter has been demonstrated on a laboratory scale (Meyers, 1981a).

This process is not developed sufficiently for an evaluation to be made although it should be pointed out that the handling of very hot caustic solution on a large scale could present severe design and operational problems. Further laboratory work to be undertaken by TRW Systems and Energy will be sponsored by the US DOE.

General Electric microwave treatment process

Development of this process by General Electric commenced in 1977 with the financial support of the US EPA (Zavitsanos *et al.*, 1978). More recent work has also been supported by the US DOE. The process is unique in that it employs microwave energy to heat the coal and remove both pyritic and organic sulphur.

There are two versions of the process. In one, dry powdered coal is subjected to microwave energy for between 20 and 60 s which selectively heats the pyrite converting it into pyrrhotite, a related mineral. Pyrrhotite has a high magnetic susceptibility and is therefore readily separated from the coal by a magnetic separator.

The second version of the process involves applying a sodium hydroxide solution to pulverised coal (590 μm × 149 μm) followed by dewatering and irradiation with microwave energy for 30 to 60 s under a nitrogen atmosphere. Both pyritic and organic forms of sulphur react with the sodium hydroxide to form soluble sodium sulphide (Na_2S) and polysulphides (Na_2S_x) during irradiation. The coal is then washed to remove the sulphur compounds and caustic. Treatment of the extract from this step might follow the procedure adopted in the Battelle hydrothermal process.

The uniqueness of microwave treatment lies in the fact that the sodium hydroxide and the sulphur species in the coal can be heated more rapidly and efficiently than the coal itself. Thus the reaction between sodium hydroxide and sulphur occurs in a short time and with such low bulk temperatures that an insignificant amount of coal degradation occurs. As a result, the heating value of the coal is either unchanged or is slightly enhanced (Contos *et al.*, 1978).

Demonstration of the process has, so far, been restricted to laboratory scale work involving pulverised coal samples of between 10 and 500 g. With microwave energy treatment alone about half of the pyritic sulphur was removed from bituminous coal. Pretreating the coal with sodium hydroxide enhanced the removal of pyritic sulphur and converted some of the organic sulphur to soluble sulphides when the coal sample was subsequently irradiated. Washing the product coal and then repeating the pretreatment and irradiation steps resulted in the removal of up to 90% of the pyritic sulphur and between 50 and 70% of the organic sulphur from several bituminous coals.

As with other desulphurisation processes using caustic solutions the extent of the retention of sodium and calcium in the coal is an important consideration. If sodium is retained in the coal to a significant extent (>0.5%) the ash resulting from subsequent combustion of the coal may cause slagging and fouling problems within the boiler. However, the major potential disadvantage of this process is the cost of the microwave energy which is, at present, unknown. Larger bench scale work and a 1 t/d pilot plant, which is planned by GE, should help to determine the economics of the process.

Aqua/refined coal process

This process, which is an offshoot of the Battelle hydrothermal process, utilises a hot caustic solution to dissolve and extract the organic matter from susceptible low ranked coals. The coal extract is filtered to remove mineral matter and the filtrate treated with acid to precipitate the organic material. The product is low in sulphur and organic material and has a submicron particle size which makes it well suited for use in coal/oil mixtures. Work on this process is continuing at Battelle, Columbus Laboratories (Wheelock, 1980).

Other processes

Magnex process

The Magnex process for beneficiating coal was developed in 1975 by Hazen Research Inc. (Kindig and Turner, 1976; Kindig and Goens, 1979). In 1977, however, the process was purchased by Roldiva, Inc. and the licensing rights transferred to Nedlog Technology Group. The basis of the process is the chemical enhancement of the magnetic susceptibility of pyrite and mineral matter which may be removed subsequently by magnetic separation. Organic sulphur is not removed.

As-received coal is crushed to less than 1.41 mm and heated to 170°C. During the heating step the coal may also be contacted with steam, at the rate of 100 kg/t coal. This removes volatile compounds and elemental sulphur which interfere with the magnetic enhancement reaction. The coal is then contacted with iron pentacarbonyl vapour which decomposes in the presence of the coal.

$$Fe(CO)_5 \rightarrow Fe + 5CO \qquad (5.24)$$

The decomposition is neither instantaneous nor as simple as represented by the above equation. Intermediates are formed by stepwise loss of carbon monoxide (Porter and Goens, 1979).

The reaction of iron pentacarbonyl with mineral matter may be represented as:

$$Fe(CO)_5 + \text{Mineral matter} \rightarrow \text{Fe mineral matter} + 5CO \qquad (5.25)$$

The atomic iron released by the decomposition of the iron pentacarbonyl combines with the mineral matter.

The reaction of iron pentacarbonyl with pyrite may be represented as:

$$FeS_2 + Fe(CO)_5 \rightarrow 2Fe_{1-x}S + 5CO \qquad (5.26)$$

The surface of the pyrite is converted into a pyrrhotite-like phase which is highly paramagnetic.

The process has been tested in both laboratory and pilot plant (91 kg/h) apparatus. About 60 different coals treated with iron carbonyl at a rate equivalent to 16 kg/t coal were evaluated in the laboratory tests. For a selected group of seven coals the removal of pyritic sulphur ranged from 57 to 92% and of mineral matter from 7 to 71%. The heating value recovered in the product coal varied between 86 and 96% of that in the feed coal. Treatment of a bituminous coal with 10 kg of iron pentacarbonyl/t coal in the pilot plant resulted in the removal of 85% of the pyritic sulphur and an 86% recovery of the heating value. Although the pilot plant did not include the regeneration of iron pentacarbonyl from the iron and CO produced in the process, this was later demonstrated separately.

The design, construction and operation of a 55 t/d demonstration plant is planned. In addition, preliminary capital and operating costs for a 7300 t/d commercial plant have been estimated.

IGT flash desulphurisation process

Work on this process was carried out by the Institute of Gas Technology (IGT), Chicago, USA between 1973 and 1977 under the sponsorship of the US EPA

(Fleming *et al.*, 1977). The process is significantly different from the other processes discussed in this chapter because it combines thermal and chemical treatments of coal. The coal is partially gasified to produce a fuel gas and a char which is not a direct substitute for coal. The lack of volatile matter would limit the use of the product in pulverised coal fired boilers. However, the product might be suitable for coal/oil mixtures or for stoker fired boilers. The uncertain application of the product is one reason the development of the process has been suspended (Wheelock, 1980).

The initial step of the process involves contacting pulverised coal (<1.41 mm) with air in a fluidised bed reactor at 400°C and near atmospheric pressure. This step, which is common to the IGT coal gasification process, destroys the caking tendency of the coal and increases the removal of organic sulphur in the subsequent hydrodesulphurisation step. Between approximately 8 and 12% of the coal is consumed during pretreatment, generating steam and a low Btu gas that can be burned on site to provide process steam or to generate power. Between 25 and 30% of the sulphur is removed, primarily as SO_2 in the low Btu gas.

The second step in the process involves contacting the pretreated coal in a second fluidised bed reactor with hydrogen at 800°C and atmospheric pressure. Sulphur is extracted as H_2S. A sulphur acceptor such as calcium oxide or iron oxide may be added to the system to limit the quantity of hydrogen consumed. The off-gases from both fluidised beds require extensive treatment to remove sulphur compounds, recover heat, and reclaim various gaseous and liquid hydrocarbons.

In a preliminary design study for a commercial plant treating a bituminous coal, the projected total energy recovery is 84% distributed as:

61.3% in char;
14.0% in fuel gas;
7.0% in light oil;
1.7% in exported power.

The process has been demonstrated in laboratory, bench and pilot plant scale apparatus. The pilot plant incorporated a 250 mm diameter fluidised bed reactor operated continuously with feed rates of between 10 and 45 kg coal/h.

Chemical comminution

The Syracuse Research Corporation has developed a process for the chemical fracturing or comminuting of coal, which is an alternative to mechanical crushing and grinding (Howard and Datta, 1977; Robinder, 1979). This is not strictly a chemical cleaning process but a precursor to the removal of pyritic sulphur and mineral matter by physical coal cleaning methods. It is claimed that the chemical comminution disrupts the natural bonding forces acting across the internal boundaries of the coal structure where the majority of the mineral matter and pyritic sulphur deposits are located. Hence, lower concentrations of sulphur in the physically cleaned product may be achieved compared with mechanical crushing or grinding. The production of fine coal is also reduced (Cullen and Quackenbush, 1978).

The Syracuse Research Corporation commenced development of this process in 1971. Part of the work was financed by the then ERDA (now US DOE) and a final report was published in 1976 (Datta *et al.*). In 1977 marketing of the process was undertaken by Catalytic, Inc., PA, USA who, in conjunction with EPRI and

TVA, prepared an economic evaluation of the process (Cullen and Quackenbush, 1978; Quackenbush *et al.*, 1979).

Coal to be processed is sized to between 380 mm and 0.149 mm. Coal of less than 0.149 mm in size passes directly to the physical coal cleaning plant. The remainder of the coal passes to a reactor where it is contacted with ammonia vapour at a pressure of 0.91 MPa for 120 min. During the reaction period the temperature in the reactor rises to between 70 and 85°C due to the heat of solution of ammonia absorbed by moisture in the coal. The ammonia does not react with the coal. The coal which is comminuted to about 10 mm top size is washed and passed to a conventional physical coal cleaning plant. The ammonia is recovered by distillation and re-used.

All work to date has been performed in laboratory or bench scale equipment. The largest tests have been with 23 kg batches of coal. The potential for removal of pyritic sulphur from chemically comminuted coal has been assessed only by coal washing data obtained on a laboratory scale. Contos *et al.* (1978) estimate that the Syracuse chemical comminution process, followed by conventional physical coal cleaning, will remove between 50 and 70% of the pyritic sulphur from coals with product recoveries of between 60 and 90%. Different coals were found to require different residence times within the reactor for adequate chemical comminution to be achieved. Hence, the economic advantage of this process over that of mechanical crushing or grinding will depend on the coal which is being treated.

Chapter 6

Evaluation of processes

Many of the chemical coal cleaning processes discussed in this review have been under development for as long as a decade. With almost total removal of pyritic sulphur and partial removal of organic sulphur, the majority of the processes were potentially capable of desulphurising a large number of coals to within the 1971 US NSPS for SO_2 emissions of 0.52 kg/GJ (1.2 lb/10^6 Btu). However development work proceeded slowly. Only now are the desulphurisation steps of some of these processes being tested at large pilot plant scale. In the meantime however, the NSPS for SO_2 emissions has been revised to include a requirement that 90% of the sulphur in a coal is prevented from being emitted to the atmosphere irrespective of the initial concentration of sulphur in the coal. None of the coal cleaning processes described in this review are capable of meeting this requirement. This restricts the product coal to industrial scale boilers (not covered by the US NSPS), to coal/oil mixtures or other specialised applications.

The majority of the processes described in Chapter 5 produce cleaned coal with a very small particle size, often less than 74 μm. Serious technical and economic constraints exist for the long distance transportation of this fine coal (Bechtel National Inc., 1980). Transport in one of the following forms may be considered:

1. as thermally dried particulates;
2. as a slurry of particulates;
3. as agglomerates.

The thermally dried particulate form may be transported by conventional methods although special handling and equipment will be required to resolve the potential problems associated with dust, fire, explosion and material handling. Whilst transport in slurry form is practicable with some modification of particle size distribution, very high capital costs are involved. It is unlikely chemical cleaning of coal will be practised at a scale sufficiently large to justify the construction of a slurry pipeline. Agglomeration of the coal particulates is also technically feasible but expensive additives are normally required. It remains to be proven whether the quality of chemically cleaned coal will be sufficient to justify the additional cost of transportation.

Several comparisons have been made of the economics of various chemical coal cleaning processes (Bechtel National Inc., 1980; Contos *et al.*, 1978; Oder *et al.*, 1977; Plants and George, 1979; Singh and Peterson, 1979). The usefulness of these is very limited. The comparisons are based on insufficient data and hence many assumptions. Most of the chemical coal cleaning processes are at different stages of development. Often only the desulphurisation step, which represents a small part of the overall cost, has been demonstrated. The only general conclusion which may be drawn from these evaluations is that the cost of chemical coal cleaning will be greater than that of physical cleaning but less than that of coal conversion processes (Friedman and Warzinski, 1977). At present, selection of a chemical cleaning process is likely to depend more on the technical merits of the process than on its potential economics.

One of the more important recommendations to emerge from the Bechtel study is:

> 'For the conceptual design and material balance, both pyritic and organic sulphur were assumed to be in the form of FeS_2. The actual forms of sulphur may be different, and in general, forms of sulphur in coal are not well understood. Future research should better define types of sulphur in the coal, with the objective of gaining a better understanding of the sulphur removal reactions of the chemical cleaning process'.

Only recently has such work been undertaken. Because of its significance to the potential of the chemical cleaning of coal this research will be discussed in detail.

Chapter 7

Current research on chemistry of desulphurisation reactions

Oxidation reactions

Squires *et al.* (1980) have determined the potential of oxidation techniques for the removal of organic sulphur from coal by subjecting model compounds to the oxydesulphurisation conditions used in the Ames process (Chang *et al.*, 1980). The model compounds used, which are shown in Figure 7.1, represent the major types of sulphur compounds commonly believed to be present in coal. They are all in the divalent or sulphide oxidation state which is consistent with the generally reduced state of coal.

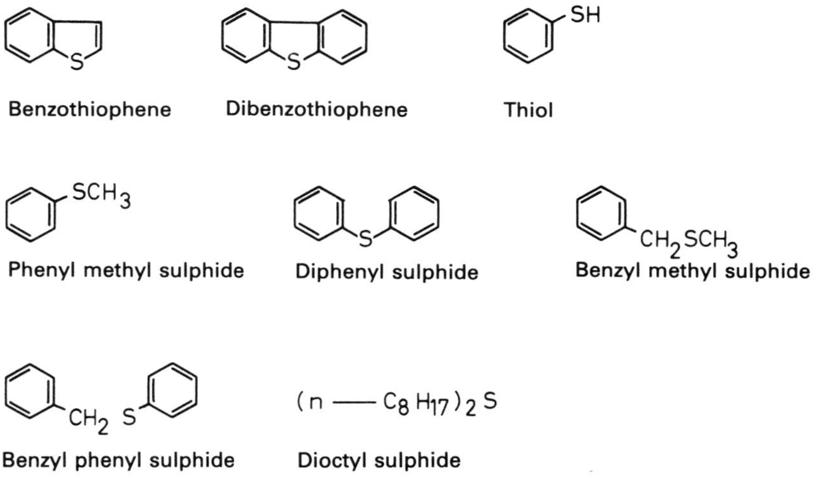

Figure 7.1 Compounds used to model the reactions of organic sulphur in coal (after Squires *et al.*, 1980)

The results of the oxydesulphurisation of the model compounds are presented in Table 7.1. The amount of starting material recovered is an indication of the degree of reaction that the compound underwent. In order to assess the possibility that coal may act catalytically in the oxidation, reactions of four compounds were carried out in the presence of coal. These results are presented in Table 7.2. To take into account the decreased recovery expected due to adsorption of the model compounds onto coal, experiments using both oxygen and nitrogen were performed. The right hand column of Table 7.2 shows the yields of recovered starting material corrected for this adsorption.

The results presented in Tables 7.1 and 7.2 show that compounds of the thiophene type and all sulphides except benzylic sulphides are inert to oxidation under the Ames oxydesulphurisation conditions. These compounds represent the

Table 7.1 Effect of the Ames process on model organic sulphur compounds (after Squires *et al.*, 1980)

Model compound	*Yields*	
	Recovered starting material %	*Products*
thiol	0	 – SO_3Na (95%)
diphenyl disulphide	25	 – SO_3Na (70%)
dibenzothiophene	96	No reaction
benzothiophene	87	No reaction
diphenyl sulphide	98	No reaction
dioctyl sulphide	90	No reaction
benzyl methyl sulphide	36	 – CHO (23%),  – COO Na (34%), CH_3SO_3Na (55%)
benzyl phenyl sulphide	29	 – CHO (31%),  – COO Na (18%),  – SO_3Na (51%)

Reaction conditions: 150°C, 1.38 MPa, 0.2 M Na_2CO_3, 3600 s

Table 7.2 Effect of the Ames process on model organic sulphur compounds in the presence of coal (after Squires *et al.*, 1980)

Model compound	*Yield of recovered starting material, %*			
	Model compound alone	*Lovilia coal added* N_2	*Lovilia coal added* O_2	*Corrected for adsorption onto coal*
Dibenzothiophene	96	83	76	93
Diphenyl sulphide	98	76	74	98
Dioctyl sulphide	90	—	82	—
Benzyl methyl sulphide	36	75	13	38

principal functional groups believed to be in coal. Thiophenol is readily oxidised to the diphenyl disulphide which under more severe conditions is oxidised to benzene sulphonic acid.

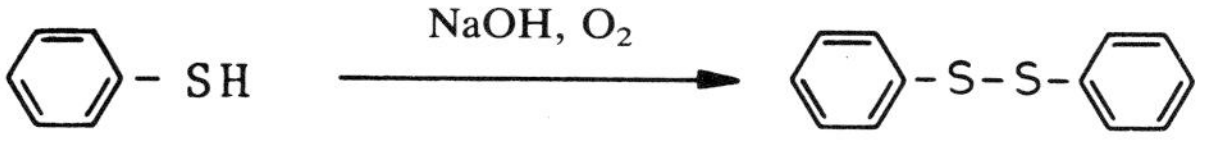

(7.1)

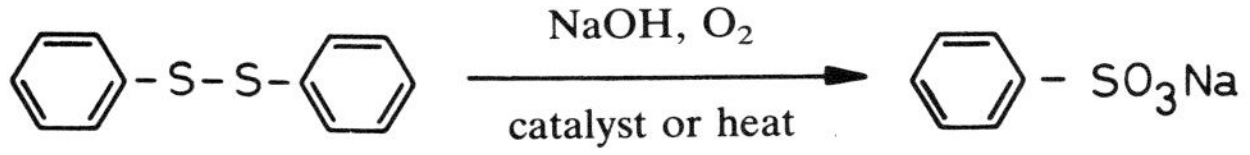

(7.2)

The oxidation of benzylic sulphides (benzyl methyl sulphide and benzyl phenyl sulphide) is believed to occur by the oxidation of the C atom in the CH_2 group and not by oxidation of the S atom. To eliminate the possibility that the benzylic sulphides are, in fact, oxidised to either the sulphide or sulphone, which may then be responsible for the observed products, both benzyl phenyl sulphoxide and benzyl phenyl sulphone were subjected to the reaction conditions. The sulphoxide was converted partially to the sulphone which did not undergo further reaction. The fact that sulphone is not a product of the oxidation of benzylic sulphides (see Table 7.1) is evidence that the oxidation of the S atom does not occur. As further evidence that the oxidation takes place at the C atom, fluorene, which also contains a CH_2 group, was found to be oxidised to fluorenone under similar reaction conditions.

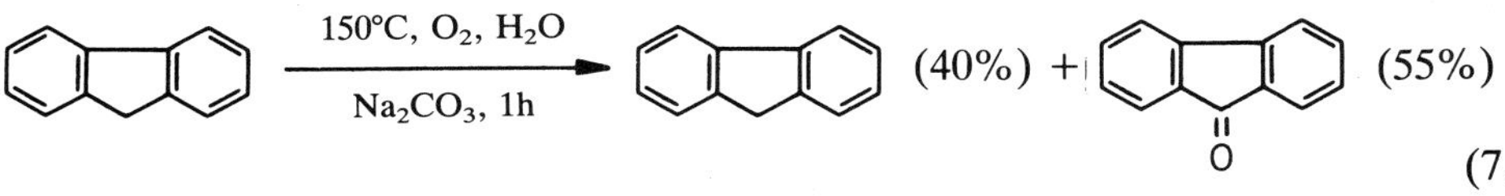

(7.3)

Dibenzothiophene, 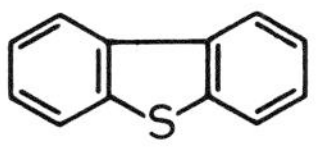 did not react. The 55% yield of fluorenone is in good agreement with the yields obtained from the oxidation of the benzylic sulphides.

More recent work has involved subjecting sulphur-containing synthetic polymers and the pyridine extract of Iowa Lovilia coal to the Ames process conditions (Squires and Venier, 1981; Squires *et al.*, 1981). The results obtained with three modified polystyrenes, shown in Table 7.3, add support to the model compound studies (Squires *et al.*, 1981). Phenylthiomethyl and tolylthiomethyl polystyrenes are analogues of the model compound, benzyl phenyl sulphide. As expected, these polymers were reactive, but less so than the corresponding model compound. Polystyrene cross-linked by dibenzothiophene was unreactive, confirming the inertness of the dibenzothiophene moiety to oxydesulphurisation.

Perhaps the most complete evidence of the unreactive nature of some organic sulphur in coal to oxydesulphurisation has been obtained with a pyridine coal extract (Squires and Venier, 1981; Squires *et al.*, 1981). All the sulphur in an extract is organically bound and hence the total sulphur analysis, which is quite accurate, is also the organic sulphur determination. The results in Table 7.4 show that oxydesulphurisation apparently does not reduce the organic sulphur content

Table 7.3 Effect of the Ames process on sulphur-containing polystyrene polymers (after Squires *et al.*, 1981)

Polymer[a]	*Sulphur removal* %
P–CH_2–S– 	10
P–CH_2–S– 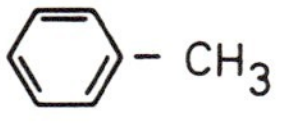	20
P–CH_2–DBT–CH_2–P	0

Reaction conditions: 150°C; 1.38 MPa; 0.2 aqueous Na_2CO_3, 3600 s
[a]P – polystyrene backbone; substituents are attached through the para position.
DBT – dibenzothiophene nucleus.

Table 7.4 Effect of the Ames process on pyridine extract of Lovilia coal[a] (after Squires *et al.*, 1981)

	Extract	*Extract processed with N_2*[b]	*Extract processed with O_2*[c]
Weight recovered:	—	93%	85%
Elemental analysis[d]:			
C	76.6	77.9	68.5
H	6.1	5.8	4.9
S	0.8	0.8	0.9
O (by difference)	14.4	12.8	22.3
S/C mole ratio	0.028	0.026	0.034
H/C mole ratio	0.95	0.90	0.85
O/C mole ratio	0.25	0.22	0.43

[a]150°C, 0.2 M aqueous Na_2CO_3, 3600 s
[b]1.38 MPa N_2 pressure
[c]1.38 MPa O_2 pressure
[d]Samples contain a small amount of ash or NaCl. Element percents are expressed ash or NaCl free

of the extract. However, some of the carbon in the extract is oxidised as evidenced by the increases in S/C ratio, the oxygen content and O/C ratio.

Warzinski *et al.* (1980) have studied the effect of the PETC oxydesulphurisation process on the following model sulphur compounds:

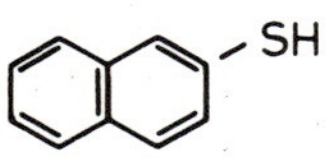

compound 1
2-naphthalenethiol

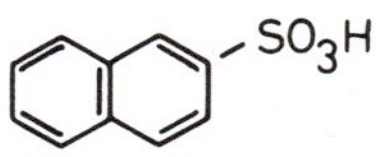

compound 2
2-naphthalene-sulphonic acid

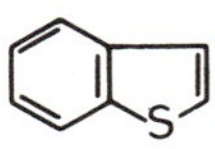

compound 3
benzothiophene

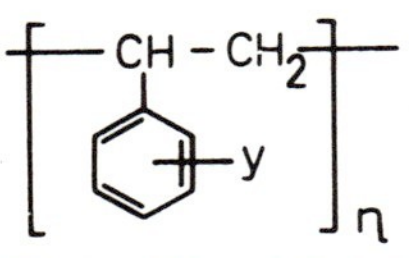

Y = –H, –SO_3H

compound 4
sulphonated polystyrene copolymer

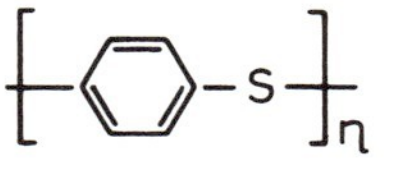

compound 4
polyphenyl sulphide

The reactions of the thiol and the sulphonic acid are similar to those reported by Squires *et al.* (1980). Compound 4, a sulphonated polystyrene copolymer also reacted in a manner similar to that reported by Squires *et al.* (1980) for organic compounds containing benzylic sulphides. Compound 5, polyphenyl sulphide, which contains no benzylic carbon was treated at 200°C with and without the presence of coal. In all cases the polymer was recovered (95%) unchanged.

The major disagreement between the work of Warzinski *et al.* (1980) and Squires *et al.* (1980) is in the reaction of benzothiophene. Squires *et al.* (1980) report an 87% recovery of the compound after subjection to the Ames oxydesulphurisation conditions. Warzinski *et al.* (1980) report 97% conversion of benzothiophene into 2-sulphobenzoic acid, benzoic acid, benzothiophene-1, 1-dioxide, and phenol, by the following proposed reaction:

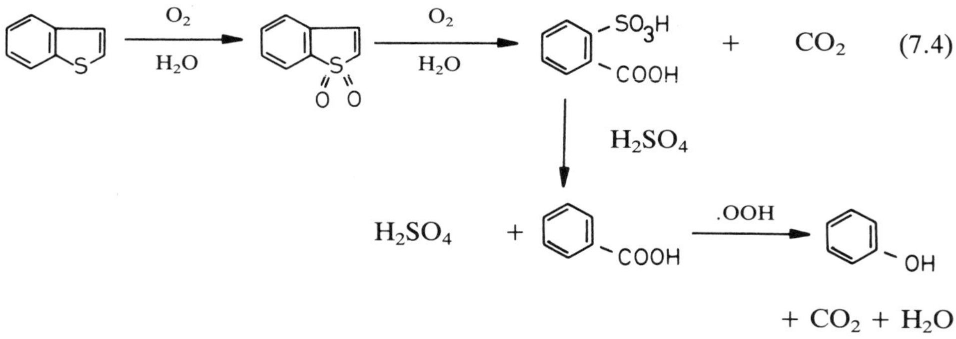

(7.4)

It is unlikely that the differences in the process conditions (200°C, acidic solution, compared with 150°C and basic solution) would account for the disagreement. Oxidation of the S atom, as proposed by Warzinski *et al.* (1980), is unlikely in view of the work at Ames, reported above.

Attar (1980) has summarised the relative reactivities of sulphur containing species in coal to oxidation and reduction (Table 7.5).

The stability of the thiophenic sulphur and sulphides under oxydesulphurisation conditions may be more simply explained in terms of bond strengths. A strong

Table 7.5 Reactivity of sulphur-containing compounds in coal (after Attar, 1980)

Compound	*Reactivity*	
	'mild' oxidizing media	*'mild' reducing media*
Smooth, perfect iron pyrites (including framboidal crystals)	non-reactive	non-reactive to 650°C
Ill-defined iron pyrites crystals	forms $FeSO_4$, $Fe(SO_4)_3$ $FeOH(SO_4)$, etc. and H_2SO_4	forms $FeS + H_2S$ at approx. 600°C
Thiols and thiophenols	forms RSO_3H	$RH + H_2S$
Alicyclic sulphides	forms RSOR, RSO_2R	forms $RH + H_2O$
Thiophenes and aryl sulphides	non-reactive	non-reactive

bond exists between a S atom and a C atom within an aromatic ring. Indeed, nowhere in the oxidation of the model compounds shown in Table 7.1 is such a bond broken. Thiophene converts to diphenyl disulphide which is oxidised by the breaking of an S–S bond. The benzylic sulphides are oxidised at the C atom in the CH_2 group which results in the breaking of the C–S bond. This is the only C–S bond to be broken in this model compound, and this is only because the C–S bond is in a benzylic position.

Squires (Ames Laboratory, 1980) is also studying the reaction of model organic sulphur compounds with chlorine gas under conditions simulating the first step of the JPL chlorinolysis process. Only the reaction with dibenzothiophene has been reported to date:

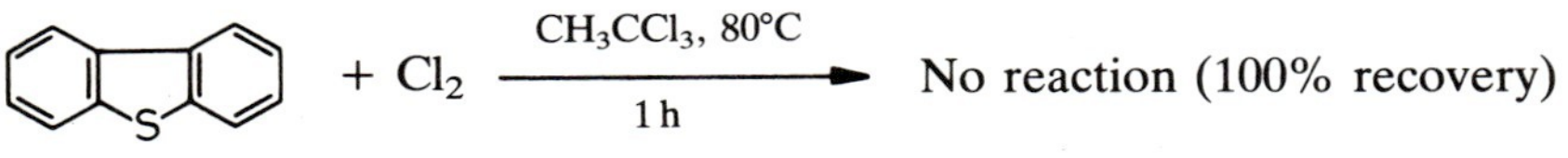

(7.5)

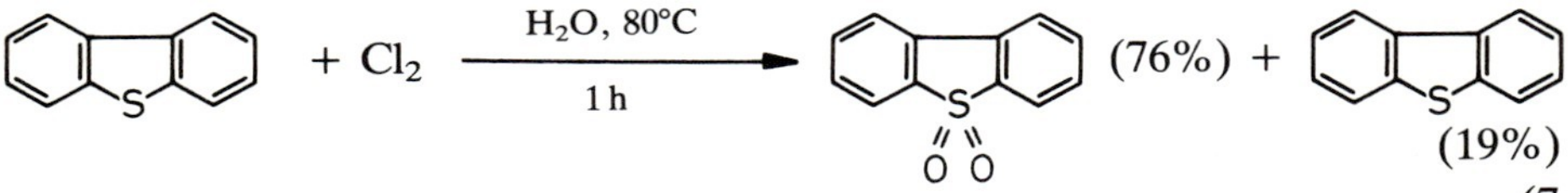

(7.6)

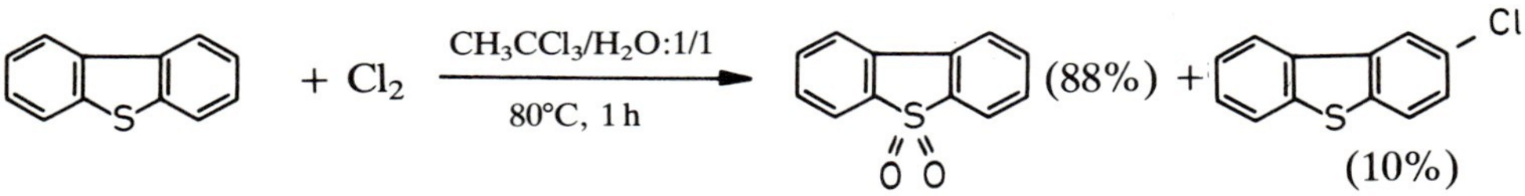

(7.7)

In 1,1,1-trichloroethane, dibenzothiophene is inert to chlorine gas (reaction 7.5). However in an aqueous medium in which dibenzothiophene is sparingly soluble, dibenzothiophene is oxidised electrophilically. Sulphone is the only product. With both solvents present the oxidation is much faster and is accompanied by some electrophilic chlorination. The authors propose the following reaction scheme:

$$Ar_2S + Cl_2 \rightleftharpoons Ar_2S^+ClCl^- \rightleftharpoons Ar_2SCl_2$$

$$Ar_2S^+ClCl^- \xrightarrow{H_2O} Ar_2SO + 2HCl$$

$$Ar_2SO + 2HCl \underset{}{\overset{Cl_2}{\rightleftharpoons}} Ar_2SO^+ClCl^-$$

$$Ar_2SO^+ClCl^- \xrightarrow{H_2O} Ar_2SO_2 + 2HCl$$

These results are apparently in contradiction to those obtained by Kalvinskas *et al.* (1980) during the bench scale testing of the JPL chlorinolysis process. They determined the removal of organic sulphur from four coals to be between 46 and 89% with 1,1,1-trichloroethane as solvent and between 0 and 24% with water as solvent. However, Kalvinskas *et al.* (1980) also report data from laboratory scale apparatus showing greater organic sulphur removal with water as a solvent for two of the coals than with 1,1,1-trichloroethane. Clearly, more development work is required on the JPL chlorinolysis process to reconcile the data from bench and laboratory scale tests. Further work on the reactions of chlorine gas with model compounds other than dibenzothiophene would also be beneficial in estimating the potential of the JPL chlorinolysis process for organic sulphur removal.

Squires (1978) determined a means of oxidising dibenzothiophene, phenyl sulphides and diphenyl sulphides under the following conditions:

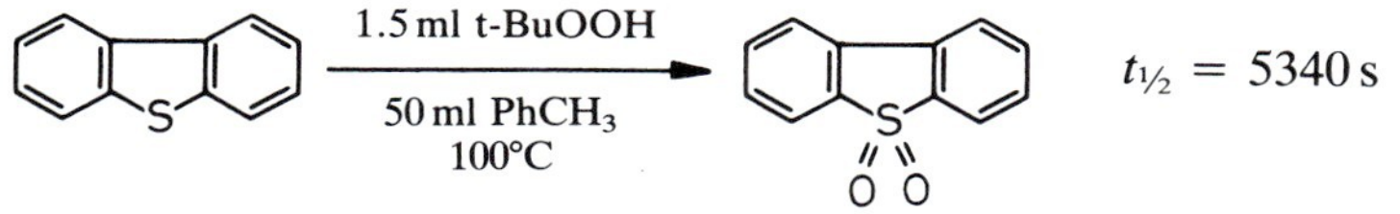

(7.8)

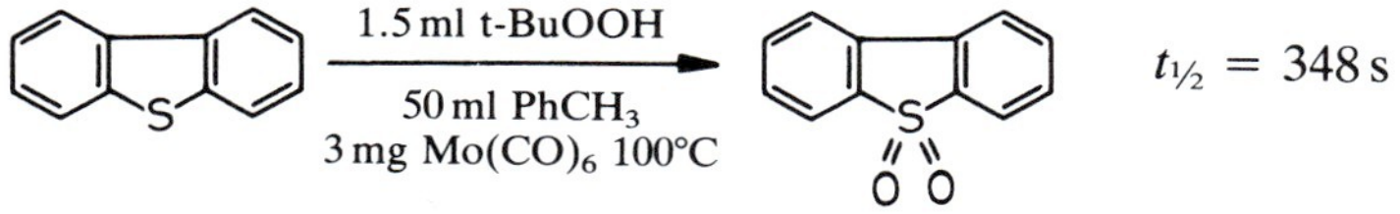

(7.9)

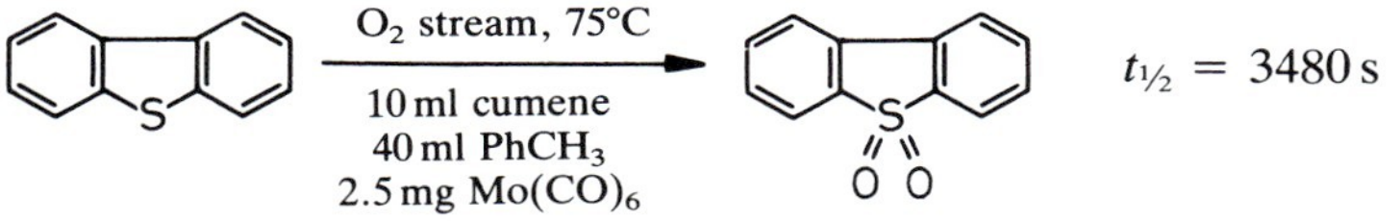

(7.10)

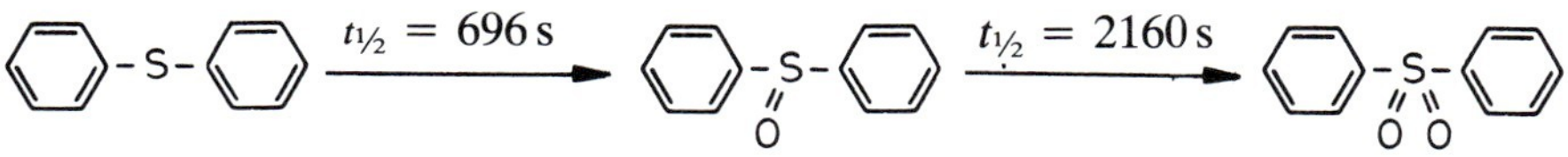

(7.11)

Reaction conditions: 3.1 ml t-BuOOH, 50 ml $PhCH_3$, 6 mg $Mo(CO)_6$, 75°C

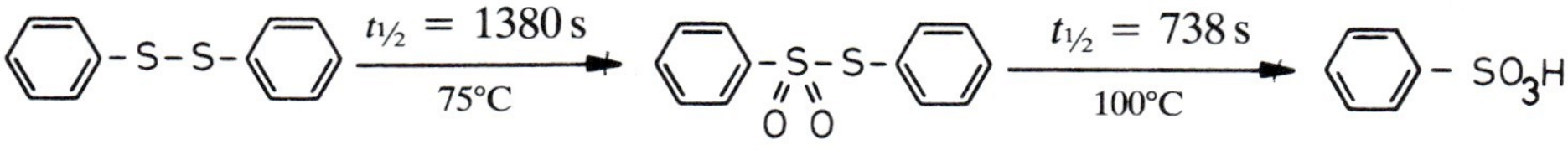

(7.12)

Reaction conditions: 13 ml t-BuOOH, 50 ml $PhCH_3$, 12 mg $Mo(CO)_6$

However, no viable method of desulphonisation was found.

Displacement reactions

As discussed in Chapter 5 displacement reactions are the main alternatives to the use of oxidation reactions for the removal of sulphur from coal. Unfortunately the chemistry of the reactions between caustic, in both the molten and aqueous solution forms, and the sulphur compounds in coal is not well understood. The only processes currently under development using caustic are the TRW Gravimelt process and the Aqua-refined coal process. Both of these are at very early stages of development. Meyers and Hart (1980) claim that thiophenic compounds in coal undergo a displacement reaction in the presence of molten caustic at temperatures of between 300 and 400°C where coal is somewhat plastic in nature:

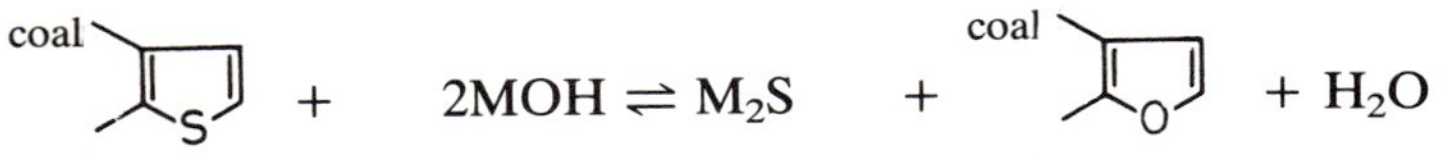

(7.13)

This reaction was found to proceed only when the competing reaction between pyrite and caustic was eliminated.

Attar (1980) suggests that the reaction of thiolic sulphur proceeds via the ion SR^-:

$$RSH + NaOH \rightarrow RS^-Na^+ + H_2O \rightarrow \text{Hydrocarbons} + NaSH \qquad (7.14)$$

and that the reaction of sulphides proceeds via the intermediate Na/RCHSR′ which subsequently decomposes to form unsaturated hydrocarbons and NaSH. In contrast to the claims of Meyers and Hart (1980), Attar (1980) postulates that aryl sulphides and thiophenes do not react with NaOH or KOH.

Several authors report erratic results of the removal of organic sulphur from coal using caustic. Also, the generally severe reaction conditions employed have resulted in large coal losses. Kasehagen (1937) studied the action of aqueous sodium hydroxide for between 20 and 30 h on a Pittsburgh seam coal over the temperature range of 250–400°C at various alkali concentrations. Decomposition of the coal into a coke-like residue and the production of phenols, acids, neutral oils, hydrocarbon gases and carbon dioxide occurred. Although Battelle report that only about 5% of the coal is dissolved by their hydrothermal treatment (Hammond, 1975) severe corrosion within the reactor and sodium retention in the coal have resulted in the development of the process being suspended. It remains to be proven whether these problems can be overcome in the TRW Gravimelt process. Whilst procedures for the handling of hot caustic are well documented the contacting of caustic with coal continuously and on a large scale may present severe operational problems.

Chapter 8

Conclusions

Despite the claims of some articles which appear in the technical press it is obvious from this review there is no simple, efficient process currently available for the chemical cleaning of coal. Many potential processes have been under active development in the USA for more than a decade. However, very few have progressed beyond pilot plant scale and, often, only the desulphurisation step of the process has been demonstrated: facts which are indicative of the process development problems involved.

Almost all chemical coal cleaning processes remove the majority of the pyritic sulphur from coal. The extent of organic sulphur removal is much less certain because of the inaccuracies involved in the determination of the organic sulphur content of coal. Standard analytical procedures may overestimate the organic sulphur content of coal by as much as 50%. The removal of organic sulphur by chemical cleaning of coal is generally reported to be between 0 and 40% although claims of up to 70% removal have been made. Consequently, it is possible that actual quantities of organic sulphur removal may be very low. Even applying the most optimistic claims for the removal of organic sulphur, the overall sulphur removal (pyritic and organic) does not enable coal to meet existing US New Source Performance Standards for SO_2 emissions when burned in large scale utility boilers. These regulations require a 90% reduction in the potential emission of SO_2. In addition, continued improvements in the reliability and performance of flue gas desulphurisation are likely to restrict further the potential applications of chemically cleaned coal in the industrial boiler market.

Little is known about the organic sulphur in coal. It is not surprising, therefore, that the chemical reactions involved in the desulphurisation steps of several chemical coal cleaning processes are not well understood. Only recently has this problem been addressed by subjecting model organic compounds, believed to be present in coal, to desulphurisation conditions. The results are not encouraging. Many of the compounds were found to be unreactive under such conditions. Increasing the severity of the desulphurisation conditions will almost certainly lead to unacceptable degradation of the product coal. Much more fundamental research is required to determine the occurrence and nature of sulphur forms in coal and their reaction under possible desulphurisation conditions.

The cost of chemically cleaning coal is not known with any certainty at present but it is likely to be considerably more than that of physically cleaning coal. At the present stage of development it is unlikely that the additional sulphur removal afforded by a chemical cleaning process would justify the additional cost.

References

Agarwal, JC, Giberti, R.A., Irminger, PF, Petrovic, LF., and Sareen, SS (1975) Chemical desulfurization of coal. *Mining Congress Journal*; **61** (3); 40–43

Ames Laboratory (1980) *Fossil Energy Annual Report, Oct 1, 1979–Sep 30, 1980*, Ames, IA, USA; Ames Laboratory, Iowa State University

Ames Laboratory (1979) *Fossil Energy Annual Report, Oct 1, 1978–Sep 30, 1979*, Ames, IA, USA; Ames Laboratory, Iowa State University; 190pp

Attar, A. (1979) Evaluate sulphur in coal. *Hydrocarbon Processing*; **58** (1); 175–179 (Jan 1979)

Attar, A. (1980) Study of precombustion methods of chemical coal cleaning. Paper presented at the *73rd Annual Meeting of the American Institute of Chemical Engineers*, Chicago, IL, USA, 16–20 Nov 1980

Attar, A. and Corcoran, WH (1977) Sulfur compounds in coal. *Industrial and Engineering Chemistry, Product Research and Development (US)*; **16** (2); 168–170

Attar, A. and Corcoran, WH (1978) Desulfurization of organic sulfur compounds by selective oxidation. 1. Regenerable and nonregenerable oxygen carriers. *Industrial Engineering Chemistry, Product Research and Development (US)*; **17** (2); 102–109

Attar, A and Dupuis, F (1978) On the distribution of organic sulfur functional groups in coal. *American Chemical Society, Division of Fuel Chemistry, Preprints*; **23** (2); 44–53

Attar, A. and Dupuis, F (1979) Data on the distribution of organic sulfur functional groups in coal. *American Chemical Society, Division of Fuel Chemistry, Preprints*; **24** (1); 166–177

Attig, RC and Duzy, AF (1969) Coal ash deposition studies and application to boiler design. In: *Proceedings of the American Power Conference*; Illinois Institute of Technology, Chicago, IL; 31; 290–300

Bechtel National Inc. (1980) *An analysis of chemical coal clearing processes*; San Francisco, CA, USA; Bechtel National Inc.; 215pp

Beckberger, LH., Burk, EH, Grossboll, MP. and Yoo, JS (1979) *Preliminary evaluation of chemical coal cleaning by promoted oxydesulfurization*; EPRI-EM-1044; Harvey, IL, USA; Atlantic Richfield Company; 128pp

Betancourt, T. and Hancock, HA (1978) Chemical desulfurization of Cape Breton coal by ferric sulphate leaching. In: *Proceedings of the coal and coke sessions. Twenty eighth Canadian Chemical Engineering Conference*; Halifax, Nova Scotia, 22–25 Oct 1978; Ottawa, Canada; Canadian Society for Chemical Engineering; p.157–167

Bettelheim, J., Halstead, WD, Lees, D.J. and Mortimer, D (1980) Combustion problems associated with high chlorine coals. Presented at VGB Conference 'Research in Power Plant Technology 1980', Essen, FRG, 21–22 May 1980. *Erdoel Kohle, Erdgas, Petrochem, Brennst.-Chem*; **33** (9); 436

Birlingmair, DH and Fisher, RW (1978) *Physical and physiochemical removal of sulfur from coal*; IS-ICP-65; Ames, IA, USA; Ames Laboratory; 16pp

Bluhm, DD, Fanslon, GE, Beck-Amontgomery, S and Nelson, SO (1980) Selective magnetic enhancement of pyrite in coal by dielectric heating at 27 and 2450 MHz. Presented at *Conference on the Chemistry and Physics of Coal Utilization*; Morgantown, WV, USA, 4 Jan 1980

Chang, LW, Goure, WF, Squires, TG. and Barton, TJ (1980) Evaluation of oxydesulfurization processes for coal. The effect of the Ames process on model organic sulfur compounds. *American Chemical Society, Division of Fuel Chemistry, Preprints*; **25** (2); 165–170

Chen, MC (1979) *Desulfurization of coal-derived pyrite using solutions containing dissolved oxygen*; Ames, IA, USA; Ames Laboratory, Iowa State University; 108pp

Chuang, K.-C., Chen, M.-C., Greer, RT, Markuszewski, R, Sun, Y. and Wheelock, TD (1980a) Pyrite desulfurization by wet oxidation in alkaline solutions. *Chemical Engineering Communications*; **7** (1–3); 79–94

Chuang, KC, Markuszewski, R. and Wheelock, TD (1980b) Desulfurization of coal by oxidation in alkaline solutions. Presented at *73rd Annual Meeting of the American Institute of Chemical Engineers*, Chicago, IL, USA, 16–20

Coal Age (1980) Magnetics attract interest. *Coal Age*; **72** (1); 72–73

Contos, GY, Frankel, IF. and McCandless, LC (1978) *Assessment of coal cleaning technology: an evaluation of chemical coal cleaning processes*; PB 289 493; Washington, DC, USA; US Environmental Protection Agency; 299pp

Cullen, PD and Quackenbush, VC (1978) Chemical comminution—an improved route to clean coal. In: *Proceedings of the coal and coke sessions. Twenty eighth Canadian Chemical Engineering Conference, Halifax, Nova Scotia, 22–25 Oct 1978*; Ottawa, Canada; Canadian Society for Chemical Engineering; pp 132-143

Datta, RS *et al.* (1976) *Feasibility study of pre-combustion coal cleaning using chemical comminution, final report*; Syracuse, NY, USA, Syracuse Research Corp.

Davidson, RM (1980) *Molecular structure of coal*; Report No ICTIS/TR08; London, UK; IEA Coal Research; 86pp

Ergun, S, Lynn, S, Petersen, EE, Vermeulen, T, Wrathall, JA, Clarry, L., Cremer, G., Mesher, J., Mixon, DA. and Smith, MC (1980) Coal desulfurisation. In: *Chemical process research and development program, FY 1979*; LBL-10319; Berkeley, CA, USA; Lawrence Berkeley Laboratory, University of California; 4.32–4.33

Fanslow, GE, Bluhm, DD and Nelson, SO (1980) *Dielectric heating in mixtures containing coal and pyrite*; Ames, IA, USA; Ames Laboratory, Iowa State University, 18pp

Fleming, DK, Smith, RD and Aquino, MRY (1977) Hydrodesulfurization of coals. In: *Coal Desulfurization, Chemical and Physical Methods*, (ed. TD Wheelock) Washington, DC, USA; American Chemical Society, *ACS Symposium Series*; (64); 267–279

Ford, CT and Boyer, JF (1979) Effects of coal cleaning on elemental distributions. In: *Proceedings of the Symposium on Coal Cleaning to Achieve Energy and Environmental Goals*; Hollywood, FL, USA, Sep 1978. Washington, DC, USA; US Environmental Protection Agency; 59–90

Friedman, S, Lacount, RB and Warzinski, RP (1977) Oxidative desulfurization of coal. In: *Coal Desulfurization, Chemical and Physical Methods*, (ed. TD Wheelock) Washington, DC, USA; American Chemical Society; *ACS Symposium Series*; (64); 164–172

Friedman, S. and Warzinski, RP (1977) Chemical cleaning of coal. *Engineering for Power*; **99** (3); 361–364

Gathen, R. von der (1979) Possibilities and limits of coal desulphurisation. *Glueckauf*; **115** (3); 112–118

George, TJ, Plants, KD and Morel, WC (1979) Chemical desulfurization: a misunderstood technology. *Coal Mining and Processing*; **16** (8); 60–63

Giammar, RD, Barnes, RH, Hopper, DR, Webb, PR and Weller, AE (1980) Evaluation of emissions and control technology for industrial stoker boilers. In: *Proceedings of the Joint Symposium on Stationary Combustion* NO_x *Control Vol III, Denver, CO, USA, 6–9 Oct 1980*; IERL-RTP-1083; Washington, DC, USA, US Environmental Protection Agency, pp. 1–38

Giberti, RA, Opalanko, RS and Sinek, JR (1979) The potential for chemical coal cleaning: reserves, technology, and economics. In: *Proceedings of the Symposium on Coal Cleaning to Achieve Energy and Environmental Goals, Hollywood, FL, USA, Sep 1978*; PB-299 384; Washington, DC, USA; US Environmental Protection Agency; pp. 1064–1095

Greer, RT (1976) Colloidal pyrite growth in coal. In: *Colloidal and Interface Science, Vol V*; New York, NY, USA; Academic Press; 422–423

Greer, RT (1977a) Coal microstructure and the significance of pyrite inclusions. *Scanning Electron Microscopy*; **1**; 79–94

Greer, RT (1977b) Coal microstructure and pyrite distribution. In: *Coal desulfurization, Chemical and Physical Methods*, (ed. TD Wheelock) Washington, DC, USA; American Chemical Society; *ACS Symposium Series*; (64); 3–15

Greer, RT (1978a) Evaluation of pyrite particle size, shape and distribution factors. *Energy Sources*; **4** (1); 23–51

Greer, RT (1978b) Pyrite distribution in coal. *Scanning Electron Microscopy*; **1**; 621–626

Greer, RT (1979) Organic and inorganic sulfur in coal. *Scanning Electron Microscopy*; **1**; 477–486

Greer, RT, Markuszewski, R. and Wheelock, TD (1980) Characterization of solid reaction products from wet oxidation of pyrite in coal using alkaline solutions. *Scanning Electron Microscopy*; **1**; 541–550

Guth, ED (1979) Oxidative coal desulfurization using nitrogen oxides: the KVB process. In: *Proceedings of the Symposium on Coal Cleaning to Achieve Energy and Environmental Goals, Hollywood, FL, USA, Sep 1978*; PB-299 384; Washington, DC, USA; US Environmental Protection Agency; 1141–1164

Hamersma, JW, Kraft, ML and Meyers, RA (1977) Applicability of the Meyers process for desulfurization of US coal (a survey of 35 coals). In: *Coal Desulfurization, Chemical and Physical*

Methods, (ed. TD Wheelock) Washington, DC, USA; American Chemical Society; *ACS Symposium Series*; (64); 143–152

Hammond, AL (1975) Cleaning up coal; a new entry in the energy sweepstakes. *Science*; **189** (4197); 128–130

Hancock, HA and Betancourt, T (1978) The effect of the addition of chemical oxidants on the desulphurization of Cape Breton coal by ferric sulphate leaching. In: *Proceedings of the coal and coke sessions, Twenty eighth Canadian Chemical Engineering Conference*, Halifax, Nova Scotia, 22–25 Oct 1978 Ottawa, Canada; Canadian Society for Chemical Engineering; 168–175

Hart, WD, McClanathan, RA, Orsini, RA, Meyers, RA. and Santy, MJ (1978a) Chemical coal desulfurization. In: *Coal Technology '78, International coal utilization convention*, Houston, TX, USA, 17–19 Oct 1978, Vol. 1; 157–170

Hart, WD, McClanathan, RA, Orsini, RA, Meyers, RA. and Santy, MJ (1978b) The Gravichem process for coal desulfurization. In: *Energy Technology Conference, ASMF Petroleum Division* New York, NY, USA; ASME; 67–73

Howard, P.H. and Datta, RS (1977) Chemical comminution: a process for liberating the mineral matter from coal. In: *Coal Desulfurization, Chemical and Physical Methods*, (ed. TD Wheelock) Washington, DC, USA; American Chemical Society; *ACS Symposium Series*; (64); 58–69

Hsu, GC, Kalvinskas, JJ, Ganguli, PS and Gavalas, GR (1977) Coal desulfurization by low-temperature chlorinolysis. In: *Coal Desulfurization, Chemical and Physical Methods*, (ed. TD Wheelock) Washington, DC, USA; American Chemical Society, *ACS Symposium Series*; (64); 206–217

Kalvinskas, J, Grohmann, K, Rohatgi, N, Ernest, J and Feller, D (1980) *Final Report. Coal desulfurization by low temperature chlorinolysis, Phase II*; Pasadena, CA, USA; Jet Propulsion Laboratory, California Institute of Technology; 150pp

Kalvinskas, JJ and Hsu, GC (1979) JPL coal desulfurization process by low temperature chlorinolysis. In: *Proceedings of the Symposium on Coal Cleaning to Achieve Energy and Environmental Goals*, Hollywood, FL, USA, 11–15 Sep 1978; PB-299 384; Washington, DC, USA; US Environmental Protection Agency; 1096–1140

Kasehagen, L (1937) Action of aqueous alkali on a bituminous coal. *Industrial Engineering Chemistry*; **29**; 600–604

Kindig, JK and Goens, DN (1979) The dry removal of pyrite and ash from coal by the Magnex process, coal properties and process variables. In: *Proceedings of the Symposium on Coal Cleaning to Achieve Energy and Environmental Goals, Hollywood, FL, USA, Sep 1978*; PB-299 384; Washington, DC, USA; US Environmental Protection Agency; 1165–1196

Kindig, JK and Turner, RL (1976) *Dry chemical process to magnetize pyrite and ash for removal from coal*; Preprint of paper presented at 1976 SME-AIME Fall Meeting and Exhibit Denver, CO, USA, 1–3 Sep 1976; Preprint No. 76-F-306; Salt Lake City, UT, USA; Society of Mining Engineers of AIME; 18pp

Land, GW (1979) *Problems of removing sulphur from coal*, Paper presented at the Eighty-Second National Western Mining Conference, Denver, USA, 31 Jan 1979 *Mining Year Book 1979*; 159–170

Levine, MD, Fullen, RE, McCarthy, WN and Foley, GJ (1976) The need for coal cleaning. *AIChE/APCA Energy and the Env., 4th Natl Conf.*, Cincinnati, OH, USA, 3–7 Oct 1976; Dayton, OH, USA; American Institute of Chemical Engineers; 273–281

Markuszewski, R, Chuang, KC and Wheelock, TD (1979) Coal desulfurization by leaching with alkaline solutions containing oxygen. In: *Symposium on Coal Cleaning to Achieve Energy and Environmental Goals, Hollywood, FL, USA, 11–15 Sep 1978*; PB-299 384; Washington, DC, USA; US Environmental Protection Agency; 22pp

Markuszewski, R, Fan, CW, Greer, RT and Wheelock, TD (1980) Evaluation of the removal of organic sulfur from coal. In: *Symposium on New Approaches in Coal Chemistry, American Chemical Society*, Pittsburgh, PA, USA, 12–14 Nov

Markuszewski, R, Wei, CK. and Wheelock, TD (1980) *Oxydesulfurization of coal treated with methyl iodide–implications for removal of organic sulfur*; CONF-800303–6; Ames, IA, USA; Ames Laboratory, Iowa State University; 9pp

Masciantonio, PX (1965) The effect of molten caustic on pyritic sulfur in bituminous coal. *Fuel*; **44**; 269–275

Mesher, JA and Petersen, EE (1979) *Coal desulfurization via chemical leaching with aqueous solutions of sodium sulfide and polysulfide*; LBL-9386; Berkeley, CA, USA; Lawrence Berkeley Laboratories, University of California; 88pp

Meyers, RA (1975) Desulfurize coal chemically. *Hydrocarbon Processing*; **54** (6); 93–96

Meyers, RA (1977) *Coal desulfurization*, New York, NY, USA; Marcel Dekker; 254pp

Meyers, RA (1979) System optimizes coal desulfurization. *Hydrocarbon Processing*; **58** (6); 123–126

Meyers, RA (1980) TRW Systems and Energy Corporation, CA, USA; Private communication

Meyers, RA (1981a) TRW Systems and Energy Corporation, CA, USA; Private communication

Meyers, RA (1981b) Coal cleaning. In: *Coal Handbook*. New York, USA; Marcel Dekker, pp. 209–302

Meyers, RA and Hart, WD (1980) Chemical removal of organic sulfur from coal. Presented at 179th national meeting of the American Chemical Society, Houston, TX, USA, 23–28 Mar 1980 CONF-800303. *American Chemical Society, Division of Petroleum Chemistry, Preprints*; **25** (2); 258–261

Meyers, RA, Hart, WD. and McClanathan, LC (1980) The Gravimelt process for chemical removal of organic and pyritic sulfur from coal. *Presented at AIChE Meeting, Symposium on Chemical Coal Cleaning*; Chicago, IL, USA, Nov 1980

Meyers, RA, Koutsoukos, EP, Santy, MJ and Orsini, R (1979c) *Bench-scale development of Meyers process for coal desulfurization. Final report*; PB-290 515; Redondo Beach, CA, USA; TRW Systems Group; 104pp

Meyers, RA, Land, JS and Flegal, CA (1971) *Chemical removal of nitrogen and organic sulfur from coal*; PB-204 863; Washington, DC, USA; US Environmental Protection Agency; 61pp

Meyers, RA, Santy, MJ, Hart, WD, McClanathan, LC and Orsini, RA (1979b) *Reactor test project for chemical removal of pyritic sulfur from coal*; Vol 1 Final Report; PB-295 211; Washington, DC, USA; US Environmental Protection Agency; 277pp

Meyers, RA, Van Nice, LJ and Santy, MJ (1979a) Coal desulfurization. *Combustion*; **51** (6); 18–24

Mixon, DA and Vermeulen, T (1979) *Oxydesulfurization of coal by acidic iron sulfate solutions*; LBL-9963; Berkeley, CA, USA; Lawrence Berkeley Laboratory, University of California; 13pp

McCandless, LC and Contos, GY (1979) Current status of chemical coal cleaning processes—an overview. In: *Proceedings of the Symposium on Coal Cleaning to Achieve Energy and Environmental Goals, Hollywood, FL, USA, Sep 1978*; PB-299 384; Washington, DC, USA; US Environmental Protection Agency; 934–959

Nelson, SO, Fanslow, GE and Bluhm, DD (1980) Frequency dependence of the dielectric properties of coal. Presented at the *15th Annual International Microwave Power Symposium, Iowa, IA, USA, 7–9 May 1980*; CONF-800501-; Springfields, VA, USA; National Technical Information Service

Oder, RR, Kulapaditharom, L, Lee, AK and Ekholm, EL (1977) Technical and cost comparisons for chemical coal cleaning processes. *Mining Congress Journal*; **63** (8); 42–49

Plants, KD and George, TJ (1979) *An economic evaluation of the TRW Gravichem process; one unit train per day of coal*; FE/EES-79/10; Morgantown, WV, USA; US DOE; 36pp

Porter, CR and Goens, DN (1979) 'Magnex' pilot plant evaluation—a dry chemical process for the removal of pyrite and ash from coal. *Mining Engineering*; **31** (2); 175-180

Prior, M. (1977) *The control of sulphur oxides emitted in coal combustion*; EAS/B1/77; London, UK; IEA Coal Research; 108pp

Quackenbush, VC, Maddocks, R R and Higginson, GW (1979) Chemical comminution: an improved route to clean coal. *Coal Mining and Processing*; **16** (5); 68–72

Rabinder, SD (1977) Chemical fragmentation of coal. In: *Proceedings of the Third Symposium on Coal Preparation, Louisville, KY, USA, 18–20 Oct 1977*; CONF-7710113; Washington, DC, USA, National Coal Association; 22–31

Reggel, L, Raymond, R, Wender, I and Blaustein, BD (1972) Preparation of ash-free, pyrite-free coal by mild chemical treatment. *American Chemical Society, Division of Fuel Chemistry, Preprints*; **17** (1); 44–48

Ruether, JA (1979a) Chemical coal cleaning. *Combustion*; **51** (6); 25

Ruether, JA (1979b) *Steady state thermal analysis of a reactor for removal of sulfur from coal by oxygen/water selective oxidation*; PETC/TR-79/6; Pittsburgh, PA, USA; Pittsburgh Energy Techology Center; 36pp

Santy, MJ and Van Nice, LJ (1979) Status of the reactor test project for chemical removal of pyritic sulfur from coal. In: *Proceedings of the Symposium on Coal Cleaning to Achieve Energy and Environmental Goals, Hollywood, FL, USA; Sep 1978*; PB-299 384; Washington, DC, USA; US Environmental Protection Agency; 960–990

Sareen, SS (1977) Sulfur removal from coals: ammonia/oxygen system. In: *Coal Desulfurization, Chemical and Physical Methods*, (ed. TD Wheelock) Washington, DC, USA; American Chemical Society; *ACS Symposium Series*; (64); 173–181

Sareen, SS, Gilberti, RA, Irminger, PF and Petrovic, LJ (1977) The use of oxygen/water for removal of sulfur from coals. Presented at 80th AIChE National Meeting, Boston, MA, USA, 7–10 Sep 1975; *AIChE Symposium Series*; **73** (165); 183–189

Singh, SPN and Peterson, GR (1979) *Survery and evaluation of current and potential coal beneficiation processes*; ORNL/TM-5953; Oak Ridge, TN, USA; Oak Ridge National Laboratory; 310pp

Slagle, D, Shah, YT and Joshi, JB (1980) Kinetics of oxydesulfurisation of Upper Freeport coal. *Industrial and Engineering Chemistry, Process Design and Development*; **19**; 294–300

Squires, TG (1978) *A systematic investigation of the organosulfur components in coal and coal derived materials. Annual Progress Report 19 Jun 1978–30 Sep 1978*; Ames, IA, USA; Ames Laboratory, Iowa State University; 12pp

Squires, TG and Venier, CG (1981) *A systematic investigation of the organosulfur components in coal. Quarterly technical progress report 1 Oct 1980–31 DEc 1980*; Ames, IA, USA; Ames Laboratory, Iowa State University; 55pp

Squires, TG, Venier, CG, Chang, LW and Schmidt, TE (1981) Chemical studies of the Ames oxydesulfurization process. Presented at the 181st National Meeting of the American Chemical Society, Atlanta, GA, USA, 29 Mar–3 Apr 1981; *American Chemical Society, Division of Fuel Chemistry, Preprints*; **26** (1); 50–59

Squires, TG, Venier, CG, Goure, WF, Chang, LW, Shei, JC, Chen, YY, Hussmann, GP and Barton, TJ (1980) The use of model organosulfur materials to investigate the chemical desulfurization of coal. Presented at the *Meeting of the American Institute of Chemical Engineers*, Chicago, IL, USA, 14pp

Stambaugh, EP (1977) Hydrothermal coal process. In: *Coal Desulfurization, Chemical and Physical Methods*, (ed. TD Wheelock) Washington, DC, USA; American Chemical Society; *ACS Symposium Series*; (64); 198–205

Stambaugh, EP, Conkle, HN, Miller, JF, Mezey, EJ and Kim, BC (1979) Status of hydrothermal processing for chemical desulfurization of coal. In: *Proceedings of the Symposium on Coal Cleaning to Achieve Energy and Environmental Goals, Hollywood, FL, USA, Sep 1978*; PB-299 384; Washington, DC, USA; US Environmental Protection Agency; 991–1015

Stambaugh, EP, Levy, A, Giammar, RD and Sekhar, KC (1976) Hydrothermal coal desulfurization with combustion results. Presented at *AIChE/APCA Energy and the Env. 4th Natl. Conf., Cincinnati, OH, USA, 3–7 Oct 1976*; Dayton OH, USA; American Institute of Chemical Engineers; 386–394

Stambaugh, EP, Miller, JF, Tam, SS, Chauhan, SP, Feldmann, HF, Carlton, HE, Foster, JF, Nack, H and Oxley, JH (1975) Hydrothermal process produces clean fuel. *Hydrocarbon Processing*; **54** (7); 115–116

Starzonski, JJ and Brothers, JA (1978) Desulphurization: research and development of Nova Scotia Research Foundation Corporation. In: *Proceedings of the coal and coke sessions. Twenty eighth Canadian Chemical Engineering Conference, Halifax, Nova Scotia, 22–25 Oct 1978*; Ottawa, Canada; Canadian Society for Chemical Engineering; 108–117

Tai, CY, Graves, GV and Wheelock, TD (1977a) Desulfurizing coal with alkaline solutions containing dissolved oxygen. In: *Coal Desulfurization, Chemical and Physical Methods*, (ed. TD Wheelock) Washington, DC, USA; American Chemical Society; *ACS Symposium Series*; (64); 182–197

Tai, CY, Graves, GV. and Wheelock, TD (1977b) Desulfurization of coal by oxidation in alkaline solutions. In: *Third Symposium on Coal Preparation, NCA/BCR Coal Conference and Expo IV, Louisville, KY, USA, 18–20 Oct 1977;* CONF 7710113; Washington, DC, USA; National Coal Association; 1–9

Van Nice, LJ, Santy, MJ, Koutsoukos, EP, Orsini, RA and Meyers, RA (1977) Coal desulfurization test plant status—July 1977. In: *Coal Desulfurization, Chemical and Physical Methods*, (ed. TD Wheelock) Washington, DC, USA, American Chemical Society; *ACS Symposium Series*; (64); 153–163

Venier, CG, Squires, TG, Chen, YY, Shei, JC, Metzler, RM and Smith, BF (1981) The fate of sulfur functions on oxidation with peroxytrifluoroacetic acid. Presented at the 181st National Meeting of the American Chemical Society, Atlanta, GA, USA, 29 Mar–3 Apr 1981; *American Chemical Society, Division of Fuel Chemistry, Preprints*; **26** (1); 20–25

Warzinski, RP, Friedman, S, Ruether, JA and LaCount, RB (1980) *Air/water oxydesulfurization of coal—laboratory investigation*. Pittsburgh Energy Technology Center; 120pp

Warzinski, RP, LaCount, RB and Friedman, S (1980) *Oxydesulfurization of coal and sulfur-containing compounds*. Presented at the Meeting of the American Institute of Chemical Engineers, Chicago, IL, USA, Nov 1980; 28pp. Report DE-83008423, DOE/NBM-3008423, NTIS, Springfield, Virginia

Warzinski, RP, Ruether, JA, Friedman, S. and Steffgen, FW (1979) Survey of coals treated by oxydesulfurization. In: *Proceedings of the Symposium on Coal Cleaning to Achieve Energy and Environmental Goals, Hollywood, FL, USA, Sep 1978*; PB-299 384; Washington, DC, USA; US Environmental Protection Agency; 1016-1037

Wheelock, TD (1980) Status of chemical coal cleaning processes. Presented at *Fifth International Conference on Coal Research, Duesseldorf, FRG, 2–5 Sep 1980*; CONF-800910-1; Springfield, VA, USA; National Technical Information Service; 21pp

Wheelock, TD (1980) Dept. of Chemical Engineering, Iowa State University, Private Communication

Wheelock, TD (1980) Oxydesulfurization of coal in alkaline solutions. Presented at: *Ruth Symposium III, Ames, IA, USA, 12 Mar 1981*; Ames, IA, USA; Ames Laboratory; 30pp

Wheelock, TD, Greer, RT, Markuszewski, R, Fisher, RW (1979) *Advanced development of fine coal desulfurization and recovery technology. Annual Technical Progress Report, 1 Oct 1977–30 Sep 1978*; Ames, IA, USA; Ames Laboratory, Iowa State University; 84pp

Wheelock, TD and Markuszewski, R (1981) Physical and chemical coal cleaning. Presented at: *Conference on the Chemistry and Physics of Coal Utilization, Morgantown, WV, USA, 2–4 Jun 1980*; CONF-8006116–; New York, NY, USA; American Institute of Physics; *AIP Conf. Proceedings*; (70); 357–387

Wrathall, JA and Petersen, EE (1979) *Desulfurization of coal model compounds and coal liquids*; LBL-8576; Berkeley, CA, USA; Lawrence Berkeley Laboratory; 101pp

Wrathall, J, Vermeulen, T. and Ergun, S (1979) *Coal desulfurization prior to combustion*; LBL-10118; Berkeley, CA, USA; Lawrence Berkeley Laboratory; 17pp

Yurovskii, AZ (1960) *Sulfur in coals*; translated from Russian original, first published in 1960 TT70-57216; Springfield, VA, USA; National Technical Information Service; 459pp

Zavitsanos, PD, Golden, JA, Bleiler, KW and Kinkead, WK (1978) *Coal desulfurization using microwave energy*; PB-285 880; Washington, DC, USA; US Environmental Protection Agency; 79pp

Part 2

Control of sulphur oxides from coal combustion

Summary

This review discusses and compares the technologies for the control of the emissions of sulphur dioxide, SO_2, from the combustion of coal at three stages of the combustion cycle: prior to combustion by coal cleaning; during combustion mainly by the use of SO_2 sorbents in fluidised bed combustors (FBC); and after combustion by flue gas desulphurisation (FGD) processes. Physical, chemical and microbiological coal desulphurisation processes are discussed. The section covering SO_2 removal during fluidised bed combustion concentrates on methods of increasing the utilisation of sorbents. These include: pretreatment of sorbent; post-sulphation treatment of sorbent; and modification of the operating parameters of a fluidised bed combustor. Both wet and dry FGD processes are described. Wet processes are subcategorized as regenerable and non-regenerable. Disposal of FGD waste is discussed. Dry processes include those based on dry injection, spray dryer and dry adsorption techniques. It is concluded that FGD processes currently provide the only means of consistently removing high percentages (>90%) of potential SO_2 emissions from coal combustion. FBC has the potential to equal these removal rates when the technology becomes available for power generation. Practical restrictions limit the amount of sulphur removed by physical desulphurisation processes to about 50 wt% of the total pyritic sulphur in coal. Organic sulphur is not removed. Chemical and microbiological desulphurisation processes are not yet commercial.

Chapter 9

Introduction

This section discusses and compares the technologies for the control of the emissions of sulphur, principally SO_2, from the combustion of coal. The economics of these technologies are not covered here but are discussed by Prior (1977). For the purpose of this review the siting of power stations and more particularly the point of emission of flue gases to the atmosphere is not considered a method of control. There is no reduction in the total quantity of SO_2 emitted to the atmosphere although its distribution may be affected. However, such options often form part of the legislation 'controlling' SO_2 emissions from coal fired power stations and are adopted by the UK, in the form of a 'tall stack policy'. Readers seeking further information about the implementation of such policies should consult the papers of Ross (1978, 1979).

Sulphur may be removed at three stages of the combustion cycle:

1. prior to combustion;
2. during combustion;
3. from flue gases.

The removal of sulphur prior to combustion as part of a coal cleaning process, if it can be performed economically and without excessive technical difficulties, offers a convenient and attractive way of reducing emissions of sulphur dioxide. A cleaned coal may result in improved boiler availability and reliability. The reduced weight of a cleaned coal also results in lower shipping costs. Cleaning processes are based on physical, chemical or microbiological principles. Chemical desulphurisation of coal was discussed in Part 1 of this volume.

The removal of sulphur during combustion is exemplified mainly by the addition of sulphur sorbents to fluidised bed combustors, although less successful attempts have been made to incorporate such sorbents in the pellets of coal fired in stoker boilers (Ban, 1980). Research on the injection of sorbents into pulverised coal-fired boilers is also in progress.

Flue gas desulphurisation is, at present, the most popular method of controlling SO_2 emissions particularly in the USA, Japan, and to a lesser extent, the Federal Republic of Germany (FRG). There are many such processes although only those which have been shown to be applicable to coal fired power stations are discussed here.

Before discussing these technologies it is necessary to review briefly the occurrence of sulphur and the regulations controlling its emission from coal fired power plants in various countries as these ultimately determine the control strategy.

Chapter 10

The occurrence of sulphur

Sulphur is a relatively abundant element and is present in the atmosphere as a result of emissions from both natural and anthropogenic sources. The main natural sources of atmospheric sulphur are geothermal activity, sea-spray and biological decay. Anthropogenic emissions of sulphur, generally in the form of SO_2, result mainly from the combustion of fossil fuels and the smelting of sulphur-containing ores. Sulphur does not accumulate in the atmosphere. It is removed by the processes of wet and dry deposition. However, the equilibrium concentration of sulphur in the atmosphere is dependent on the total emissions of sulphur.

Natural emissions of sulphur remain essentially constant whereas those from anthropogenic sources are increasing. In 1956 global anthropogenic emissions were less than 33% of those from natural sources. By 1974 this figure has risen to more than 66%. Anthropogenic emissions are increasing at a rate of 2.2%/year (Cullis and Hirschler, 1980). If present rates of growth are maintained and emission factors are not reduced significantly anthropogenic emissions will equal and exceed natural emissions well before the end of the century. The relationship between anthropogenic and natural sources is region dependent. In NW Europe for example, anthropogenic sources contribute more than 80% of total sulphur emissions to the atmosphere (Granat *et al.*, 1976).

As mentioned below the most important anthropogenic source of sulphur dioxide is the combustion of fossil fuels. In the late 19th century, coal, the predominant energy source, accounted for almost all anthropogenic emissions of SO_2. However, the increasing use of petroleum and petroleum products in the 20th century reduced the significance of coal although it still remained the major source of anthropogenic emissions. By the mid-1960s coal accounted for less than 70% of anthropogenic emissions and oil accounted for 20%. However, in the 1970s the balance between the use of oil and coal was again changed by political events and the realisation of the finite nature of oil reserves. It is now generally accepted that coal will again become the dominant energy source for industry.

Figure 10.1 shows scenarios of industrial and utility coal use to the year 2000 from IEA and WOCOL studies for several industrial countries of North America, Europe and the Pacific (International Energy Agency, 1978; Wilson, 1980). The IEA scenario includes all coal not used for coking. This includes domestic and commercial uses of coal plus synthetic fuels processes which are not included in the WOCOL scenario for industry. The major conclusion from these studies is that by the year 2000 industrial and utility coal use will increase by between 200 and 300% over that of the present.

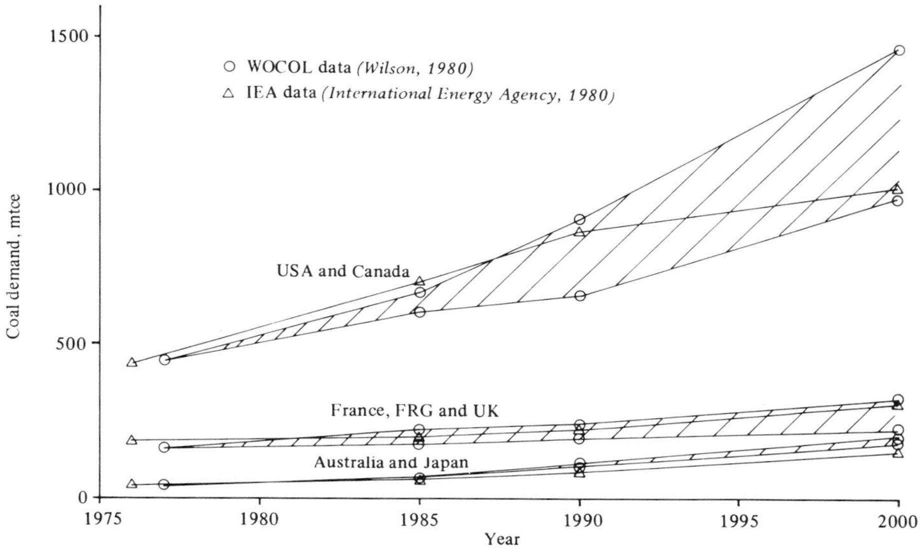

Figure 10.1 Scenarios of future industrial and utility coal demand

Table 10.1 Sulphur dioxide emission regulations and guidelines applicable to coal fired boilers (after Rubin, 1981; Zabel, 1979)

Country	*Regulation or guideline*
EEC	None
Federal Republic of Germany	$\leqslant$4 TJ/h thermal input (1100 MW): $\leqslant$1%S. $>$4 TJ/h thermal input (1100 MW): 0.73 kg/GJ. Specified stack height to limit ambient concentration
France	Specified stack height Rhone zone: $\leqslant$1%S in fuel. Paris and North zone: 0.95 kg/GJ. Power plants: lowest %S of ambient SO_2 $>$1000 $\mu g/m^3$
Japan	Specified stack height according to the formula $q = K \cdot He^2 \cdot 10^{-3}$ *where* q = SO_2 emission, m^3/h corrected to 0°C and atmospheric pressure; *He* is the sum of the stack height and plume ascent height, m; *K* is a constant for the geographic region ranging from 3.0 to 17.5 for existing plants and from 1.17 to 2.34 for new plants
Spain	Bituminous coal or anthracite: power plant: 2400 mg/m^3 other combustors: 2400 mg/m^3 Lignite: power plant: 9000 mg/m^3 other combustors: 6000 mg/m^3
UK	Specified stack height to limit short term ambient concentration (nominal 3 minute average $\leqslant$0.17 ppm SO_2)
USA	New steam generators $>$73.3 MW_t: 0.52 kg/GJ. All other boilers: applicable state regulations. New steam-electric power plants $<$73.3 MW_t; SO_2 reduction of 70% below 0.26 kg/GJ; SO_2 reduction of 90% below 0.52 kg/GJ

The effects on man and the environment of the potential increase in the concentration of sulphur in the atmosphere are the subjects of controversy. That there is public concern, however, cannot be denied. It is reflected in the increasing amount of legislation directed at controlling the emissions of sulphur from various sources, particularly those from coal combustion. This legislation, much of which has been introduced within the last 10 years, is summarised in Table 10.1.

Chapter 11

Physical desulphurisation of coal

Sulphur occurs in coal in essentially two forms:

1. Organic sulphur or sulphur which is chemically associated with the organic matter in coal.
2. Inorganic sulphur, mainly in the form of pyrite.

Pyrite occurs in coal as narrow veins, nodular and discrete crystals (Morrison, 1981). Physical coal desulphurisation techniques are applicable only to the removal

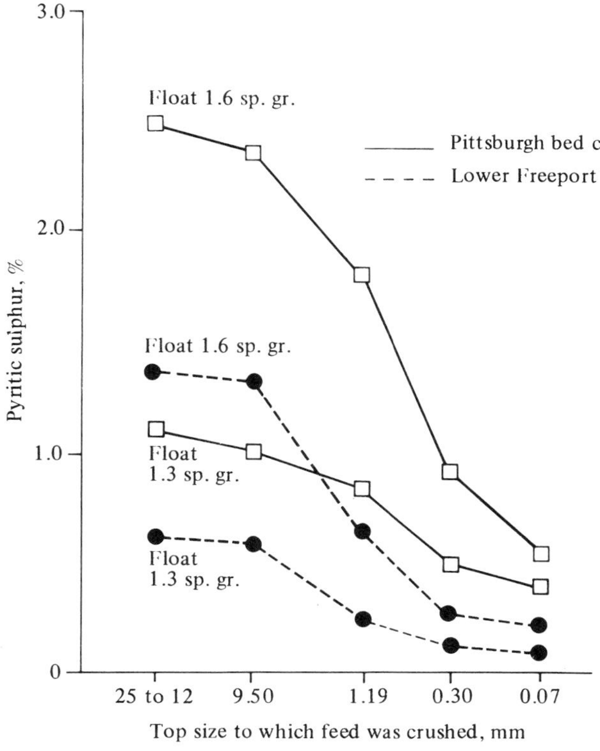

Figure 11.1 Removal of pyritic sulphur by a gravity separation process as a function of coal particle size (after Cavallaro and Deurbrouck, 1977)

of inorganic sulphur. Organic sulphur is not removed. Although the physical cleaning of coal has been practised for many years the majority of coal preparation plants were commissioned when coal prices were low and substantial losses of combustible material were accepted. With few exceptions plants were designed primarily for the removal of mineral matter from coal, any removal of sulphur being of secondary importance. Only in recent years has consideration been given to processes designed especially for the removal of sulphur.

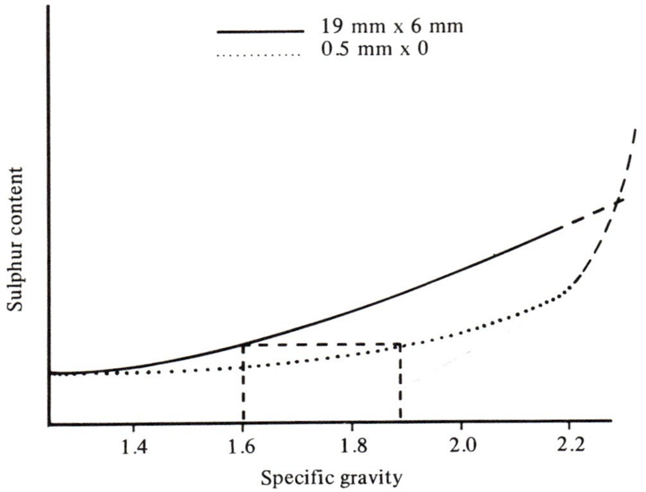

Figure 11.2 Sulphur distribution for two different size fractions of coal (after Huettenhain, 1978)

For any physical technique to be most effective it is first necessary to liberate the pyrite by crushing the coal. The more closely the particle size of the raw coal approaches that of finely disseminated pyrite crystals the greater is the potential for the removal of sulphur as illustrated in Figure 11.1. Substantially greater reductions in the sulphur contents of the two coals are obtained at a size of 0.30 mm compared with those obtained at 25.4 to 12.7 mm. The increased separation of coal and pyrite with decreasing particle size is also illustrated in Figure 11.2. The sulphur content of the smaller size fraction is concentrated in the higher relative density ranges. However, crushing coal to this size order is an expensive operation and may result in a product with a high moisture content and poor handling characteristics. Figure 11.3 shows the effect of particle size on the handling characteristics of coal expressed as the maximum angle of the walls of a storage bin with the vertical to maintain controllable flow. In practice a maximum angle of 20° is required to ensure controlled flow of a smalls/filter cake ratio of 65:35 whilst maintaining acceptable storage capacity (Jenkinson and Cammack, 1981). Figure 11.3 shows that this can be achieved for smalls/filter cake mixtures only at very low total moisture contents.

It is disputed whether the benefits of obtaining low sulphur products justify the additional costs and problems encountered (Cavallaro and Deurbrouck, 1977). It is likely that the amount of sulphur removed will involve a balance between the desired sulphur content of the cleaned coal, the loss of combustible material that

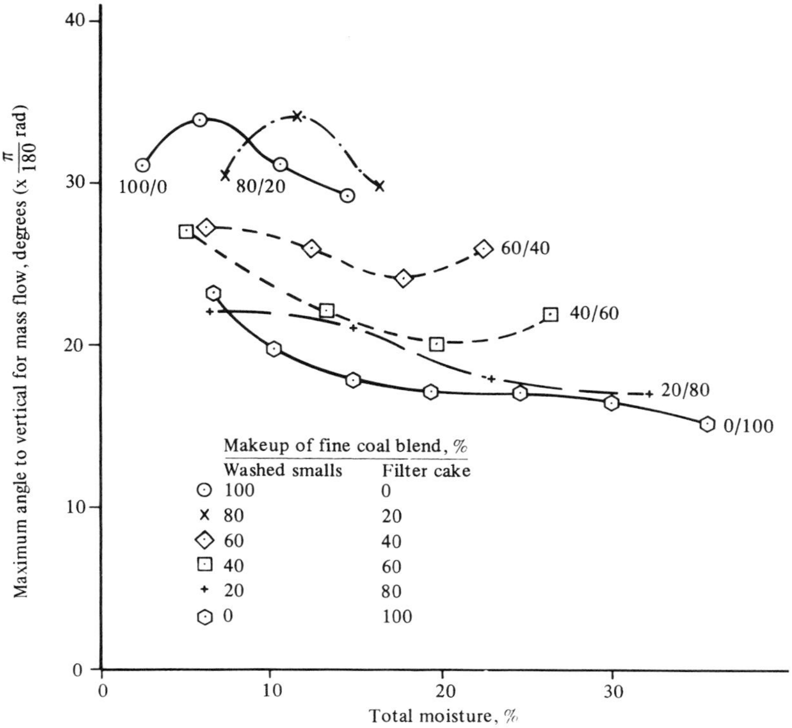

Figure 11.3 Handling characteristics of a washed smalls/filter cake blend (after Jenkinson and Cammack, 1981)

can be tolerated and the handling characteristics of the product. These factors will depend on the coal being cleaned and the use for which it is required. Studies of several hundred samples of US coals revealed that even with a 50% rejection of coal during preparation, 68% of the coals could not be cleaned sufficiently to comply with the then US SO_2 emission standard of 0.52 kg/GJ (Land, 1979).

Whilst it is impractical for coal to be crushed to the optimum size for maximum sulphur removal virtually all coal preparation plants incorporate an initial size reduction or comminution of coal. This is achieved by breakers which essentially drop raw lumps of coal. The coal shatters while the relatively hard rock remains intact and is discarded. Some fine coal inevitably results from this process. For some plants this is the extent of coal preparation: coal is screened and crushed. At other plants some degree of cleaning of the coarse coal is performed. It may then be mixed with raw fines. Modern plants may separate coal into as many as four size fractions (Pruce, 1978; McConnell, 1979). Each fraction is cleaned with the most suitable equipment (Lihach, 1979a).

The majority of commercial physical coal cleaning processes depend on the difference in the relative densities of the constituent coal and associated impurities although differences in surface properties are also exploited to a lesser extent. More recently, sulphur removal processes based on electrostatic and magnetic properties have been studied.

Differences in relative densities

The relative density of pyrite is 5.0, that of shale is 2.4 and that of bituminous coal is approximately 1.2. The differences in these relative densities form the operating principle of the majority of today's coal preparation equipment. The technique is often termed 'gravity separation'. For maximum sulphur removal the separation should be carried out in a medium with a relatively density only slightly above that of clean coal. However, coal losses increase sharply as this figure is approached because the rejected material contains progressively more combustible material. In practice, it is difficult to achieve an acceptable separation in media of relative densities of less than 1.3 or 1.4. It has been shown that for three typical, high sulphur content UK coals little reduction in sulphur content is achieved with media of greater than 1.6 to 1.65 relative densities (Jenkinson and Cammack, 1981).

One of the simplest types of equipment operating on the principle of gravity separation is the jig washer. Coal rests on a screen plate through which a pulsating current of water is passed. Stratification of the coal and contaminant minerals occurs such that clean coal is moved towards the top of the mixture while the relatively dense impurities sink. Jigs, particularly Baum jigs, are widely used in coal preparation plants. Although they are suitable for a wide range of coal sizes they are limited to a minimum separating gravity of about 1.5, resulting in little effective reduction of the sulphur content. A more recent development of jig technology is the Batac jig (Killmeyer, 1980). Close control of the compressed air which produces the pulsating action of the water enables separations to be carried out at specific gravities as low as 1.45. This increases the removal of pyritic sulphur particularly from 12.7 to 0 mm fine coal (Cavallaro and Deurbrouck, 1977).

Concentrating tables consist of tilted, ribbed vibrating surfaces. Coal/water slurry is fed to one corner at the top of the table. As the slurry moves down and across the table the mixture stratifies in a predictable manner. The relatively dense refuse moves towards the bottom edge of the table and the less dense coal remains nearer the top. Tables are capable of cleaning intermediate sizes of coal, below 10 mm, but in common with other equipment which use water as the separation medium, they have relatively low separation efficiencies.

A more effective separation can be achieved in a dense media vessel. This is a tank through which move streams of raw coal and a heavy medium, usually magnetite and water. As the raw coal travels slowly through the medium the relatively dense refuse particles settle out and leave the vessel through a low-level exit. The less dense coal particles exit at a higher level. The magnetite is later recovered from the clean coal and refuse fractions by screening and magnetic separation. Controlled size grinding of the magnetite and/or the use of stabilizers such as bentonite enables such systems to operate at relative densities as low as 1.32 (Bettelheim and Morris, 1979). However, such systems are relatively expensive to install and, because of losses of magnetite, are relatively expensive to operate. They are also limited to the cleaning of coarse coals, normally above 6 mm.

Dense media cyclones operate on the same principle as dense media vessels but the centrifugal motion imparted by the cyclone produces greatly increased separation rates. This enables fine coal down to a minimum of between 74 μm and 149 μm (Hise, 1982) to be cleaned: this minimum being set not by the limitations of the cyclones but by the difficulties of recovering the magnetite medium from smaller sizes. Dobby and Kelland (1982) describe the use of HGMS to separate fine coal and magnetite. Although dense media cyclones share the higher capital and operating costs of dense media vessels they are well suited to the removal of pyrite.

Application of dense media vessels and cyclones in conjunction with concentrating tables and froth flotation has been shown to remove between 30 and 36% of total sulphur from a coal initially containing 5.0% sulphur. Mass and heat recoveries of 84% and 91% respectively were achieved (Tarkington *et al.*, 1979).

Separation using surface properties

It is theoretically possible to effect a separation of inorganic mineral impurities from coal by using the differences in their respective surface properties, particularly their affinity to water. Coal is generally hydrophobic whereas most of the mineral impurities are hydrophilic. Froth flotation and oil agglomeration are two methods of cleaning fine coal particles based on this principle. However, these techniques are not well suited to the separation of pyrite from coal because of the similarities in the surface properties of coal and pyrite, particularly pyrite in a reduced or unweathered state (Wheelock and Ho, 1979). Research projects aimed at enhancing the differences in surface properties of coal and pyrite are described below.

Froth flotation

Froth flotation was introduced into coal preparation from the metallurgical industries where it was employed in mineral separations. It is now the most common technique for the cleaning of fine coal (<0.5 mm). Numerous small air bubbles are generated within an aqueous suspension of the coal to be cleaned. Some of the bubbles become attached to the hydrophobic coal particles and buoy them to the surface from where they are recovered as a froth. The hydrophilic minerals remain in the suspension. A frothing agent or 'frother' is added to promote froth formation. The most widely used is methyl isobutyl carbinol (MIBC).

Aplan (1977) studied the kinetics of coal flotation and the effects of different frothers, frother concentrations and operating conditions in order to optimize the separation of pyrite from coal. He concluded that the rate of coal flotation is considerably greater than that of pyrite or other mineral matter (Figure 11.4) and that above certain threshold values, increasing concentrations of frother, aeration rates and impeller speeds increase the rate of flotation of pyrite more than they increase the rate of flotation of coal. A relationship was found between the yield of coal and the amount of pyrite which floats. Figure 11.4 shows that less than 10% of the pyrite floats at coal yields of less than 80%. Hence, to effect the maximum separation of coal and pyrite it is necessary to operate the froth flotation cell under 'gentle' conditions: i.e. short residence times, low concentrations of frother, low air flow rates and low impeller speeds.

These conclusions were used to suggest a method of pyrite removal using multiple froth flotation cells. Between 50 and 65% of the coal yield is floated in froth flotation cells operating under gentle conditions. The coal is very low in pyrite. The balance of the recoverable coal is removed in a scavenger circuit under more intense operating conditions. The scavenger concentrate, which is relatively

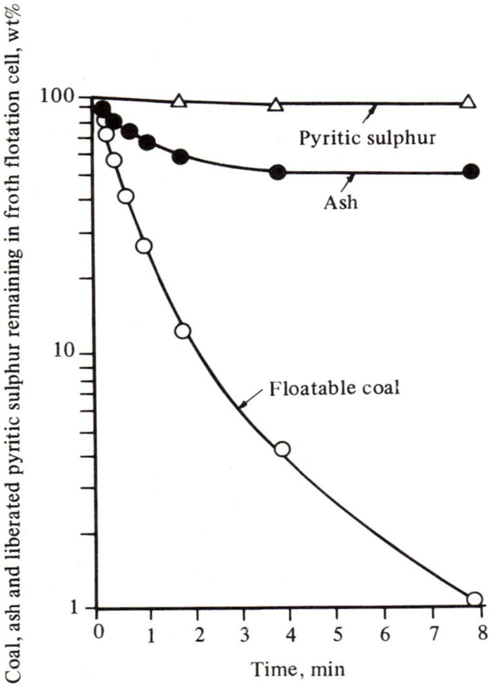

Figure 11.4 Rates of flotation of coal, pyrite and other mineral matter (after Aplan, 1977)

high in pyritic sulphur, is passed to additional flotation cells in which separation is achieved by selectively depressing either the pyrite or the coal (Aplan, 1977) as described below.

Research has also been conducted into the use of additives which either diminish or enhance the affinity to water of either coal or pyrite. Additives which increase the hydrophobicity of either coal or pyrite are known as collectors; those which increase their hydrophilicity are known as depressants. Several authors have investigated pyrite depressants with little success (Min, 1977; Le, 1977; Laros, 1977). However, in laboratory scale work Wheelock and Ho (1979) demonstrated that basic compounds, in particular calcium hydroxide, $Ca(OH)_2$, depress pyrite preferentially. It is suggested that the strongly depressant action of calcium hydroxide is due to the adsorption of both calcium ions and hydroxyl ions. Calcium combines with various pyrite oxidation products forming hydrated films of calcium sulphate and/or carbonate on the pyrite surface. These films are hydrophilic. Wheelock (1982) has also shown that wet oxidation of pyrite in alkaline solutions has an even greater depressant effect on pyrite than the use of the alkali alone.

In 1974 the US DOE (then Bureau of Mines) patented a two stage coal flotation process based on the use of a coal depressant and a pyrite collector. Because the process reverses the natural hydrophobic property of coal and the hydrophilic property of pyrite the separation of pyrite must be achieved in two stages. The first stage is a conventional froth flotation in which the majority of the mineral matter and the coarse pyrite or pyrite which is associated with shale are removed as tailings. The first stage coal froth concentrate is then repulped and a coal depressant, a pyrite collector and a frother are added. A high sulphur content product is floated. The second stage underflow is collected as the final product. The pyrite collector is a xanthate, a sulphydryl collector that is adsorbed selectively onto the pyrite rendering the surface hydrophobic (Burger, 1980). Reductions of up to 70% in the pyritic sulphur content of some US coals have been obtained in laboratory scale tests (Miller, 1978). A full scale application of this process is in operation in Pennsylvania, USA.

Oil agglomeration

Oil agglomeration involves the addition of an oil-based liquid to a suspension of the coal particles to be cleaned. The oil is adsorbed selectively onto the surface of the hydrophobic coal particles which are agglomerated by agitating the suspension. The agglomerates of coal are removed by screening or flotation (Capes, 1979).

Research into adapting oil agglomeration to the removal of pyrite from coal is similar to that applied to froth flotation: i.e. the investigation of chemical additives which enhance the difference in the surface properties of fine pyrite and coal particles. The technique of modifying the surface properties of pyrite by oxidation in alkaline solutions, shown to be effective as a pyrite depressant during froth flotation, is also applicable to oil agglomeration (Wheelock, 1982). Coal fines are treated with an aerated solution of between 1 and 2% sodium carbonate at temperatures of between 50 and 80°C for 15 min. Application of this treatment to coal with a particle size of less than 37 μm from the Iowa State University demonstration mine followed by agglomeration with a mixture of oils reduced the inorganic sulphur content of the coal by 88%. The total mineral matter of the coal was reduced by 63% while 93% of the combustible matter of the coal was preserved (Wheelock, 1982). When the same coal was agglomerated without pretreatment, the inorganic sulphur content was reduced by 40% and the total mineral matter by 35% while the recovery of combustible matter was the same.

The microorganism *Thiobacillus ferrooxidans* has been used to pretreat coal prior to agglomeration by a process termed bioadsorption (Kempton *et al.*, 1980). This is a physical adsorption process in which the *T. ferrooxidans* preferentially render the pyrite particles hydrophilic. The treatment time of less than 30 min contrasts with the several days required for microbiological desulphurisation of coal (see Chapter 12).

The effectiveness of oil agglomeration for the removal of pyrite is dependent on

Table 11.1 The removal of pyrite from two Canadian coals by oil agglomeration (after Capes *et al.*, 1979)

	New Brunswick coal		*Nova Scotia coal*	
	Starting material	*Agglomerated product*	*Starting material*	*Agglomerated product*
Ash, wt%, dry basis	23.2	8.6	18.4	3.8
Sulphur: total	9.6	7.0	4.5	1.8
(dry) sulphate	0.7	0.2	0.5	0.2
pyritic	7.3	5.1	2.8	0.6
organic	1.7	1.7	1.2	1.0
Recovery of combustible material		95.0		94.0

coal type and in particular on the form and distribution of pyrite within the coal. This is emphasised by the work of Capes *et al.* (1979) who noted a marked difference in the extent to which oil agglomeration removed pyrite from a Nova Scotia (NS) coal and a New Brunswick (NB) coal (Table 11.1). Both coals were comminuted to the same fineness prior to beneficiation. Scanning electron micrographs revealed a difference in the crystalline structure of the pyrite in the two coals. Whilst the pyrite crystals in NB coal were well-formed (euhedral) those in the NS coal were imperfect (subhedral). Capes *et al.* (1979) postulated that the poor quality pyrite crystals were more susceptible to reaction with oxygen during comminution than the well-formed crystals of the NB coal and would thus be more readily separated from coal by oil agglomeration.

Electrostatic separation

Inculet *et al.* (1980) describe experiments for the separation of pyrite and ash from coal by electrostatics. Coal particles of between 100 and 200 μm in size are charged in a fluidised bed by a technique known as triboelectrification. When two particles of dissimilar composition or structure are in physical contact an electrical charge is transferred. If at least one of the particles is a semiconductor or an insulator, upon separation one particle remains positively charged and the other negatively charged. Coal macerals are, in general, insulators when dry. Vitrinite charges positively and fusinite, pyrite and ash charge negatively. After charging the coal particles fall by gravity between a pair of electrodes held respectively at high positive and negative electric potentials thus generating a horizontal electric field. Positively charged particles move in the direction of the electric field, and negatively charged particles in the opposite direction.

Experiments so far have been directed at reducing the ash contents of very high ash coals (~50%). It is not clear to what extent pyrite is removed.

Separation using magnetic properties

A substance which is paramagnetic possesses a positive magnetic susceptibility and is, therefore, attracted to a magnet. A diamagnetic substance possesses a negative magnetic susceptibility and is repelled by a magnet. The mineral matter in coal, including pyritic sulphur, is paramagnetic whilst coal is diamagnetic. It is theoretically possible to use these properties to effect a separation of the mineral matter from coal. However, it is only in recent years that developments in magnet technology, particularly the introduction of high gradient magnetic separation (HGMS), have rendered such a process practicable (Murray, 1977). However, the separation is difficult, requiring powerful magnets and large field gradients.

The magnetic separation of pyrite from coal is enhanced if the weakly magnetic pyrite particles are converted to a material such as pyrrhotite, the magnetic susceptibility of which is almost four orders of magnitude greater than that of pyrite (Bluhm *et al.*, 1981). This can be achieved by heating the pyrite. However, if the coal is also heated this represents not only a waste of energy but can also result in pyrolysis of the coal. The sulphur released might also react with the organic portion and be more difficult to remove. One solution to this problem currently being investigated is the selective dielectric heating of pyrite in coal with minimal heating of the coal. Dielectric heating is essentially the heating of a substance by the absorption of electromagnetic energy. The amount of heating depends on the intensity and frequency of the electromagnetic energy and on the following properties of the material being heated: the specific heat, the density, and the relative dielectric loss factor. The exact relationship is discussed in detail by Bluhm *et al.* (1981). These authors calculate that pyrite should be heated approximately three times faster than coal. In practice selective dielectric heating of pyrite has proven difficult to implement with less than the theoretical increase in the magnetic susceptibility of the pyrite being achieved (Bluhm *et al.*, 1981).

High gradient magnetic separation (HGMS)

Conventional magnetic separators are largely confined to the separation or filtration of relatively large particles of strongly magnetic materials. They employ a single surface for separation or collection of magnetic particles. A variety of transport mechanisms are employed to carry the feed past the magnet and separate the magnetic products. The active separation volume for each of these separators is approximately the product of the area of the magnetised surface and the extent of the magnetic field. In order for the separators to have practical throughputs, the magnetic field must extend several centimetres. Such an extent implies a relatively low magnetic field gradient and weak magnetic forces.

To overcome these disadvantages HGMS has been developed. Matrices of ferromagnetic material are used to produce much stronger but shorter range magnetic forces over large surface areas. When the matrices are placed in a magnetic field, strong magnetic forces are developed adjacent to the filaments of the matrix in approximately inverse proportion to their diameter. Since the extent of the magnetic field is approximately equal to the diameter of the filaments the magnetic fields are relatively short range. However, the magnetic field produced is intense and permits the separation and trapping of very fine, weakly magnetic particles (Oberteuffer, 1979).

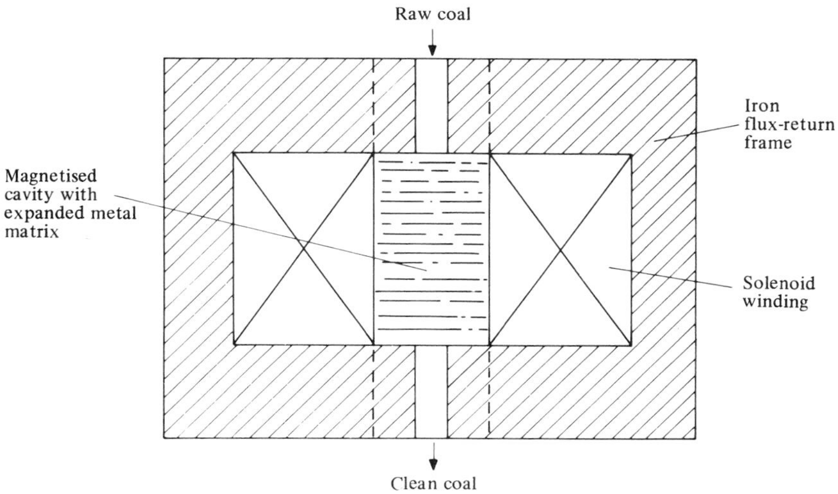

Figure 11.5 Equipment for the batch processing of coal by HGMS (after Hise, 1979)

The transport medium for HGMS can be either liquid or gaseous. Dry HGMS processing has the advantage of a dry product although classification of the pulverised coal is required to ensure proper separation. Small particles tend to agglomerate and pass through the separator. It has been shown that individual particles of coal in the discharge of a power plant pulveriser flow freely and hence separate well only if the material below about 10 μm is removed (Eissenberg *et al.*, 1979). Even then drying of that part of run of mine coal to be treated by HGMS may be required to ensure good flow characteristics.

A schematic representation of a batch HGMS process is shown in Figure 11.5 (Hise, 1979, 1980; Hise *et al.*, 1979). It consists of a solenoid, the core cavity of which is filled with an expanded metal mesh. Crushed coal is fed to the top of the separator. Clean coal passes through while much of the inorganic material is trapped to be released when the solenoid is later deactivated.

Data from a batch HGMS process of one size fraction of one coal are plotted in Figure 11.6 as weight per cent of material trapped in the magnetic matrix, the product sulphur and the product ash versus the independent variable of superficial transport velocity. At low superficial transport velocities the amount of material removed from the coal is high partly due to mechanical entrapment. As the velocity is increased the importance of this factor diminishes but hydrodynamic forces on the particles increase. These hydrodynamic forces oppose the magnetic force and the amount of material removed from the coal decreases (Hise, 1979).

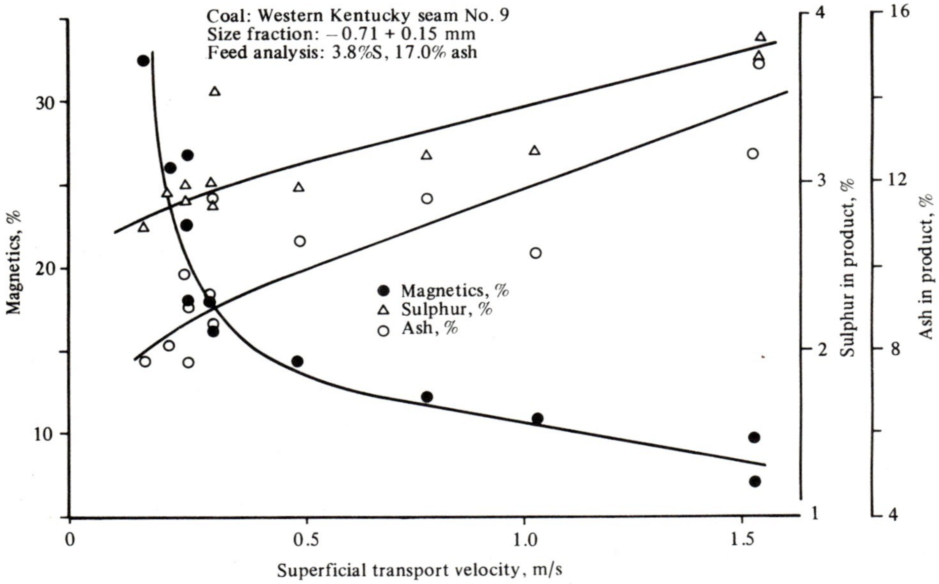

Figure 11.6 Product sulphur and ash content of coal cleaned by HGMS (after Hise, 1979)

For comparison, Figure 11.7 shows data from a specific gravity separation of the same size fraction of the same coal. While the sulphur contents of the products from the two separation processes are similar the ash content of the HGMS product is considerably higher than that of the specific gravity product. It should be emphasised that this comparison was made for one size fraction of one coal.

More recently dry HGMS has been demonstrated at a scale of 1 t/h on carousel type equipment which processes coal continuously (Figure 11.8; Hise *et al.*, 1981). A metal mesh passes continuously through the magnetised cavity so that the product coal passes through while the trapped inorganics are carried out of the field and released separately.

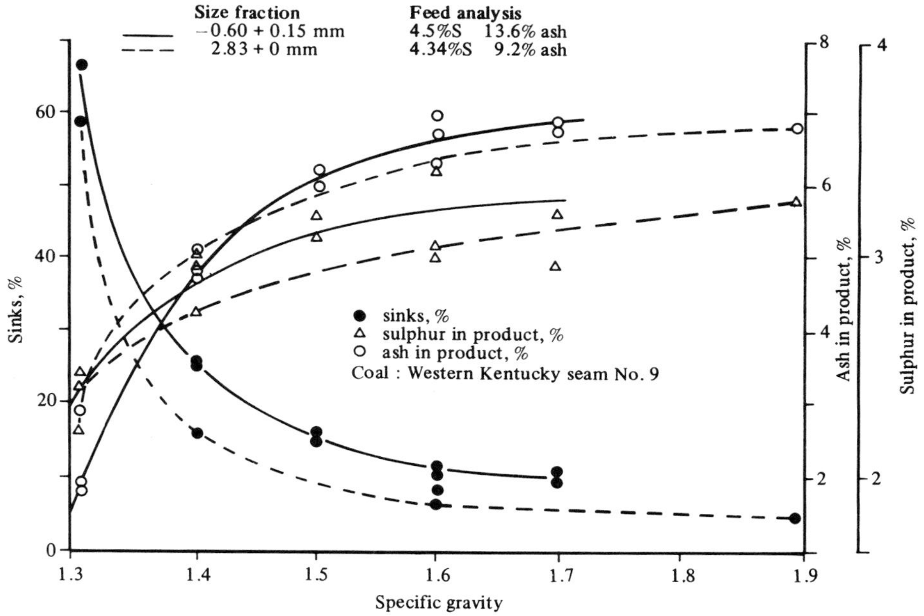

Figure 11.7 Product sulphur and ash content of coal cleaned by a gravity separation process (after Hise, 1979)

Wet HGMS is able to treat a much wider range of coal particle sizes than dry HGMS. The efficiency of separation increases with decreasing particle size. However, depending on the end use a considerable quantity of energy may have to be expended in drying the wet, fine coal product. Wet HGMS may find particular application to the precleaning of coal for use in preparing coal water mixtures for subsequent combustion as both pulverising the coal to a fine particle size and transporting the coal in a water slurry are operations common to both processes.

Work at Bruceton, PA, USA has compared the pyrite reduction potential of froth flotation followed by wet HGMS with that of a two stage froth flotation process (Hucko and Miller, 1980). Typical results are shown in Figures 11.9 and 11.10. The reduction in pyritic sulphur is similar in each case although a greater reduction in ash content is achieved by froth flotation followed by HGMS than by two stage froth flotation. However, Hucko (1979) concludes that it is highly unlikely that HGMS would be used for coal preparation independently of other beneficiation processes. As with froth flotation there is considerable variation in the amenability of various coals to magnetic beneficiation.

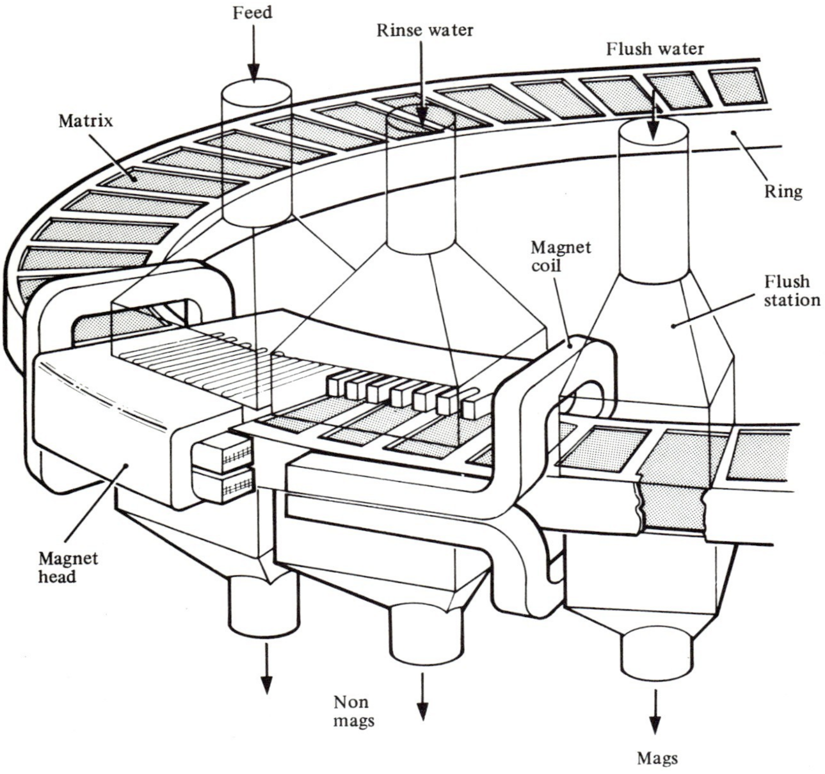

Figure 11.8 Equipment for the continuous processing of coal by HGMS (after Hise *et al.*, 1981)

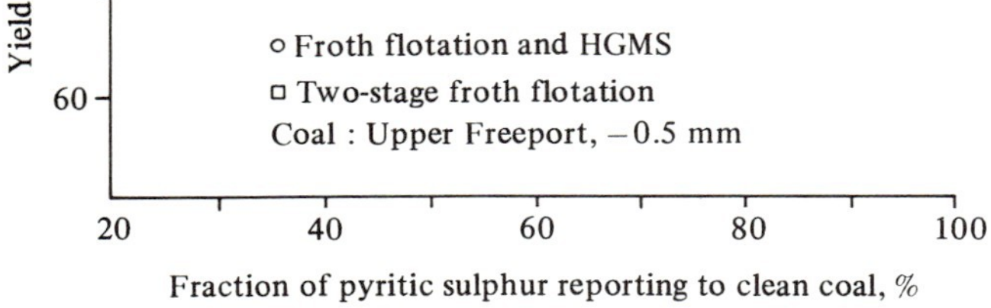

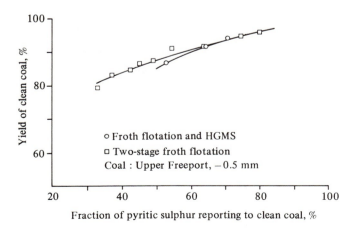

Figure 11.9 Sulphur content of coal cleaned by froth flotation followed by HGMS and of coal cleaned by two-stage froth flotation (after Hucko and Miller, 1980)

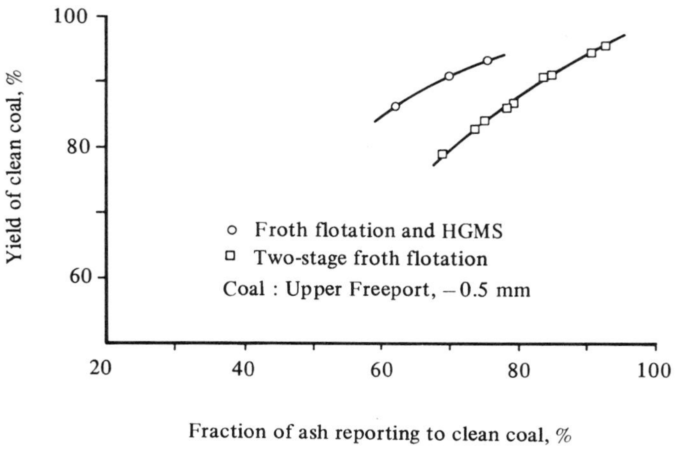

Figure 11.10 Ash content of coal cleaned by froth flotation followed by HGMS and of coal cleaned by two-stage froth flotation (after Hucko and Miller, 1980)

Open gradient magnetic separation (OGMS)

Although virtually all current research into the magnetic separation of minerals from coal is concerned with HGMS the possibility of using open gradient magnetic separation (OGMS) has also been investigated (Hise, 1979; Holman *et al.*, 1982).

Dry coal particles are passed, either mechanically or by free-fall, through a magnetic field that exhibits a gradient normal to the direction of travel of the coal.

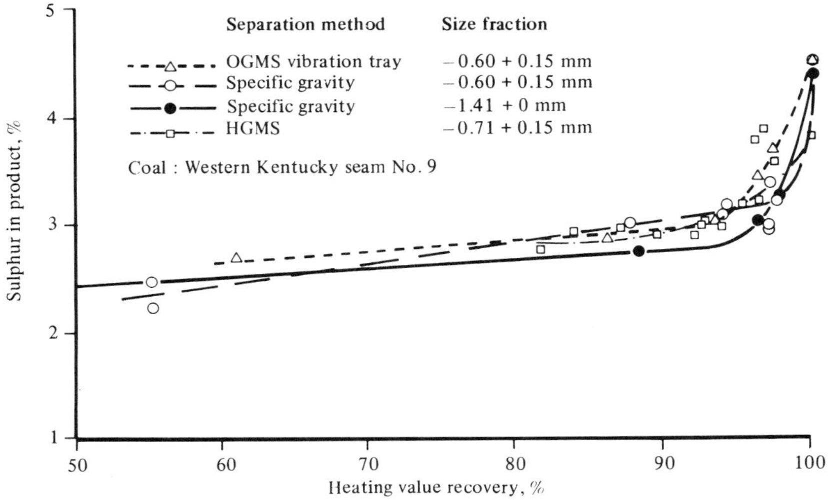

Figure 11.11 Product sulphur versus heating value recovery for OGMS and HGMS (after Hise, 1979)

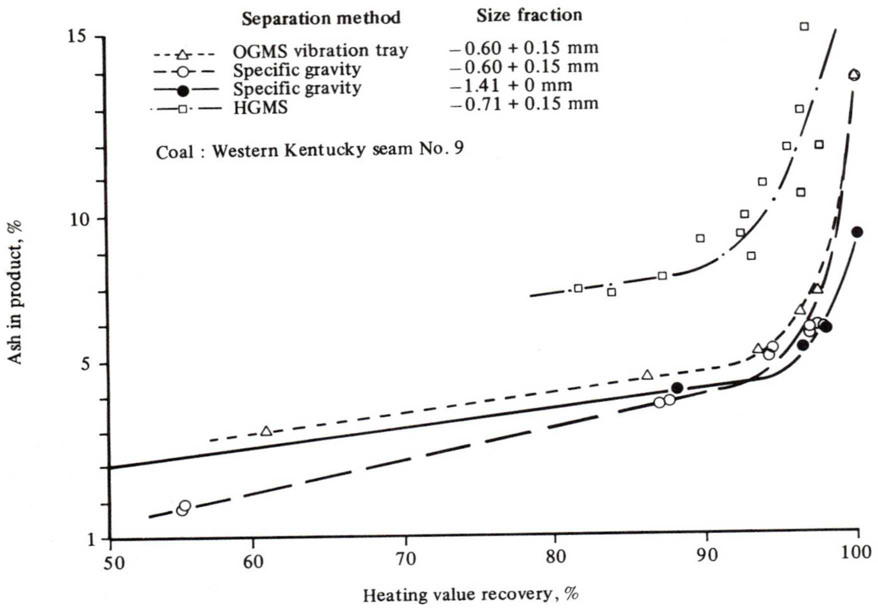

Figure 11.12 Product ash versus heating value recovery of OGMS and HGMS (after Hise, 1979)

The paramagnetic material, mainly mineral matter, is deflected up gradient whilst the diamagnetic material, mainly coal, is deflected down gradient. OGMS is therefore capable of using:

1. The magnetic attractive force exerted on particles having positive susceptibility.
2. The magnetic repulsive force exerted on particles having negative susceptibility.
3. The gravitational force exerted on all particles.

In contrast, HGMS uses only the first of these forces.

In theory therefore OGMS has the potential of performing the more precise separation. This has been confirmed in practice by the work of Hise (1979; Figures 11.11 and 11.12). The sulphur reductions achieved by HGMS and OGMS are virtually identical to those obtained by gravity separation at a given recovery of heating value. The inferior ash reductions obtained by HGMS may be attributed to the differences in the positive magnetic susceptibilities of the different mineral impurities in coal (Hise, 1979). However, it must be remembered that the results for OGMS and HGMS presented in Figures 11.11 and 11.12 were obtained from different size fractions of different coal samples. It is unlikely that the magnetic field will be generated over a large enough volume to make OGMS competitive with HGMS for the cleaning of fine coal. Little work has been reported on OGMS This may be due, in part, to the absence of suitable large scale equipment.

Chapter 12

Microbiological desulphurisation of coal

Atmospheric oxidation or weathering of coal can result in a significant reduction in the sulphur content (Figure 12.1; Chandra *et al.*, 1982). It is theoretically possible to catalyse this oxidative dissolution of pyrite, FeS_2, in coal by the use of microorganisms. Two genera of bacteria have been identified for this purpose; *Thiobacillus ferrooxidans* and *Sulfolobus acidocaldarius*. *T. ferrooxidans* functions at a temperature of about 20°C whereas *S. acidocaldarius* functions at temperatures between 60 and 80°C. Both these microorganisms are aerobic and acidophilic, thriving in acid environments with a pH of between 1.0 and 4.0.

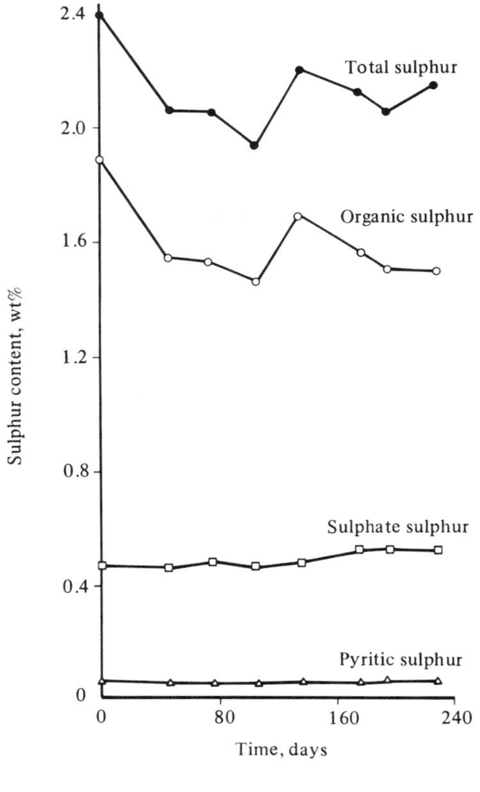

Figure 12.1 Effect of weathering on the sulphur content of coal (after Chandra *et al.*, 1982)

The dissolution of pyrite from coal using these organisms can be explained by the direct attack on pyrite by ferric ions in solution (Detz and Barvinchak, 1979):

$$FeS_2 + 14Fe^{3+} + 8H_2O \rightarrow 15Fe^{2+} + 2SO_4^{2-} + 16H^+ \quad (12.1)$$

The ferric ion is regenerated by the action of the organisms according to the reaction:

$$O_2 + 4H^+ + 4Fe^{2+} \rightarrow 4Fe^{3+} + 2H_2O \quad (12.2)$$

Any reduced forms of sulphur which have not been oxidised completely are converted to sulphate by the microorganisms:

$$2S_2^{2-} + 7O_2 + H_2O \rightarrow 2SO_4^{2-} + 2H^+ \quad (12.3)$$

The overall oxidation and solubilisation of pyrite may be expressed as:

$$4FeS_2 + 15O_2 + 2H_2O \rightarrow 2Fe_2(SO_4)_3 + 2H_2SO_4 \quad (12.4)$$

The rate of pyrite solubilisation depends on the concentration of Fe^{3+} in solution. In the presence of the microorganisms the rate of reaction (12.2) is very rapid and the rate of solubilisation is controlled by the rate of reaction (12.1). The rate of reaction (12.3) is temperature dependent. Hence, faster solubilisation of pyrite from coal can be achieved by using *S. acidocaldarius* rather than *T. ferrooxidans* at the expense of providing additional energy to maintain the higher temperature required by this microorganism.

Table 12.1 Process conditions for a conceptual plant for the microbiological desulphurisation of coal (after Detz and Barvinchak, 1979)

Throughput, t/d	7300
Slurry concentration, wt%	20
Coal size, μm	<74
Pyritic sulphur in feed coal, %	2
Temperature, °C	28
Residence time, days	18[a]
Reactors	Lined and covered lagoons 6 m deep × 10 ha
Coal recovery, %	99
Product coal moisture content, %	10[b]
Lime requirement	1 kg per kg pyrite solubilised (37.5 kg lime per tonne of coal)

[a]corresponds to removal of 93% of the pyritic sulphur
[b]vacuum filtration to 23% moisture content followed by thermal drying to 10% moisture content

Process conditions for a conceptual 7300 t/day plant for the desulphurisation of coal with *T. ferrooxidans* are given in Table 12.1. A large storage volume, in the form of lagoons, is required to provide a residence time of 18 days. These lagoons must be lined, covered, agitated and aerated. The residence time and hence the required storage volume may be reduced by using *S. acidocaldarius* but at the additional expense of heating the lagoons to between 60 and 80°C. The requirement for lagoons may also be reduced by using slurry pipelines as plug flow, mixed culture reactors (Hoffmann *et al.*, 1981). However, the economics of the microbiological desulphurisation of coal have yet to be demonstrated.

Neither *T. ferrooxidans* nor *S. acidocaldarius* reduce the organic sulphur content of coal. In fact, certain organic compounds such as short chain carboxylic acids may poison *T. ferrooxidans*. Under such conditions it may be necessary to use

Table 12.2 Effect of bacterial treatment on the removal of organic sulphur from coal (after Chandra *et al.*, 1979)

Sample	*Total sulphur* wt%	*Organic sulphur* wt%	*Removal of organic sulphur* wt% of organic sulphur
Original coal	6.68	5.53	—
Unsterile coal without bacteria (control sample)	6.56	5.44	1.6
Sterilized coal without bacteria (control sample)	6.54	5.42	2.0
Sterilized coal treated with bacterial culture	5.40	4.43	19.9
Unsterile coal treated with the bacterial culture	5.62	4.62	16.5

additional organisms such as *T. acidophilus* or *T. perometabolis*. Although these organisms do not oxidise iron or sulphur they metabolise the organic compounds detrimental to *T. ferrooxidans*. *S. acidocaldarius* is unaffected by organic compounds. Chandra *et al.* (1979) have shown the possibility of removing organic sulphur from coal by using a culture enriched in dibenzothiophene (DBT). Preliminary results are shown in Table 12.2. The removal of less organic sulphur in the case of the unsterile coal suggests there is an interaction of the DBT-enriched culture with the indigenous *Thiobacillus* present in coal. Work is continuing to identify the bacteria and to optimise the conditions for the removal of organic sulphur from coal.

Chapter 13

Sulphur removal during combustion

As discussed in Chapter 9, there exists the possibility of capturing SO_2 as it is formed during the combustion process if an SO_2 acceptor such as calcium carbonate is present. The latter calcines under combustion conditions; the reaction for which may be expressed by the endothermic reaction:

$$CaCO_3 \rightarrow CaO + CO_2 \quad (13.1)$$

The calcium oxide is porous with a relatively large internal surface area. This promotes the sulphation of the calcium oxide which may be represented by:

$$CaO + SO_2 + 1/2O_2 \rightarrow CaSO_4 \quad (13.2)$$

The calcium sulphate eventually forms an impervious skin around the particle which stops any further reaction (Dutkiewicz and Naude, 1981).

The exact mechanism of the sulphation reaction is disputed. It is generally accepted that SO_2 and O_2 present in the gaseous phase diffuse into the CaO particles to form $CaSO_4$. However Lallai *et al.* (1979) suggest that oxidation of SO_2 to SO_3 occurs outside the particle due to the catalytic action of trace elements in the coal such as V_2O_5:

$$2SO_2 + O_2 \rightarrow 2SO_3 \quad (13.3)$$

The SO_3 then diffuses into the pores of the CaO where it reacts producing calcium sulphate:

$$SO_3 + CaO \rightarrow CaSO_4 \quad (13.4)$$

Ban (1980) describes the processing of high sulphur coal with limestone into carbonised-composite pellets known as Helifuel. Limestone and coal are mixed in a molar ratio from about one to six parts of CaO to one part S. The mixture is ground to a particle size of approximately 0.589 mm before being balled and compacted. The pellets are carbonised and pyrolysed at temperatures of between 649 to 1316°C. Helifuel has been tested in commercial stokers and gasifiers.

The possibility of in-flame desulphurisation by injection of calcium-based sorbents during the combustion of pulverised coal is being studied at the International Flame Research Foundation (IFRF), the Netherlands, in conjunction with the Steinmueller Company of the FRG (Flament, 1982). The main limitation on the application of this technique is temperature. Calcium sulphate, $CaSO_4$, is unstable above 1200°C and the reactivity of calcined limestone is reduced by sintering at relatively high temperatures. A further limitation is the short time available for contact between the solid CaO and the combustion gases in pulverised coal fired boilers. Whilst encouraging results have been obtained with the firing of brown coal, in which the flame temperature is about 1150°C, low sulphur capture has been reported for the combustion of bituminous coal in which flame temperatures can reach 1500°C (Flament, 1982). However, the recent development of low NO_x staged-combustion burners has revived interest in the application of the injection of calcium based sorbents during the combustion of pulverised bituminous coals (Morrison, 1980). Staged combustion shifts the position of the peak flame temperature to an off axis position and may also lower the peak flame temperature. Staged combustion also produces a reducing atmosphere close to the burner mouth in which sulphur may be captured in accordance with the reaction:

$$CaO + H_2S \rightarrow CaS + H_2O \qquad (13.5)$$

Sulphur removal rates of the same order of magnitude as those currently achieved with limestone addition in fluidised bed combustors have been achieved experimentally (Flament, 1982). However there are several potential problems which may arise from the injection of calcium-based sorbents into pulverised coal flames:

1. The increased calcium content may lower the ash melting point with the associated risk of increased slagging and fouling (Morrison, 1978).
2. The handling of increased quantities of ash.
3. The possible effect of calcium addition on downstream equipment such as electrostatic precipitators.
4. The possible influence of sorbent injection on the radiative properties of the flame.

These aspects, as well as demonstrating the process on commercial scale equipment, must be investigated before the injection of sorbents into pulverised coal flames can be considered a viable SO_2 control technique.

The main application for the addition of calcium based sorbents during coal combustion is in the fluidised bed combustion of coal. The thorough mixing within the bed ensures good contact between sorbent and combustion gases. Also, the temperature of the bed is uniform and usually between 750 and 950°C. The remainder of this chapter discusses the removal of S during the fluidised bed combustion of coal.

Selection and utilisation of sorbent

Limestones and dolomites are used extensively as sorbent materials. Oil shales have been studied but the relatively low calcium content of these materials makes them less suitable for use in FBCs than limestone (Wilson *et al.*, 1980). Limestones of different types and from different sources exhibit varying sulphur capture efficiencies (Meserole *et al.*, 1979). These variations are the result of differences in impurities, pore size distributions and grain or particle sizes of the limestones. The purity or calcium carbonate content of the limestone is important from the operator's point of view because it influences both the requirement for raw limestone and the amount of waste for disposal.

Porosity and pore size distribution influence strongly the availability of the active constituent, calcium, in the limestone. Hartman *et al.* (1978) have shown that natural porosity persists during the calcination process and that to be suitable for use in FBCs carbonate rocks should have an initial porosity prior to calcination of more than 30%. In general, the older the limestone the less porous it is. The geological age of a limestone may therefore be taken as a very approximate index of its sulphur capturing efficiency (Munzner and Bonn, 1980). Criteria for the selection of SO_2 sorbents for FBC are discussed by Ulerich *et al.* (1979) and by Fee *et al.* (1980). The effect of pore size distribution is illustrated in Figure 13.1. The variations in the average pore diameters shown in this figure were obtained partly by the use of different limestones and partly by the addition of sodium chloride. There is an optimum pore diameter of approximately 0.3 μm. Pores smaller than this are rapidly blocked by the formation of $CaSO_4$ which has a larger molar volume than the original $CaCO_3$. As the average pore diameter is increased beyond 0.3 μm gaseous permeability increases but the effective surface area decreases. Surface area eventually becomes the controlling factor over permeability and the overall sorbent reactivity decreases with further pore enlargement (Shearer *et al.*, 1979).

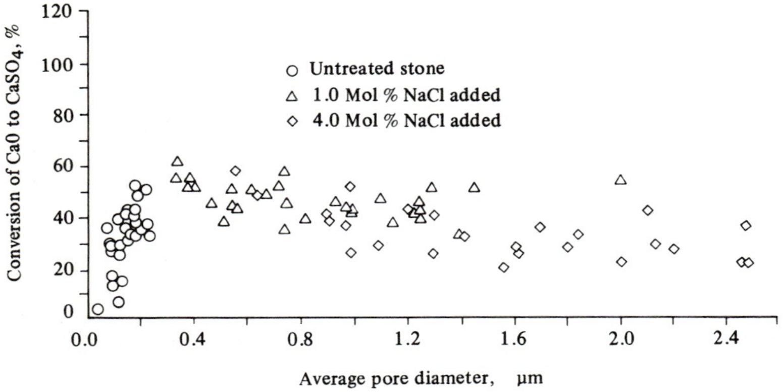

Figure 13.1 Effect of pore size on the sulphur capture efficiency of limestone sorbent (after Shearer *et al.*, 1979)

Increasing the amount of limestone added and hence the Ca/S mole ratio increases the amount of sulphur retained within a FBC. From the stoichiometry of Equation (13.2) it can be predicted that a Ca/S mole ratio of 1.0 is required to reduce the emission of SO_2 to zero. This theoretical requirement is equivalent to the addition of 3.15 kg of limestone or 5.75 kg of dolomite per kg of sulphur in coal (Gibson, 1981). In practice the relationship between the Ca/S mole ratio and the amount of SO_2 removed varies according to the properties of the limestone and the conditions within the bed, but a typical relationship is shown in Figure 13.2 (National Coal Board, 1980). The addition of the theoretical requirement of limestone results in the retention of less than 50% of the sulphur. An addition of limestone of between 6 and 12 kg/kg coal (Ca/S mole ratio of between 2 and 4) is generally required to meet the current US EPA sulphur emission standards (Grimm, 1981). Newby *et al.* (1978) estimate that 90% desulphurisation of medium

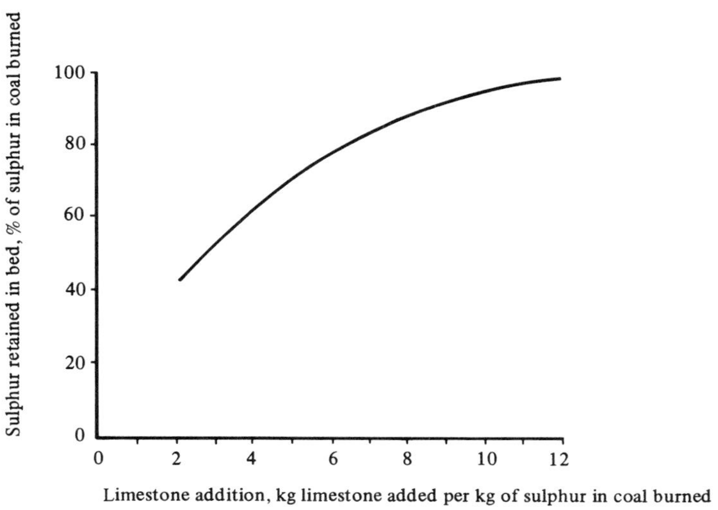

Figure 13.2 Sulphur retention in a fluidised bed as a function of limestone addition (after National Coal Board, 1980)

sulphur British coal (1.5% S by weight) will require an addition of limestone of at least 10 to 12% by weight of that of the coal. Highton and Webb (1980) give a figure of 15% by weight of the coal input. This scale of limestone utilisation increases the operating costs of FBC and detracts from its environmental acceptability. Not only must limestone be quarried, transported and crushed but large quantities of spent sorbent, often equal in weight to the ash produced from the combustion of the coal, must be disposed of (Burdett *et al.*, 1981).

Much of the current R and D effort on sulphur removal in FBCs is directed at increasing the efficiency of limestone utilisation, thereby decreasing the quantities required. Ultimately, the goal must be a sorbent which can be regenerated and reused indefinitely. However, limited success in this approach has led to investigation of processes in which spent sorbent is reactivated and reused for a limited number of passes through the FBC. The efficiency of limestone utilisation can also be increased by careful control of the operating parameters in a FBC and by pretreatment of the limestone (Gasner and Setesak, 1978). These techniques are discussed below.

Pretreatment of limestone

Pretreatment of limestone includes precalcination, heat treatment and the use of additives.

Precalcination

There is evidence that significantly improved sorbent utilisation can be achieved by controlled calcination of the sorbent prior to sulphation (Ulerich *et al.*, 1977). When untreated limestone is fed to a fluidised bed combustor calcination and sulphation reactions occur simultaneously. However, calcination occurs more slowly under these conditions than during calcination alone. The lower calcination rate during simultaneous calcination and sulphation produces a material having smaller pores and thus a low degree of utilisation (Smyk *et al.*, 1980).

Newby *et al.* (1980) list the following ranges of parameters governing the conditions for calcination:

1. The temperature of calcination should be in the range of 815 to 935°C.
2. The concentration of CO_2 in the calciner should be between 15 and 60% at atmospheric pressure.
3. A retention time at calcination conditions of up to 4 hours.
4. Higher heating rates produce sorbents of low porosity and hence lower reactivity.

These parameters can act individually or collectively to change the degree of utilisation of the calcined sorbents. Sorbent utilisation generally improves with increasing calcination temperature within the range given in condition 1, increasing CO_2 partial pressure, and increasing retention time. The optimum calcination conditions remain specific to each sorbent and must be determined experimentally.

Heat treatment

If a precalcined limestone is further heated, crystal growth occurs. Although total porosity decreases as the pores grow the surface area of the small pores, which become blocked during sulphation, is decreased and the area of the larger pores is increased. This results in a nett increase in active area for sulphation. Argonne National Laboratory is currently studying this method of improving the utilisation of limestone (Newby *et al.*, 1980). Five limestones have been heat treated at temperatures between 1000 and 1200°C for between 5 and 180 minutes under an atmosphere of air containing 20% CO_2. Three of the limestones exhibited increased utilisation after treatment whilst there was no change for the two remaining limestones.

Combined precalcination and heat treatment as a means of controlling pore structure has the potential of doubling the utilisation of a moderately reactive limestone (Smyk *et al.*, 1980).

Chemical treatment

The sulphur capture efficiency of limestone sorbents is enhanced by the addition of small quantities of chlorides, notably sodium chloride, NaCl (Shearer *et al.*, 1979, 1980a,b; Newby *et al.*, 1980). Whilst this fact has been known for many years there has been a reluctance on the part of developers of FBCs to introduce salt into large scale boilers because of the possibility of corrosion. Recent research at the University of Maryland, USA and the Argonne National Laboratory, USA has been directed at understanding the mechanism by which salts affect sorbents in order to develop other additives with the same effect. The results of this research indicate that salt induces structural rearrangement in limestones that can lead to an optimum pore distribution for reaction with SO_2/O_2 mixtures. This is primarily due to the presence of a liquid phase that increases the ionic diffusion and mobility of the system, enhancing calcination and crystallisation of CaO and creating a pore structure permeable to reactant gas diffusion (Shearer *et al.*, 1979, see Figure 13.3).

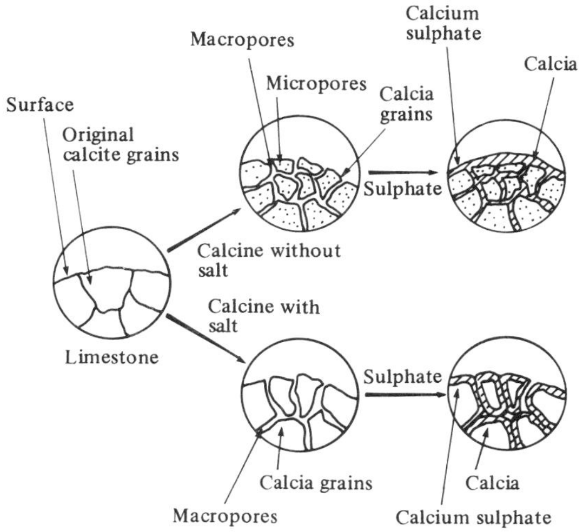

Figure 13.3 Effect of sodium chloride addition on the calcination and sulphation of limestone sorbent (after Shearer *et al.*, 1979)

Four chemically treated sorbents, developed by the University of Tennessee under an agreement with TVA, have been evaluated in the FBC facility at Morgantown Energy Technology Center (METC), USA (Grimm, 1981). These sorbents consisted of limestone and hopper ash in combination with a binder. A limestone was also tested as a control in addition to a mixture of this limestone and a power plant fly ash. The results are presented in Table 13.1. The percentage sulphur retention of the sorbents is shown in the first line. The second line shows the equivalent Ca/S ratio of limestone required to achieve the retention of the chemically treated sorbents and the limestone/fly ash. The third and fourth lines express the relative effectiveness of the sorbents normalised to limestone on a molar and a weight basis. The sorbents consisting of limestone, fly ash and binder show improved sulphur capture performance over the limestone alone. However, difficulties were experienced with the apparent separation of sorbent pellets in the FBC. It was recommended that further attention be given to the attrition resistance of chemically treated sorbents.

Snyder *et al.* (1977) have studied the use of alkali and alkaline earth metal oxides as sorbents (see also Newby and Keairns, 1978). Although several solid metal oxides react with SO_2 and O_2 to form stable sulphates most are suitable only for

Table 13.1 Sulphur capture efficiencies of chemically treated sorbents (after Grimm, 1981)

	Limestone	*Mixtures of limestone, fly ash and binder*				*Mixtures of limestone and fly ash*	
Sulphur retention, %	71	84	78	80	86	62	75
Limestone Ca/S for equivalent retention	2	3.48	2.56	2.78	4.00	1.43	2.43
Relative effectiveness, molar basis	1	1.74	1.28	1.39	2.00	0.71	1.21
Relative effectiveness, wt. basis	1	1.46	1.14	0.96	1.71	0.50	0.86

use at temperatures between 130 and 700°C, becoming unstable at the temperatures of 850–950°C experienced in FBCs. However, CaO on a support of αAl_2O_3 was found to be suitable at these temperatures. The sulphation rate of this sorbent was found to be highly dependent on its physical properties, particularly its pore size distribution. Porosity decreased with increasing concentrations of CaO thus limiting the effectiveness of the sorbent.

Operating parameters

Whilst changes in the operating parameters of a FBC are not as effective as other methods discussed in this review in improving limestone utilisation they should be considered during the design of a system (Smyk *et al.*, 1980). Those parameters having significant effects on the utilisation of limestone are limestone particle size, bed temperature, fines recycle, and those parameters which influence the gas residence time such as fluidising velocity, bed height and pressure.

Limestone particle size

The effect of limestone particle size on sulphur capture efficiency is illustrated in Figure 12.5. In general, below a particle size of 0.1 mm, the smaller the particle the greater is the percentage of sulphur captured. However, the minimum practical particle size may be determined by the fluidising velocity. At relatively high fluidising velocities fine limestone particles are elutriated from the bed before they

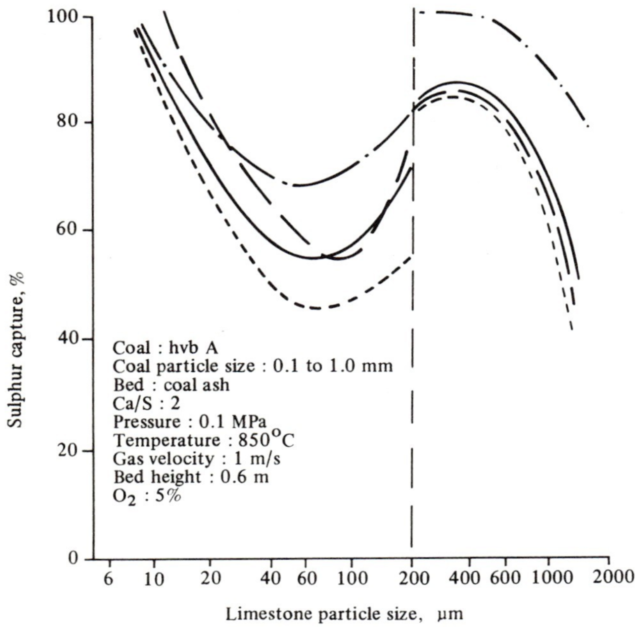

Figure 13.4 Sulphur efficiency versus limestone particle size for various limestones (after Munzner and Bonn, 1980)

have sufficient time to react with sulphur. This effect was noted by Highley (1975) who reports 20% greater SO_2 removal using limestone of less than 150 μm compared with limestone of 3 mm at a fluidising velocity of 0.9 m/s, but a corresponding 15% decrease in SO_2 removal at a fluidising velocity of 2.4 m/s. Figure 13.4 also indicates an optimum particle size of between 0.3 and 0.4 mm for particle sizes greater than 0.1 mm. This is confirmed by the work of Haque *et al.* (1979) who obtained maximum desulphurisation rates for two limestones at a particle size of 0.31 mm.

Bed temperature

The optimum bed temperature at which the sulphur removal efficiency is a maximum lies between 820 and 840°C. The value of this optimum is in disagreement with the theoretical temperature of 1100°C. The reason for this is still not fully understood. Highley (1975) reports that the effect of temperature on limestone reactivity is dependent on the degree of sulphation as shown in Figure 13.5. With fresh limestone the reactivity increases with temperature at least up to 950°C. However, with limestone sulphated to about 20% the optimum temperature is about 850°C. At higher percentages of sulphation the optimum becomes more pronounced and occurs at a lower temperature. This optimum temperature does not exist with dolomite in pressurised fluidised bed combustors (PFBCs). In fact the reactivity tends to increase slightly with increasing temperature.

Highley (1975) reports the effect of temperature on sorbent reactivity to be reversible: i.e. SO_2 emissions increase as the temperature is raised but revert to a low value when the optimum bed temperature is restored. The effect of temperature cannot therefore be explained in terms of sintering, slag formation or alteration of the pore structures of particles. Highley (1975) speculates that an

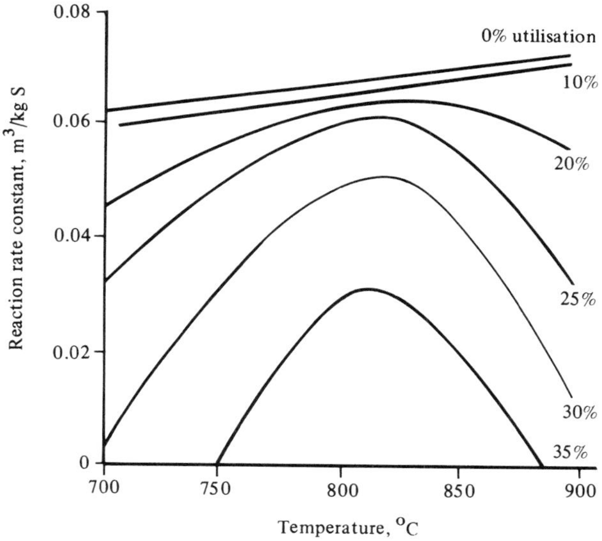

Figure 13.5 Effect of temperature on the reactivity of limestone sorbent at different degrees of sulphation (after Highley, 1975)

oxygen-containing species, originating from the residual water in limestone, participates in the sulphation reaction. Relatively high temperatures might cause this species to become mobile, lowering the sulphation reaction rate. The reduced sulphation rate might also be attributed to the lower SO_3 partial pressure at high temperatures.

Fluidising velocity

Increasing the fluidising velocity decreases the rate of desulphurisation. Highley (1975) correlates this effect by:

$$\frac{\text{sulphur emitted}}{\text{sulphur retained}} = XV$$

where V is the fluidising velocity and X is a constant which depends on the other operating conditions.

Bed height

An increase in bed height increases the residence time of both gases and solids and would therefore be expected to improve the efficiency of SO_2 removal. However, the efficiency of gas/solids contacting may reduce with increasing bed height because of increased bubble coalescence rates. Increases in sulphur removal efficiency with increasing bed height reported in the literature vary considerably up to about 15%. This variation may be due to the influence of immersed heat transfer equipment on the rate of bubble coalescence (Haque *et al.*, 1979).

Fines recycle

Fines elutriated from a bed may contain unreacted limestone as well as unburnt carbon and fly ash. The amount of unreacted limestone depends on the operating conditions within the bed, in particular the fluidising velocity and the Ca/S ratio. Hence, any increase in the efficiency of sulphur removal achieved by the use of fines recycle will depend on the original operating conditions. Highley (1975) gives an example of a 0.9 m square combustor operating at a fluidising velocity of 0.6 m/s and a Ca/S mole ratio of 1.6 in which total recycle of fines larger than 10 μm increases the SO_2 removal from 73 and to 99%.

Pressure

Dolomite is the preferred sorbent for use in pressurised fluidised bed combustors. With limestone, sulphur retention is reduced with increasing pressure. Limestone does not calcine under these conditions thus limiting sulphation to the outer layers of the limestone particle. However, the magnesium carbonate content of dolomite calcines at low temperature and generates a large internal reaction surface. Highley (1975) reports increasing sulphur retention rates for dolomites with increasing pressure up to 0.6 MPa which he attributes to better gas/solid contacting. However, more recent work at higher pressures of up to 2.0 MPa has shown a marked decrease in the sulphur sorption properties of dolomites with increasing pressure (Ulerich *et al.*, 1981). A Ca/S ratio of 1.6 was required to achieve removal of 90% of the sulphur from a fluidised bed combustor operating at 0.6 MPa. This ratio was increased to 2.7 for similar sulphur removal from a bed operating at 1.6 MPa. The authors attribute this increased sorbent requirement to the reduction in bed volume and hence residence time of the gas in the bed.

Post-sulphation treatment

Limestone which has passed through a FBC as a sulphur sorbent is generally only partially sulphated. A shell of $CaSO_4$ forms on the limestone particles effectively isolating the remaining CaO from SO_2. Rarely is more than 50% of the CaO present in limestone or dolomite converted to $CaSO_4$ and often this figure is as low as 10% (Shearer *et al.*, 1980c). Increased utilisation of limestone may be achieved by either converting the $CaSO_4$ back to CaO or by rearranging the physical structure of the partially sulphated particles to expose fresh CaO.

Regeneration

At temperatures of between 1000 and 1100°C limestone or dolomite may be regenerated by the following reductive decomposition reactions:

$$CaSO_4 + CO \rightarrow CaO + SO_2 + CO_2 \qquad (13.6)$$

$$CaSO_4 + H_2 \rightarrow CaO + SO_2 + H_2O \qquad (13.7)$$

The reducing gas, CO + H_2, can be generated by the partial combustion of coal. The feasibility of such a process depends on:

1. The reactivity and resistance to attrition of the sorbent during subsequent sulphation cycles.
2. The production of a gas stream sufficiently rich in SO_2 to enable it to be treated by a sulphur recovery process.

The process has been studied at Argonne National Laboratory, USA (Montagna *et al.*, 1977a,b; 1978a,b; Jonke *et al.*, 1977). A flowsheet is shown in Figure 13.6. Tests were conducted over a temperature range of 1000 to 1100°C with a particle size of about 1.1 mm. Between 15 and 89% regeneration of sulphated Tymochtee dolomite was achieved using methane as the fuel. In subsequent tests, Tymochtee dolomite and Greer limestone were regenerated using coal in a 10.8 cm regenerator operating at 152 kPa. Regeneration approached 95%. Tymochtee dolomite was also subjected to cyclic combustion–regeneration tests. The dolomite was fed to a 15.2 cm diameter combustor operating at 810 kPa. At the end of each test the sulphated dolomite was fed to the above mentioned regenerator at 152 kPa and 1100°C. This sulphation followed by regeneration of sorbent was continued for 10 cycles. Losses of both combustion and regeneration activity were recorded. Similar tests were made with Greer limestone using a combustor operating at 300 kPa. Again, declining activity was apparent.

At lower temperatures (≈800°C) and under more strongly reducing conditions the formation of CaS is favoured:

$$CaSO_4 + 4CO \rightarrow CaS + 4CO_2 \qquad (13.8)$$

$$CaSO_4 + 4H_2 \rightarrow CaS + 4H_2O \qquad (13.9)$$

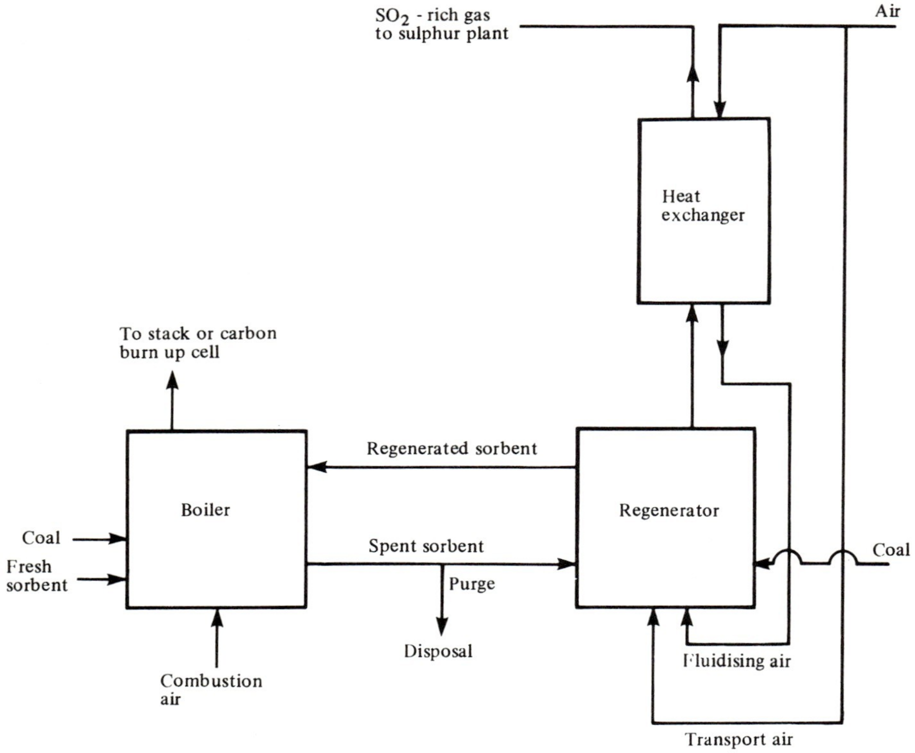

Figure 13.6 Regeneration of spent sorbent from an FBC by reductive decomposition (after Newby *et al.*, 1980)

The CaS can be reacted with carbon dioxide and steam to produce calcium carbonate and hydrogen sulphide, from which sulphur may be recovered. This route is not suitable for atmospheric pressure FBC (AFBC) as it would introduce an unnecessary heat load to regenerate to carbonate and then recalcine to the oxide. However, in pressurised FBC (PFBC) the carbonate is not calcined and this route might therefore be acceptable.

A combined process incorporating both regeneration and sorbent disposal is illustrated in Figure 13.7 (Newby *et al.*, 1980). The main advantage claimed for this process is the elimination of the expensive processing of the SO_2 stream, described above, by the disposal of highly sulphated limestone.

Several authors have evaluated critically sorbent regeneration (e.g. Highley, 1975; Newby *et al.*, 1980). Highley (1975) concludes that overall plant efficiency, capital investment and operating costs favour sorbent disposal and that regeneration can only be justified as a means of reducing the environmental impact of solid wastes. Newby *et al.* (1980) list the following technical uncertainties for the combined regeneration/sorbent disposal process: limited small scale data; ash sintering; sensitivity of process to sorbent type; ash accumulation; sulphur removal requirement; unknown environmental impact; and high temperature solids circulation. The regeneration process incorporating sulphur recovery is claimed to have the following additional uncertainties: sulphur recovery efficiency; quality of recovered sulphur; impact of SO_2 recycle to combustor; plant control integrated with sulphur recovery; and sensitivity to attrition.

Hydration

Whilst calcium hydroxide has long been recognised as a highly reactive species for SO_2 sorption (Shearer *et al.*, 1980c,d) the economics of producing $Ca(OH)_2$, the problems of handling hydrated lime, and the attrition and elutriation of soft $Ca(OH)_2$ have made its use as a sorbent in FBC application impracticable. However, work at Argonne National Laboratory, USA involves the hydration of spent or unreacted sorbent that on subsequent dehydration reacts readily with SO_2. The formation of $Ca(OH)_2$ in a partially sulphated sorbent results in a material that has sufficient physical integrity, due to the outer sulphate layer, and high reactivity to make it suitable as a sorbent feed material in FBCs.

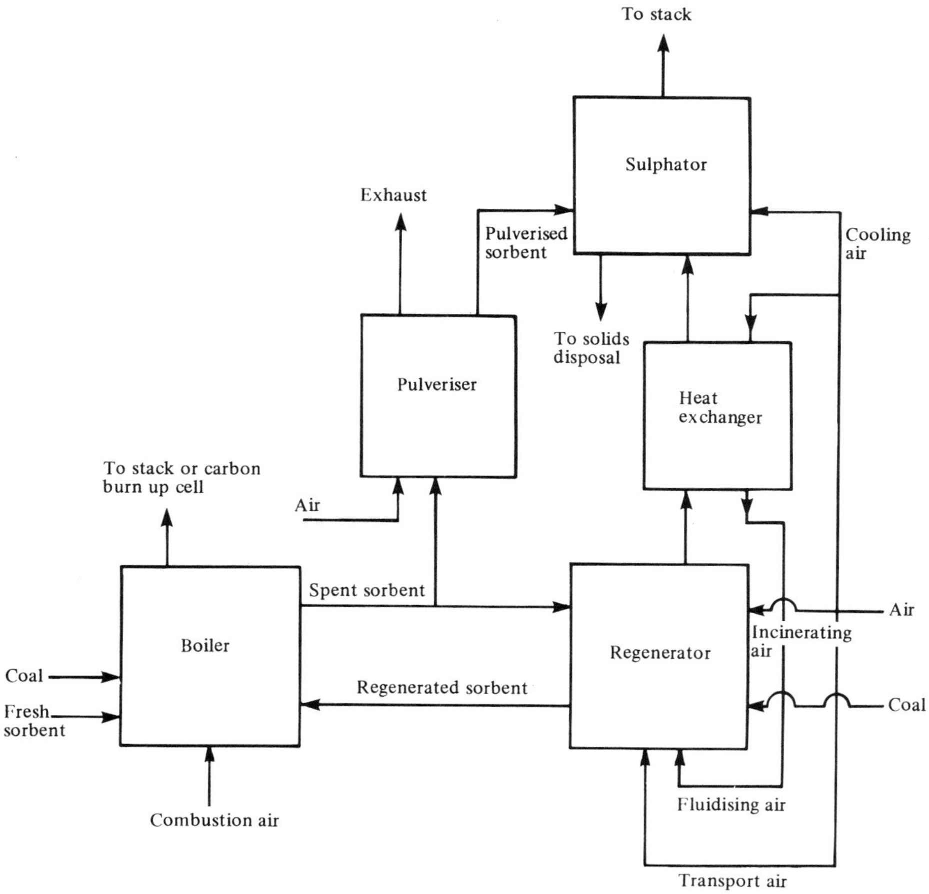

Figure 13.7 Combined process incorporating regeneration and disposal of spent sorbent from an FBC (after Newby *et al.*, 1980)

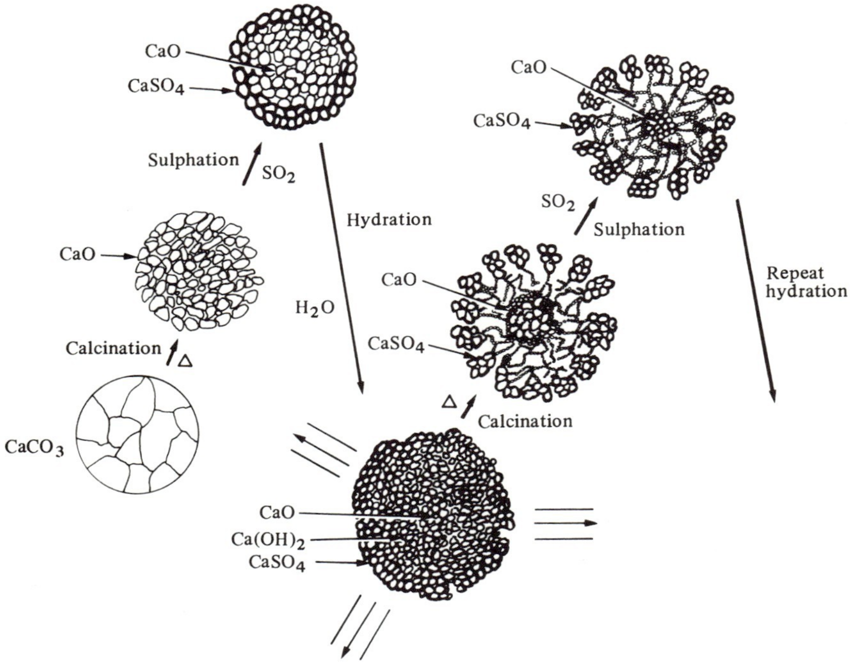

Figure 13.8 Schematic model of the enhancement of limestone sulphation by hydration (after Shearer *et al.*, 1980c)

Figure 13.8 is a schematic illustration of the mechanism of the enhancement of limestone sulphation by hydration (Shearer *et al.*, 1980c). The mechanism is based on laboratory experiments, porosity measurements and scanning electron microscope photography. Unreacted sorbent enters the FBC where calcination and sulphation occur simultaneously producing a dense layer of $CaSO_4$ around a core of residual CaO. The pore structure is that of the original sorbent. Upon partial hydration of the unreacted CaO, particle expansion occurs due to the formation of $Ca(OH)_2$, the molar volume of which is larger than that of CaO. The surface layer of $CaSO_4$ is unaltered other than being subjected to cracking from the expansion of the particle interior. The partially hydrated sorbent is reintroduced into the FBC where dehydration and sulphation occur simultaneously. The gas permeability of the treated material increases greatly due to an increase in average pore size, an increase in total porosity and an increase in surface area due to recrystallisation of the CaO. The sulphation reaction proceeds until sufficient $CaSO_4$ has formed to seal off any remaining CaO from further reaction. The hydration reaction may then be repeated.

Figures 13.9 and 13.10 show the percentage conversion of CaO to $CaSO_4$ for untreated and hydrated/resulphated limestones and dolomites respectively (Shearer *et al.*, 1980c; Smith *et al.*, 1981). There is a marked increase in the total sulphation capacity for the sorbents studied. Sorbents having a lower initial calcium conversion exhibit the greatest increases in conversion due to a single hydration/resulphation step. In three of the experiments a limestone and two dolomites were

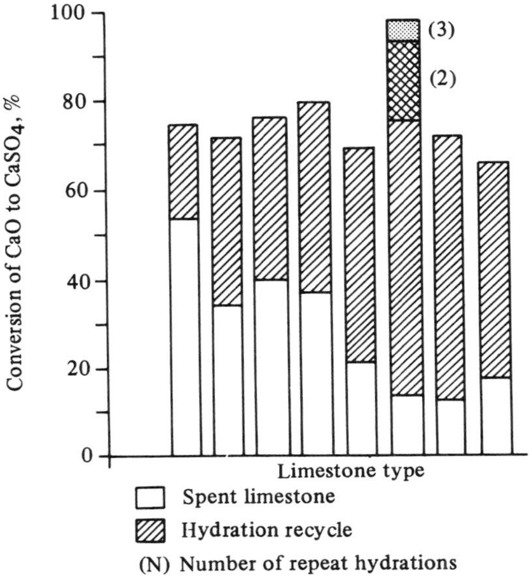

Figure 13.9 Increased sulphation of several limestones by hydration of spent sorbent (after Shearer *et al.*, 1980c)

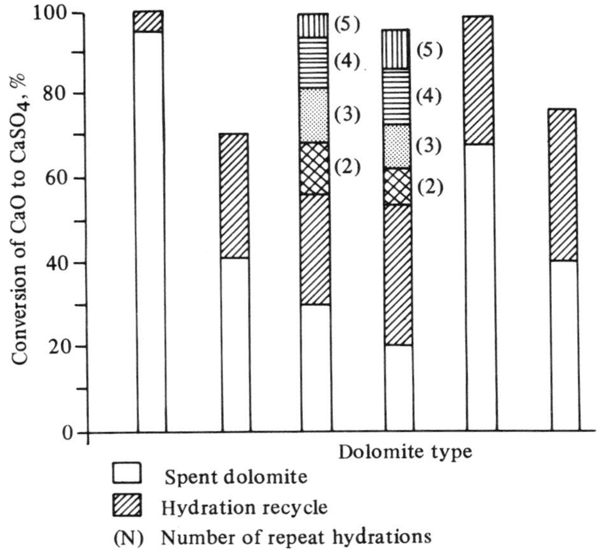

Figure 13.10 Increased sulphation of several dolomites by hydration of spent sorbent (after Shearer *et al.*, 1980c)

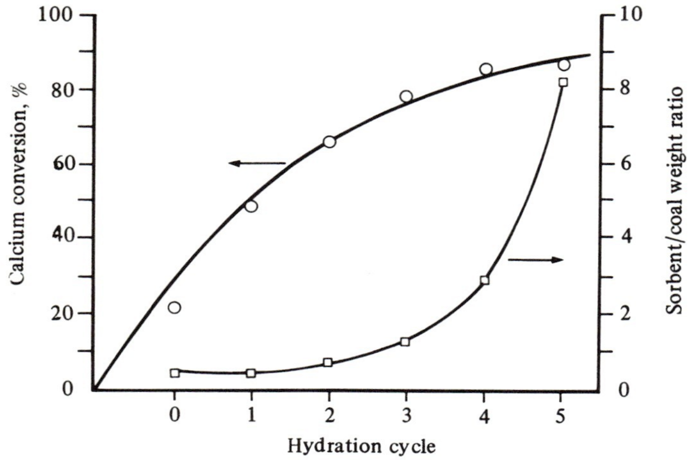

Figure 13.11 Calcium conversion as a function of hydration cycle for a limestone sorbent (after Shearer *et al.*, 1980a)

subjected to several successive hydration–sulphation cycles to determine the maximum utilisation of the available calcium oxide. The results indicate the potential for complete sulphation even in sorbents with the lowest initial reactivity.

Figure 13.11 shows the results of a series of cyclic hydration–sulphation experiments (Shearer *et al.*, 1980a; Smith *et al.*, 1981). After cycle 2, 68% of the calcium is utilised. This is three times the conversion of 22% experienced in the initial sulphation of the sorbent represented by hydration cycle in Figure 13.11. Although after five cycles the sulphated sorbent approaches a maximum utilisation of approximately 90% the amount of sorbent required increases significantly. The authors conclude that it is not advantageous to perform more than three hydration cycles because the rapid increase in the sorbent/coal weight ratio more than offsets the moderate increase in calcium conversion. After the fifth cycle a significant loss of reactivity of the available calcium occurs due to a buildup of an impermeable layer of $CaSO_4$ and, possibly, to a sintering of the residual CaO.

Agglomeration

Agglomeration of both fresh and spent sorbent has been suggested as a means of enhancing sorbent utilisation (Dunne and Gasner, 1980). Agglomerates can retain much of the chemical reactivity of their component fine particles. In particular, the high reactivity due to the high surface to volume ratio of fine particles can be retained. Agglomerated particles have inherently large macropores. Less reactive limestones could thus have agglomerates tailored to a more favourable internal pore structure and reactive internal surface. However, the main application of agglomeration for increasing the efficiency of sorbent utilisation is for reactivating spent sorbent.

Two spent sorbent agglomerates have been studied at the University of Maryland (Dunne and Gasner, 1980). One was produced from a spent bed with a high iron content and the other from a bed with a low iron content. The agglomerates were identified as RSB and GSB, respectively. The spent bed materials were pulverised

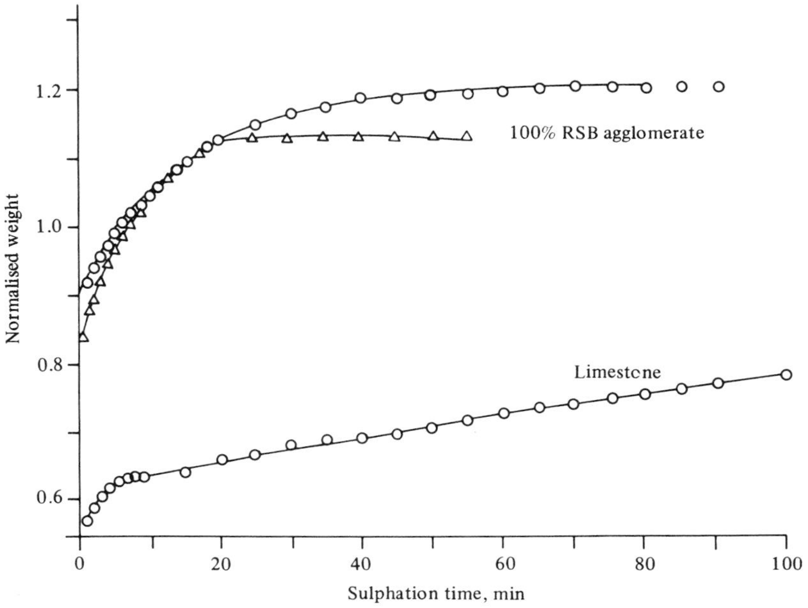

Figure 13.12 Rate and capacity of sulphur capture for RSB agglomerate material (Dunne and Gasner, 1980)

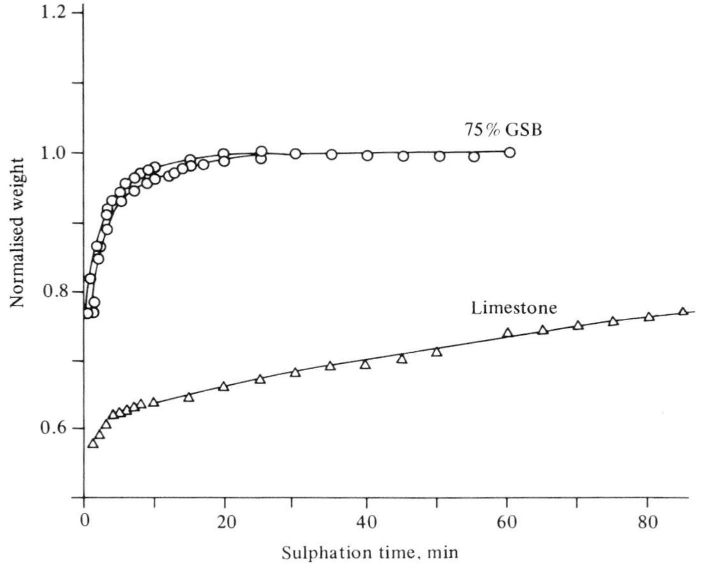

Figure 13.13 Rate and capacity of sulphur capture for GSB agglomerate material (Dunne and Gasner, 1980)

to a mean size of about 320 μm prior to agglomeration in a rotary drum. As a fine spray of water was used in the drum to assist agglomeration, the agglomerates were effectively hydrated. The agglomerates were calcined and sulphated at 850°C. The sulphation capacities of the agglomerates were determined from the difference in weight of a sample prior to calcination and after sulphation. Figures 13.12 and 13.13 are graphs of the increased weight of a sample versus time for both the RSB and GSB materials and also, for comparison purposes, for Greer limestone. The agglomerated materials not only capture more sulphur but do so more rapidly than does the Greer limestone. However, the approximate twofold increase in sulphur capture is similar to the results obtained by Shearer *et al.* (1980c,d) from experiments on the hydration of spent sorbent. It is not clear how much of the increased sulphur capture of agglomerates is due to changes in their physical properties and how much is due to the hydration of the sorbent during the agglomeration process.

Chapter 14

Wet scrubbing flue gas desulphurisation processes

There are in excess of fifty processes in various stages of development and commercial application for the removal of SO_2 from industrial waste gases and boiler flue gases (Leivo, 1978). This review considers the processes which have been applied or are potentially applicable to coal-fired utility boilers.

There are many ways in which these processes can be categorised: e.g. reagent chemistry; form of product; equipment; chronology of development; country of development; degree of development or commercial application. In this review a major distinction is made between those processes which result in a wet product and those which result in a dry product. The physical absorption of SO_2 in water has been studied for many years. In the 1930s a number of flue gas desulphurisation (FGD) applications were based on this technique (Yarze and Beiersdorf, 1979). The solubility of SO_2 in water is low. Hence, unless very high liquid to gas ratios are used the rate of removal of SO_2 from flue gases is also low. The solubility of SO_2 is increased by increasing the alkalinity of the scrubbing media (increasing the pH value). Contacting flue gas with an alkaline solution or slurry in a gas absorption vessel forms the basis of many of the FGD processes in existence today. The vessel normally takes the form of a spray tower with either countercurrent or cocurrent flow (Hollinden *et al.*, 1979b; Jackson, 1981) although the dispersal of gas bubbles through the liquid has been investigated (Idemura *et al.*, 1978). Many of the processes also remove some of the particulate matter and trace elements from flue gas (Jahnig and Shaw, 1981b). During the scrubbing process evaporation of water cools the flue gas to slightly above its dew point. This necessitates two further processing steps. Mist eliminators are required to remove entrained droplets of scrubbing media. These generally take the form of baffles or vanes upon which the desulphurised flue gases impinge. It is also necessary to reheat the flue gas to about 130°C to prevent condensation, fouling and corrosion in ducts, fans and stacks (Verhoff and Choi, 1979), to avoid a visible plume, and to improve plume dispersion from the stack. Dry SO_2 removal processes minimise or eliminate the requirements for mist elimination and stack gas reheat.

There are both non-regenerable and regenerable wet scrubbing processes. Non-regenerable processes are defined as those in which the scrubbing media are disposed of as solid or liquid wastes. The liquid, or sludge, form of waste is the more common and requires storage in ponds. Further processing to make the waste suitable for landfill is sometimes practised. This is a major disadvantage of non-regenerable processes. Because of its importance a separate section of this review is allocated to the problem of sludge disposal. In regenerable processes the scrubbing medium is regenerated and reused whilst SO_2 is recovered in a form which may be processed further to sulphuric acid or elemental sulphur. Such processes are generally more complex than non-regenerable processes.

Processes are further classified within the above groups according to the form of the chemical reagent used. Wet scrubbing processes may use either an aqueous slurry or an aqueous solution. When scrubbing with an aqueous slurry some of the absorbent and some of the reaction products are present as suspensions of solids in water. During the absorption step of the scrubbing process the suspended alkali dissolves to react with the absorbed SO_2. Waste solids are precipitated out of solution to be disposed of in non-regenerable processes or treated to regenerate active alkali in regenerable processes. Control of the process chemistry is very important to prevent the deposition of solids in the form of scale in the scrubber and associated equipment. Scrubbing with an aqueous solution, in which the absorbent and the reaction products remain in solution, minimises the occurrence of scale formation.

Table 14.1 shows the classification of wet scrubbing processes used in this review.

Table 14.1 Classification of wet FGD processes

Non-regenerable		Regenerable	
Slurry	*Solution*	*Slurry*	*Solution*
limestone	Chiyoda Thoroughbred 121	Magnesium oxide	Wellman Lord
lime	Saarberg-Hoelter		Citrate
alkaline fly ash	Sodium carbonate		Ammonia
	Dual alkali		

Non-regenerable processes

The list of non-regenerable processes is dominated by lime and limestone scrubbing or variations of these calcium based processes. The main requirements for these processes are that the scrubbing media be readily available and relatively inexpensive. Lime and limestone processes have the lowest energy requirements of all FGD processes (Thomas, 1978a,b). In the USA almost 50% of present and planned FGD capacity is of these types (Jahnig and Shaw, 1981a). Because of their predominance lime or limestone scrubbing processes are often considered as 'first generation' or 'base' processes when comparison between FGD processes is made. However, the list of non-regenerable processes also encompasses several other alkali scrubbing processes.

Lime and limestone scrubbing processes

Lime and limestone scrubbing processes involve the contacting of solutions or suspensions of lime or limestone with flue gases in gas absorption vessels. In the FRG the lime scrubbing process is known as the Bischoff process (Davids, 1979). A much simplified form of the chemistry of these processes is represented by:

$$SO_2 + CaCO_3 \rightarrow CaSO_3 + CO_2 \qquad (14.1)$$

$$SO_2 + CaO \rightarrow CaSO_3 + H_2O \qquad (14.2)$$

Because oxygen is present in the flue gas some of the calcium sulphite is oxidised to calcium sulphate, $CaSO_4$, also known as gypsum:

$$CaSO_3 + 1/2O_2 \rightarrow CaSO_4 \qquad (14.3)$$

Reactions (14.1) and (14.2) proceed more rapidly at high values of pH However, the solubilities of $CaCO_3$ and CaO decrease with increasing values of pH. In practice it is necessary to maintain the pH of lime systems between 6.5 and 8.9 to achieve a compromise between the concentrations of lime in solution and acceptable reaction rates. It is practically impossible to achieve a value of pH greater than 7 when using limestone.

Even when the above conditions are imposed the rate of absorption of SO_2 is low. Hence, in order to avoid excessively high rates of recirculation of the scrubbing media the dissolution of calcium salts *in situ* is often practised (Toprac and Rochelle, 1982). This ensures that the scrubbing solution remains saturated with respect to calcium as the reaction proceeds. The pH of the scrubbing medium decreases as it passes through the absorption vessel increasing the solubility of calcium compounds. However, the circulation through the scrubber equipment of solid $CaCO_3$ or CaO and possibly $CaSO_3$, which may be precipitated from solution, increases the possibility of scaling and blocking. Chang and Dempsey (1982) discuss the effect of limestone type and particle size on scrubber efficiency.

Scaling

Scaling is one of the major problems associated with lime/limestone scrubbing particularly in early applications of the processes when the importance of the control of the process chemistry was not fully appreciated (Jones *et al.*, 1978; Corbett *et al.*, 1977). Scale formation can result from carbonate deposition, sulphite deposition or sulphate deposition. As discussed above, careful control of the concentration of calcium carbonate in the scrubbing media and of the pH of the scrubbing media minimises the formation of the two former deposits (Mesich and Jones, 1979). Sulphate deposition is by far the most difficult to control (Rosenberg and Grotta, 1979). Calcium sulphate is virtually insoluble and, unlike calcium sulphite, its solubility does not increase with decreasing values of pH. In addition, the solubility of calcium sulphate is decreased by the presence of chloride ions. Chloride ions may enter the unit operation from the limestone, the make-up water or from the coal. There have been several approaches to the control of sulphate scaling. These include: limiting the extent of oxidation of sulphite to sulphate, prevention of deposition, and preferential deposition (Slack, 1978).

Calcium sulphate is formed continuously in the absorber loop. It may be removed from the system as part of the calcium sulphite/sulphate slurry. When the degree of oxidation of sulphite to sulphate is high, more calcium sulphate is formed in the slurry than is able to leave the system. This causes the system to operate supersaturated with respect to calcium sulphate. It has been found that calcium sulphate can supersaturate to a factor of 1.4 before massive nucleation occurs but deposition can still occur below this value (Karlsson and Rosenberg, 1980a). If the system is operated so that the maximum oxidation in the slurry loop is about 16% the scrubbing medium remains unsaturated with respect to calcium sulphate and scale deposition is prevented (Laseke and Devitt, 1979a). Such operation is, however, difficult to achieve, especially when the ratio of oxygen to SO_2 in the flue gas is high. This occurs during the combustion of low sulphur coals or in older boilers with high volumes of excess air. The use of calcium hydroxide (also known as carbide lime because it was a byproduct of acetylene manufacture) has the effect of inhibiting the oxidation rate (Van Ness *et al.*, 1980) although the exact mechanism by which this occurs is not known (Karlsson and Rosenberg, 1980a,b).

The use of additives has been successful in the prevention of sulphate deposition. The addition of small quantities of magnesium sulphite to the scrubbing medium permits a much higher concentration of sulphate ions in solution before supersaturation occurs. The relatively high solubility of magnesium sulphite has the additional benefit of increasing the rate of reaction (14.1) and hence the removal of SO_2. A magnesium additive forms the basis of the Pullman Kellogg magnesium promoted limestone slurry process (Raymond *et al.*, 1978; Granger *et al.*, 1979; Yarze and Beiersdorf, 1979). A lime-magnesium FGD process incorporating oxidation of the $CaSO_3$ to gypsum has been installed on a 700 MW power station in the FRG (Atzger, 1979). The use of organic acids as additives has also been studied in the USA by the Tennessee Valley Authority (TVA) and EPA (Head *et al.*, 1979). Both benzoic acid and adipic acid have the effect of buffering the pH of the scrubbing medium during contact with flue gas (Maxwell, 1979; Mobley and Chang, 1981). This increases the mass transfer of SO_2 from the flue gas to the liquid. In addition, because sulphurous acid is the stronger acid, the benzoate or adipate ions act as bases during the absorption of SO_2 thus increasing the total alkalinity of the scrubbing medium (Devitt *et al.*, 1980). Preliminary tests have shown adipic acid to be effective when used in conjunction with forced oxidation and in the presence of chlorides; conditions which adversely affect magnesium additives. However, deterioration or decomposition of adipic acid occurred in the scrubber (Chang and Borgwardt, 1980; Meserole, 1980; Burbank *et al.*, 1981).

The third method of controlling sulphate deposition is that of preferential deposition. This is normally achieved by allowing large amounts of gypsum to recirculate so that deposition starts on these seed crystals instead of on scrubber surfaces. A concentration of solid gypsum of 5% is sufficient for this purpose. Forced oxidation whereby sulphite is deliberately oxidised to the sulphate within the FGD unit may be used to provide nucleation sites. In Japan, gypsum seed crystals are sometimes added to the process. Fly ash may also be used to provide nucleation sites. A total solids content of greater than 10% causes crystallisation of the sulphate on the solids (Karlsson and Rosenberg, 1980a). However, an increased solids content in the slurry loop makes it difficult to pump the high volumes of slurry required for high rates of SO_2 removal. An upper limit of 15% is generally accepted (Karlsson and Rosenberg, 1980a).

Prescrubbing

Prescrubbing of flue gas prior to lime/limestone scrubbing is not generally required unless the chloride content of the coal being burned is high. In fact, the lime/limestone absorption vessel provides additional removal of particulates so that less efficient removal by the particulate removal system up-stream of the scrubber can be tolerated. Jahnig and Shaw (1981a) predict a particulate removal efficiency of greater than 97% assuming 95% removal of fly ash in an upstream electrostatic precipitator.

A prescrubber may, however, be required to reduce the number of chloride ions reaching the absorption vessel. As well as increasing the possibility of scaling, chloride ions decrease the efficiency of SO_2 removal and cause corrosion of process equipment (Paul, 1978; Hollinden *et al.*, 1979b). The use of a prescrubber or 'quencher' in indirect or double loop operations is described below.

Open, closed and double loop systems

The term 'loop' is often used to refer to the recirculation of lime/limestone slurry through the absorption vessel and recirculation tank. An 'open' loop describes a unit operation in which fresh make-up water is used to replace the water discarded with the waste sludge. However, waste water streams are a potential source of pollution. Selenium in FGD process water has been identified as a potential pollutant of surface and ground water (Sayre, 1980). Waste water streams for power plants with and without SO_2 control processes are characterised by Sugarek and Sipes (1978a,b). In most countries there exist regulations relating to the quality of water which may be discharged from a plant and in specific circumstances the discharge of all water is prohibited. This necessitates the use of 'closed loop' systems in which water is recirculated.

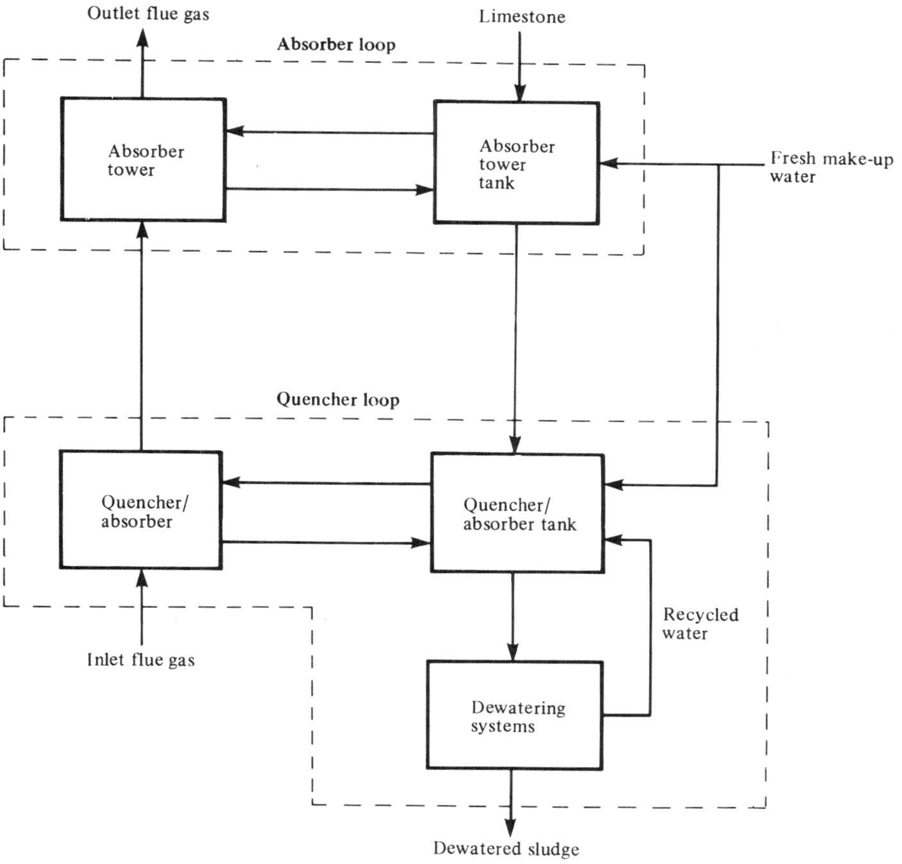

Figure 14.1 Double-loop limestone FGD process (Braden, 1978)

Operating a lime/limestone scrubbing system with a closed loop creates two major operational problems:

1. The return water is saturated with respect to calcium sulphate.
2. Inorganic salts, particularly chlorides, accumulate within the recirculated water.

Relatively high quality wash water is required to flush the demisters within a lime/limestone scrubbing system. It may be possible to dilute the return water with sufficient make-up water to achieve this quality. However, in certain operating modes the return water must be used undiluted, creating conditions which can result in serious scaling of the demister (Braden, 1978). The accumulation of chlorides is equally serious. In open loop operation the chloride concentration usually remains relatively low, seldom exceeding 3000 ppm (Braden, 1978). However, in closed loops chloride concentration may reach as high as 50 000 ppm necessitating the use of high molybdenum content steels, nickel carbide alloys or organic linings within the absorption vessel. In addition, a high concentration of chlorides has a detrimental effect on the relative solubilities of the reagents used in the scrubbing media, as discussed above. The problems of water management in closed loop systems are discussed by Borgwardt (1980) and Domahidy (1979).

Braden (1978) describes a double loop system, operated by Research-Cottrell which, it is claimed, obviates the need to use recirculated water within the absorber whilst not discharging water from the plant. The two loops, shown in Figure 14.1, operate at discrete process conditions with different values of pH and different concentrations of solids in the slurries. The absorption vessel operates as an open loop. By maintaining a pH value of 6.3 and a solids concentration in the slurry of 10% the oxidation of calcium sulphite to sulphate is suppressed and hence the formation of sulphate scale prevented. The quench vessel is operated as a closed loop. The quench solution is supersaturated with respect to calcium sulphate. Scale formation is prevented by providing a solids concentration of 15% to act as nuclei for sulphate deposition. The pH of the quench solution is maintained at about 5.0 to ensure that any solid limestone entering the quench loop from the absorber vessel is rapidly dissolved. Although the amount of SO_2 removed from the flue gas by the quench loop is low, about 40%, when combined with the additional removal achieved in the absorber loop overall removal rates can approach 95%. High concentrations of chlorides are confined to the quench loop.

Selection and utilisation of reagent

Lime is produced by calcining limestone at 825°C in a lime kiln. The cost of this processing is reflected in the cost of the lime produced which is considerably more than that of limestone. The cost of transporting lime is also higher than that of limestone as lime must be protected from moisture. Karlsson and Rosenberg (1980a) estimate the cost ratio of lime to limestone on a molar basis to be between 2 and 4 depending on the transportation distance.

Lime has certain advantages over limestone in FGD applications. Cases of pH instability have been reported for limestone due to its relatively slow rate of dissolution. Limestone has a greater liquid-side resistance to mass transfer. It is also claimed that an unsaturated mode is more readily attained with lime (Karlsson and Rosenberg, 1980a). Despite these relative advantages FGD systems based on lime are becoming less popular than those based on limestone. Recent advances in the reliabilities of both lime and limestone systems have resulted in the cost of reagent being the overriding consideration. Of the plants under construction or contracted in the USA twice as many are based on limestone than on lime. Very few, if any, lime systems are under construction for projected plants (Karlsson and Rosenberg, 1980a).

Whether lime or limestone is employed, efficient utilisation of reagent is important from an economic point of view. Single loop, calcium based FGD systems must operate at pH values of about 6 to 6.5 to obtain adequate removal of SO_2. The concentration of reagent, particularly when using limestone, in the solution at these values of pH can be as low as 2 to 4%. At these low solubility values part of the reagent may be discharged with the slurry. When 90% or greater removal of SO_2 is required limestone utilisation may be as low as 70% (Braden, 1978) but is typically between 75 and 90% (Karlsson and Rosenberg, 1980a). With double loop operation reagent utilisation approaches 100% because conditions are such that any unused reagent discharged from the absorber loop is dissolved in the quench loop.

As of January 1980 there was a total of 35 680 MW of plant in operation or committed to limestone or lime scrubbing processes in the USA. This represented over 70% of the total FGD capacity in the USA. Operating and maintenance experience is well documented (e.g. Spring, 1980; Hewitt and Saleem, 1980).

Scrubbing with alkaline fly ash

Under certain circumstances fly ash can be used to supplement or even replace lime or limestone as the principal scrubbing agent in FGD systems. Many of the early scrubber plants installed in the western states of the USA for particulate control were discovered to be removing appreciable amounts of SO_2 (30 to 40%) and developing sulphate scale without the use of external reagents (Laseke and Devitt, 1979a). This was found to be due to the inherent alkalinity of the fly ash associated with western USA lignite and subbituminous coals. These coals may contain concentrations of sodium oxide, magnesium oxide and calcium oxide amounting to between 25 and 30% of the total fly ash. Davis and Fielder (1982) have correlated sulphur retention in fly ash as a function of the alkaline metal content of the fly ash. Research into the feasibility of alkaline fly ash scrubbing for SO_2 removal was undertaken by the Grand Forks Energy Research Center of the US DOE in 1971 (Laseke and Devitt, 1979a). As of January 1980, 1480 MW of capacity were in operation.

Three main factors need to be considered in establishing the potential for using fly ash as a source of alkali: the quantity of SO_2 to be removed per unit weight of coal, the quantity of fly ash produced per unit weight of coal, and the availability of alkali contained within the fly ash. Alkali availability increases significantly at values of pH of 4 or below. With a conventional lime/limestone scrubbing process SO_2 removal efficiencies would be extremely low at these low values of pH necessitating the use of high liquid to gas ratios. This is not the case when fly ash is present (Grimm *et al.*, 1978; Johnson, 1979).

Figure 14.2 shows typical performance curves for SO_2 removal efficiency as a function of liquid to gas ratio for both a limestone system and an alkali fly ash system. At SO_2 removal efficiencies of greater than about 40% the liquid to gas ratio is lower for the fly ash system than for the limestone system for the same removal efficiency. Hence, substantial savings in pumping costs can be realised (Johnson, 1979). Further cost advantages can be effected by integrating the

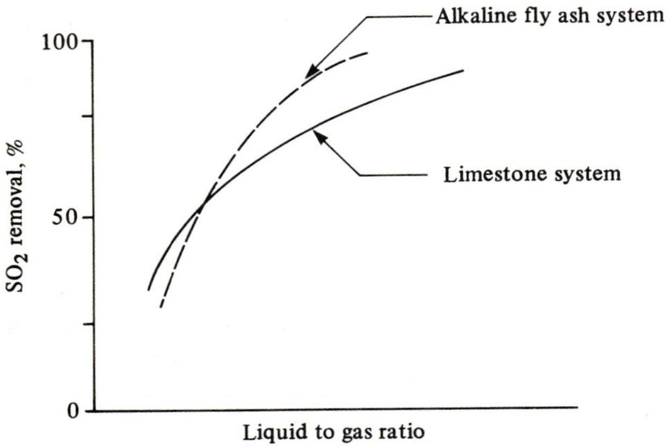

Figure 14.2 SO_2 removal efficiency as a function of liquid to gas ratio for a limestone system and an alkaline fly ash system (after Johnson, 1979)

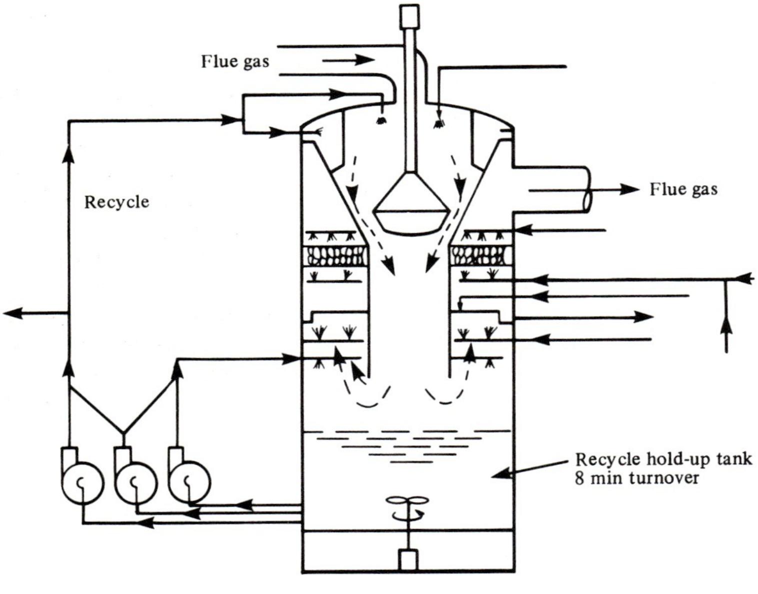

Figure 14.3 Venturi scrubber and SO_2 absorber (after McCain, 1979)

particulate removal equipment with the SO_2 removal equipment. This normally takes the form of a venturi scrubber combined with the SO_2 absorber as shown in Figure 14.3 (McCain, 1979).

An additional advantage of using fly ash is claimed to be improved waste solids characteristics. The chemistry of the alkali fly ash system results in all the waste solids being oxidised to calcium sulphate. Calcium sulphate has improved settling and dewatering properties compared with calcium sulphite/sulphate sludges generated by conventional lime/limestone scrubbing processes.

Whilst the use of the alkaline properties of fly ash clearly has both operational and economic advantages over lime/limestone scrubbing, its application is limited to those coals which produce fly ashes with suitable chemical properties.

Chiyoda Thoroughbred 121 process

This process is based on the use of a limestone slurry as sorbent but with forced oxidation of the $CaCO_3$ to produce marketable gypsum. Whilst the chemical reactions involved are essentially those described in the section on non-regenerable processes, the process differs from others described in this chapter in that absorption, oxidation and gypsum formation take place in a single sparged tank of limestone slurry known as a jet bubbling reactor (Idemura, 1979). Flue gas is bubbled through a solution of limestone supersaturated with respect to gypsum. Air is supplied to the absorber to ensure that the $CaSO_3$ formed by reaction (14.1) is oxidised to $CaSO_4$ by reaction (14.3). The crystals of gypsum grow as they are recirculated in the slurry until they reach sufficient size to precipitate to the bottom of the vessel. Gypsum slurry is continually withdrawn from the absorber. Part of this slurry is recycled to provide seed crystals. The balance is pumped to centrifuges for dewatering (Leivo, 1978).

Prescrubbing of the flue gas is required to prevent excessive contamination from fly ash or chlorides which impair crystallisation and filtration of the gypsum product (Jahnig and Shaw, 1981a). The prescrubbers are spray towers in which the flue gas is adiabatically cooled by sprays of recirculating slurry. Part of the slurry is sent to the absorber where it is neutralised. A bleed stream is maintained to control the concentration of suspended solids in the recirculating slurry at about 5%. Make-up water is added to the prescrubber to offset evaporation and bleed losses.

The Chiyoda process has been demonstrated on a 500 MW oil-fired boiler at the Hokuriku Electric Company, Japan and on a 20 MW coal-fired boiler at the Scholz station of Gulf Power, USA (Rush and Edwards, 1978a,b,c; Morasky *et al.*, 1981). The latter plant commenced operation in 1975 and has operated reliably and efficiently. The performance of the plant is summarised by the following results (Behrens and Hargrove, 1980a,b):

SO_2 removal efficiency with SO_2 inlet concentration of between 1200 and 3000 ppm:	90–95%
Plant reliability:	99.5%
Limestone utilisation in the absorber:	>98%
Gypsum purity:	>98%
Total system pressure drop:	4.7–5.2 kPa

In view of these encouraging results the Electric Power Research Institute (EPRI) is arranging a 100 MW demonstration of this process.

Saarberg–Hoelter process

Although based on the reaction between SO_2 and calcium compounds the Saarberg–Hoelter process offers significant advantages over conventional lime/limestone scrubbing processes. SO_2 is scrubbed with a solution containing no suspended solids. This minimises the possibility of scale formation and deposition within the absorber. Forced oxidation of calcium sulphite to calcium sulphate is integrated into the process resulting in a product of gypsum rather than a calcium sulphite/sulphate sludge (Glamser *et al.*, 1979).

The process was developed by Saarberg–Hoelter of the FRG and Kobe Steel of Japan. A 40 MW demonstration plant has been in operation since 1974. In May 1979 a 175 MW FGD unit was installed at the 700 MW Weiher station in the FRG It was proposed to increase the capacity of this to 350 MW at a later date (Reeves, 1979). The process is available in the USA under license by Davy Powergas (Kirkby, 1979).

The scrubbing medium is a solution of slaked lime, $Ca(OH)_2$, to which small amounts of hydrochloric acid, HCl, and formic acid, HCOOH, are added. The chloride ions form calcium chloride, $CaCl_2$, whose high rate of dissolution increases the concentration of available calcium ions in the scrubber solution. This eliminates the need for recycle of the solution and enables a relatively low liquid to gas ratio to be used. The relatively high solubility of $CaCl_2$ enables the pH value of the solution to be adjusted to between 10 and 11 without fear of scaling. This high pH value favours the reaction between calcium and SO_2. A controlled drop in pH value is permitted within the scrubber to maximise the production of calcium bisulphite, $Ca(HSO_3)_2$, the only calcium–sulphur compound with significant solubility in water.

Within the range of pH between 8 and 11 the following reaction takes place:

$$Ca^{2+} + 2OH^- + 2SO_2 + 2Cl^- \rightarrow Ca^{2+} + SO_3^{2-} + HSO_3^- + H^+ + 2Cl^- \qquad (14.4)$$

As the reaction proceeds the concentration of OH^- ions decreases while the concentration of H^+ ions increases. The pH value of the solution drops rapidly to between 5 and 4.5 at which it is temporarily held by the buffering action of the formic acid which absorbs H^+ ions to form molecular HCOOH.

$$COOH^- + H^+ \rightarrow HCOOH \qquad (14.5)$$

This value of pH favours the formation of calcium bisulphite while maintaining acceptable SO_2 removal rates.

After the formate ions are consumed, the buffering effect is lost and the pH value of the solution falls further to about 4, the optimum value for the oxidation of calcium bisulphite to gypsum.

$$Ca(HSO_3)_2 + O_2 + 2H_2O \rightarrow CaSO_2{\cdot}2H_2O + H_2SO_4 \qquad (14.6)$$

Oxidation is carried out in a separate oxidiser vessel. Lime is then added to the scrubbing solution to replenish the calcium ions consumed by the formation of gypsum and to adjust the pH value to that required for SO_2 absorption. The sulphuric acid, formed in the oxidiser by reaction (14.6) is neutralised, producing additional gypsum:

$$H_2SO_4 + Ca(OH)_2 \rightarrow CaSO_4{\cdot}2H_2O \qquad (14.7)$$

The concentration of formic acid is adjusted prior to the separation of the gypsum in a thickener. The scrubbing solution is returned to the absorber.

Prescrubbing of the flue gas is not required. The presence of HCl in the flue gas reduces the requirement for the addition of hydrochloric acid. Chloride corrosion, potentially serious at low pH values, is not seen as a problem in this process because the chlorides are in solution, not in suspension (Midkiff, 1979a). Not all of the process is carried out below a pH value of 6.9 where chloride corrosion occurs. Even so, careful consideration must be given to the materials of construction for equipment which comes into contact with desulphurised flue gas. The reliability of the demonstration unit in the FRG has exceeded 95% over 20 000 h of service, achieving SO_2 removal efficiencies of between 90 and 95% (Reeves, 1979).

Sodium carbonate scrubbing process

This process was developed to overcome the problems of scaling and deposition associated with early lime/limestone applications. Both the reactant, Na_2CO_3, and the reaction products, Na_2SO_3 and $NaHSO_3$, have relatively high solubilities in water so that there are no suspended solids. The sodium alkali is also very reactive enabling low liquid to gas ratios to be employed within the scrubber. The system also adapts rapidly to changes in SO_2 load (Leivo, 1978).

A purge stream of spent alkali is continuously withdrawn to maintain the chemical balance. This stream which is slightly acidic is neutralised with sodium carbonate before disposal. Process water make-up is required to compensate for the water evaporated in the flue gas and lost in the purge stream. Scaling can occur if this water contains a high concentration of calcium (Leivo, 1978). The spent alkali purge stream is normally discharged to a lined evaporation pond for drying although precipitation of sodium salt and the recovery of sodium sulphate (salt cake) for sale are feasible.

The major disadvantage of this process is that it consumes a relatively expensive, premium chemical, Na_2CO_3, without alleviating the problem of liquid waste disposal. This limits the application of the process to geographical areas having a source of low grade carbonate. A prototype unit, serving two industrial coal fired boilers equivalent to 25 MW at the General Motors assembly plant in St Louis, USA have been in operation since 1972. Three 125 MW units are in operation at the Reid Gardner Station of the Nevada Power Company, USA. Two of the units were commissioned in 1974 and the third in 1976. A 520 MW unit was constructed in 1979 at the Jim Bridger Station in Wyoming, USA (Leivo, 1978). All the units burn low sulphur content coals and are designed for 90% removal of SO_2. No further units have been ordered (Jahnig and Shaw, 1981a).

Dual alkali process

The dual (sometimes called 'double') alkali flue gas desulphurisation process removes SO_2 from flue gas by wet scrubbing with sodium sulphite. In a second stage the sodium sulphite is regenerated with lime or limestone and a waste sludge of calcium sulphite and calcium sulphate is formed. The use of two alkalis in separate stages provides advantages over direct lime/limestone scrubbing processes. The higher solubility of sodium compared with calcium salts means that SO_2 absorption is limited by gas-phase mass transfer rather than solid dissolution rates (Leivo, 1978). This permits the use of lower liquid to gas ratios in the dual alkali process compared with lime/limestone processes. In addition, the amount of soluble and undissolved calcium in the scrubber is minimised thus reducing the potential for scale formation.

The main disadvantages of the process concern the loss of the relatively expensive sodium salts. Prescrubbing of the flue gas may be required to remove

HCl which would otherwise react with the alkaline sodium solution to form sodium chloride. In addition, despite the fact that calcium compounds precipitated from the regeneration step are washed, the filter cake may contain up to 2 wt% sodium (Jahnig and Shaw, 1981a).

The dual alkali process can be operated in either a 'concentrated' or 'dilute' mode depending on the conditions of operation. These terms refer to the concentration of active alkali, sometimes misleadingly called 'active sodium', in the scrubbing medium. SO_2 is actually absorbed by or reacts with the sulphite, hydroxide, or carbonate ions rather than the sodium ion. The concentrated mode is the simpler and more widely used of the two systems. A total capacity of 1170 MW is either operational or under construction in the USA (Devitt and Laseke, 1980).

Van Ness *et al.* (1979, 1981) describe the application of a demonstration dual alkali FGD process, operating in the concentrated mode, to a 300 MW utility boiler. Operating experience with a dual alkali process treating flue gas from a 265 MW utility boiler is described by Durkin *et al.* (1981). The chemistry of the removal of SO_2 may be presented as:

$$Na_2SO_3 + SO_2 + H_2O \rightarrow 2NaHSO_3 \quad (14.8)$$

$$2Na_2O_3 + SO_2 + H_2O \rightarrow Na_2\,SO_3 + 2NaHCO_3 \quad (14.9)$$

$$NaHCO_3 + SO_2 \rightarrow NaHSO_3 + CO_2 \quad (14.10)$$

$$2NaOH + SO_2 \rightarrow Na_2SO_3 + H_2O \quad (14.11)$$

Sodium sulphite is the major reagent used. Only minor quantities of sodium hydroxide and sodium carbonate are present.

An important reaction which occurs simultaneously is the oxidation of sodium sulphite to sodium sulphate by the absorption of oxygen from the flue gas:

$$Na_2SO_3 + 1/2O_2 \rightarrow Na_2SO_4 \quad (14.12)$$

This reaction effectively reduces the amount of active alkali present in the scrubbing solution. The rate of oxidation is proportional to the mass transfer of oxygen which is a function of the absorber design, oxygen concentration in the flue gas, flue gas temperature and the nature and concentration of the species in the scrubbing solution. For a given set of conditions the oxidation rate expresed as moles of sulphite oxidised per unit of time is virtually independent of the SO_2 removal rate. However, for convenience, the degree of oxidation is frequently expressed, on an equivalent basis, as a percentage of the SO_2 removed. Van Ness *et al.* (1979) estimate that when burning a high sulphur content coal with 4 to 5% O_2 and 2500 ppm SO_2 in the flue gas between 5 and 10% of the SO_2 removed from the flue gas would be oxidised and appear as sulphate in the spent scrubbing solution. The sulphate must leave the system either as calcium sulphate or as a purge of sodium sulphate at the rate at which it is being formed in the system. This process is explained below.

The spent scrubbing solution is regenerated with slaked lime, $Ca(OH)_2$ by a two step process:

$$Ca(OH)_2 + 2NaHSO_3 \rightarrow CaSO_3 \cdot 1/2H_2O + Na_2SO_3 + 3/2H_2O \quad (14.13)$$

$$Ca(OH)_2 + Na_2SO_3 + 1/2H_2O \rightarrow CaSO_3 \cdot 1/2H_2O + 2NaOH \quad (14.14)$$

Reaction (14.13), the neutralisation of bisulphite to sulphite, proceeds to completion. Reaction (14.14) is a precipitation reaction in which the equilibrium hydroxide concentration is limited by the relative solubility products for calcium sulphite and calcium hydroxide, and the concentrations of hydroxide and sulphite in solution. Calcium sulphite is precipitated as the hemihydrate salt in both reactions.

Simultaneously with reactions (14.13) and (14.14) a limited amount of calcium sulphate is also precipitated:

$$Ca^{2+} + SO_4^{2-} + xH_2O \rightarrow CaSO_4{\cdot}xH_2O \qquad (14.15)$$

The sulphate co-precipitates with the sulphite resulting in a mixed crystal, or solid solution, of calcium-sulphur salts. The amount of sulphate co-precipitated with calcium sulphite is a function of the concentrations of sulphate and sulphite and the pH of the solution. The system is, to some extent, self-regulating with respect to sulphate concentration in solution. As the rate of oxidation of sulphite to sulphate increases, the ratio of sulphate to sulphite in solution increases. This continues until the rate of precipitation is equivalent to the rate of formation of sulphate. Hence, provided the active sodium concentration does not fall below about 0.15 M, the solution remains unsaturated with respect to calcium sulphate, thereby avoiding high calcium concentrations and attendant scaling problems (Van Ness *et al.*, 1979). An oxidation rate equivalent to 25% of the SO_2 absorbed can be accommodated before it is necessary to purge sodium sulphate (Leivo, 1978; Van Ness *et al.*, 1979).

When the degree of oxidation cannot be controlled to below 25%, as with low sulphur content coal and/or high O_2/SO_2 ratios, the dilute mode of operation must be employed. This involves the dilution of the scrubbing medium with water to less than 10% by weight of sulphate. In the absence of sulphite, sulphate reacts rapidly with lime:

$$Na_2SO_4 + Ca(OH)_2 \rightarrow 2NaOH + CaSO_4 \qquad (14.16)$$

The active alkali is, in this case, in the form of NaOH so that the reaction described by equation (14.11) predominates within the scrubber.

However, the dilution involves recirculating large volumes of scrubbing solution which contains large amounts of dissolved calcium. To avoid problems of scaling, the regenerated absorbent is 'softened' by the addition of sodium carbonate (Leivo, 1978). Calcium carbonate is precipitated:

$$CaSO_4 + Na_2CO_3 \rightarrow CaCO_3 + Na_2SO_4 \qquad (14.17)$$

$$Ca(OH)_2 + Na_2CO_3 \rightarrow CaCO_3 + 2NaOH \qquad (14.18)$$

Regenerable processes

Regenerable processes are those in which SO_2 is recovered in a form which may be processed to sulphuric acid or elemental sulphur. The regeneration of $CaSO_3$/$CaSO_4$ sludges by a reductive roasting process to form CaS followed by carbonation to $CaCO_3$ has been shown to be technically feasible (Mozes, 1978) but is not used in practice for economic reasons. The regenerable processes described below are based on the use of substances such as Mg, Na and NH_3.

Magnesium oxide process

There are several alternative magnesium based FGD processes the most common of which is based on scrubbing with a slurry of slaked magnesium oxide (Selmeczi and Stewart, 1978).

$$Mg(OH)_2 + SO_2 \rightarrow MgSO_3 + H_2O \quad (14.19)$$

The resulting magnesium sulphite is converted back to the oxide releasing concentrated SO_2 gas.

$$MgSO_3 \rightarrow MgO + SO_2 \quad (14.20)$$

The main advantage of this process is that there is no waste sludge to be disposed of. Despite the fact that a slurry of magnesium oxide is used as the scrubbing medium scaling of absorbers is not reported as a problem as in lime/limestone scrubbing processes. This is because the solubilities of magnesium sulphite and magnesium sulphate are many orders of magnitude higher than those of the corresponding calcium salts (Leivo, 1978).

Prescrubbing of flue gases is required to remove HCl, HF and fly ash the concentrations of which would build up in the recycled magnesium oxide. Even so, a small purge stream of scrubbing medium may be required to maintain the concentration of inerts in the recirculating solids to below 20% (Jahnig and Shaw, 1981a).

In the absorption vessel magnesium oxide reacts with SO_2 to form magnesium sulphite hydrates, $MgSO_3{\cdot}6H_2O$ and $MgSO_3{\cdot}7H_2O$. Some of the magnesium sulphite is oxidised to magnesium sulphate, $MgSO_4{\cdot}7H_2O$. The magnesium sulphite-sulphate crystals are separated in a centrifuge and pass to a dryer where the large amount of water of crystallisation is removed. The dried crystals are conveyed to a calciner where they are mixed with a small amount of reducing agent (coke or carbon) and calcined under a reducing atmosphere at about 760 to 870°C to produce SO_2 and MgO (Leivo, 1978). The reducing agent is required to reduce the sulphate only. After dust removal, the SO_2-rich gas stream may be used for sulphuric acid or sulphur production. The MgO, together with any required make-up, is recycled to the scrubber system.

There is limited long term performance data on this process. The first application in the USA was on a 155 MW oil-fired boiler at Mystic Station No. 6 of the Boston Edison Co.. However, several operating difficulties were experienced limiting the availability of the plant to 80% (Pruce, 1981b). These include:

1. The formation of fine trihydrate sulphite crystals in the scrubber which are more difficult to handle than those of the hexahydrate.
2. Dust emissions from the dryer.
3. Erosion of pumps, valves and piping.
4. Excessive wear on the internal parts of the centrifuge.

Because the plant was oil-fired it did not provide information on the effects of fly ash on the system. The unit was operated from 1972 to 1974. The SO_2 removal efficiency was greater than 90%. MgO losses averaged about 10% per cycle.

The first application in the USA to a coal-fired boiler was at the Dickerson No. 3 unit of the Potomac Electric and Power Co. which operated intermittently from 1973 to 1975. Particulate removal was in excess of 99% and SO_2 removal efficiency ranged from 88 to 96%, depending on the gas flow rate (Leivo, 1978). However, operational difficulties limited the availability of the plant to 64% (Pruce, 1981b).

Following the initiation of a 120 MW prototype test programme, which is still in progress, magnesium oxide scrubbers representing 696 MW capacity are being retrofitted to the Eddystone station of the Philadelphia Electric Co. Initial results of the prototype test proved disappointing with several operating problems and an availability of the SO_2 removal system of only 32% (Gille and MacKenzie, 1978). A further 150 MW capacity is to be installed at the Cromby station of the same Utility (Pruce, 1981b). However, TVA recently retracted a proposal to install magnesium oxide scrubbers at its 1400 MW coal-fired Johnsonville plant but is to continue investigation of the process at a scale of 1 MW (Marcus *et al.*, 1981).

Wellman–Lord process

The Wellman–Lord process is based on scrubbing with sodium sulphite to absorb SO_2 followed by regeneration which releases a concentrated stream of SO_2. The chemistry of the absorption stage is similar to that of the dual alkali process. Most of the sodium sulphite is converted to sodium bisulphite by reaction with SO_2 although some sodium sulphite is oxidised to sodium sulphate.

Prescrubbing of the flue gases is required to saturate and cool the flue gas to about 55°C (Sugarex and Sipes, 1978b). This avoids excessive evaporation in the absorber which would result in deposition of solids. Prescrubbing also removes chlorides and any remaining fly ash.

Advantages of the Wellman–Lord Process include scrubbing with a solution rather than slurry, which prevents scaling, and the production of a marketable material (Pedrose and Press, 1979). The disadvantages of high energy consumption and maintenance are due to the relative complexity of the process. A further disadvantage is that a purge stream of about 15% of the scrubbing solution is required to prevent the build up of sodium sulphate (Sugarex and Sipes, 1978b). This represents a stream of about 1 t/h for a 500 MW station (Highton and Webb, 1980). The stream is cooled to about 0°C in a chiller/crystalliser. A mixture of sodium sulphate and sodium sulphite is crystallised out and centrifuged to produce a cake of 40% solids. Development work is in progress to obviate the need for this purge stream by the processing of sodium sulphate to sodium carbonate for subsequent reaction to the sulphite required by the process (Bechtel National, 1980). Regeneration of the remainder of the scrubbing solution is carried out in a two stage evaporator according to the following reaction:

$$2NaHSO_3 \xrightarrow{\Delta} Na_2SO_3 + H_2O + SO_2 \qquad (14.21)$$

Vapours from the first stage at 94°C are condensed to supply heat to the second stage at 77°C (Jahnig and Shaw, 1981a). At these temperatures there is an increased formation of thiosulphate by the following disproportionation reactions:

$$6NaHSO_3 \rightarrow 2Na_2SO_4 + Na_2S_2O_3 + 2SO_2 + 3H_2O \qquad (14.22)$$

$$2NaHSO_3 + 2Na_2SO_3 \rightarrow 2Na_2SO_4 + Na_2S_2O_3 + H_2O \qquad (14.23)$$

Thiosulphate must be purged from the regenerated sodium sulphite. The concentrated SO_2 stream may be compressed, liquefied and catalytically oxidised to produce sulphuric acid or reduced to elemental sulphur (Pruce, 1981b). The regenerated sodium sulphite crystals are dissolved and returned to the absorber.

Over thirty Wellman–Lord installations are in operation although many of these treat tail gases from Claus plants and sulphuric acid plants. The process has been in operation in Japan on two 35 MW oil-fired boilers since 1971. Close to 100% availability is reported (Leivo, 1978b). In 1973 two larger oil-fired systems including a 220 MW utility boiler were commissioned in Japan and have subsequently operated successfully (Highton and Webb, 1980). In the USA a Wellman–Lord FGD unit was retrofitted to the 115 MW boiler at the Mitchell station of the Northern Indiana Public Service Company in 1977 (Mann and Adams, 1981). Coal with a sulphur content of 2.9% is burned. The facility combines the Wellman–Lord process with a chemical process, which converts SO_2 to elemental sulphur, marketed by the Allied Chemical Company. During two years of extensive testing SO_2 removal efficiency exceeded 90% but availability of the FGD unit averaged only 61% due mainly to condensation and corrosion problems with the absorber and problems of sulphur deposition within the sulphur plant (Wood, 1978). The Wellman–Lord FGD process has also been installed at the San Juan station of the Public Service Company of New Mexico. The units treat flue gas from four boilers with a total capacity of 1700 MW. Low sulphur content coal is burned. Units 1 and 2 incorporate the Allied Chemical Company's process for the production of sulphur. Commissioned in 1981 these units are experiencing mechanical problems due to the highly corrosive acidic environment and the lack of a steady supply of steam. The FGD plants on units 3 and 4 will supply SO_2 to a sulphuric acid plant due to be operational by the end of 1982 (Pruce, 1981b).

The Citrate process

The Citrate FGD process removes SO_2 from flue gas by absorption in an aqueous solution of sodium citrate. The absorbent is stripped with steam to recover concentrated SO_2 (about 95%). A prescrubber is required to remove fly ash and chlorides from the flue gas prior to scrubbing with citrate solution. A purge stream of absorbent is passed to a crystalliser where sodium sulphate is removed as $Na_2SO_4 \cdot 10H_2O$ (Jahnig and Shaw, 1981a).

Sodium citrate is present as a 'buffering agent' enhancing the solubility of SO_2 in water. As SO_2 is absorbed and dissolved in water it dissociates into HSO_3^- ions and H^+ ions:

$$SO_2 + 2H_2O \rightarrow HSO_3^{2-} + H_3O^+ \qquad (14.24)$$

Sodium citrate reacts with the H^+ ions effectively removing them from the above reaction which therefore proceeds further to the right hand side.

The absorbent solution passes through a heat exchanger to a steam heated stripping tower where SO_2 is distilled out of solution. The mixture of SO_2 and water vapour is condensed forming two immiscible liquid phases: a wet SO_2 phase and a water phase. The water phase is returned to the stripping tower. The SO_2 phase may be contacted with H_2S in a separate reactor to form sulphur by a liquid phase Claus reaction. Alternatively the SO_2 phase may be processed in a Resox unit. The Resox process is a proprietary development of the Foster Wheeler Energy Corporation. Gaseous SO_2 is reduced to sulphur with coal as the reductant (Beychok, 1980; Steiner *et al.*, 1980).

Initial development of the Citrate process was undertaken by the US DOE (then the USBM; Crocker *et al.*, 1979; Nissen and Madenburg, 1979; Madenburg and Seesee, 1980) although several companies in the USA are now developing the process (Pruce, 1981b). A prototype unit treating 60 m^3/s of flue gas from a coal-fired industrial boiler was commissioned in 1981. The boiler is part of the 120 MW GF Weaton power station of the St. Joe Zinc Company at Monaca, PA, USA Preliminary results indicate an SO_2 removal efficiency of 90%. Sulphur of 99% purity is produced by the liquid phase Claus reaction (Pruce, 1981b).

Ammonia scrubbing

Ammonia scrubbing to control sulphur emissions in applications other than regenerable FGD processes has been in commercial use for many years. However, they have been characterised by the formation of plumes of ammonia salt aerosols (Jahnig and Shaw, 1981a). A process is under development jointly by Catalytic of the USA and Institut Francais du Petrole (IFP) of France which avoids the generation of vapour plumes by control of concentrations and temperatures within the scrubber. The process also produces elemental sulphur. The initial development of this process is described in detail by Blanc (1978).

Prescrubbing of the flue gas is required to remove chlorides, some SO_3 and residual fly ash. The flue gas then enters the absorber which incorporates multiple scrubbing stages mounted horizontally. Each stage contains proprietary packing material and a liquid storage sump. The circulation of the scrubbing solution within each stage is controlled to achieve the desired liquid to gas ratio. The final scrubbing stage is a spray chamber in which flue gas is scrubbed with water to remove and recover any residual ammonia. Mist eliminators are included after the venturi and each stage. The flow of scrubbing liquid through the absorption stages is countercurrent to the flue gas flow except that ammonia solution is fed to the individual stages to control the scrubbing liquor concentration.

SO_2 is removed by the following primary reactions:

$$SO_2 + 2NH_4OH \rightarrow (NH_4)_2SO_3 + H_2O \quad (14.25)$$

$$(NH_4)_2SO_3 + H_2O \rightarrow 2NH_4HSO_3 \quad (14.26)$$

$$SO_3 + (NH_4)_2SO_3 \rightarrow (NH_4)_2SO_4 + SO_2 \quad (14.27)$$

$$2(NH_4)_2SO_2 + O_2 \rightarrow 2(NH_4)_2SO_4 \quad (14.28)$$

The amount of sulphate formed is small, usually less than 10%.

Regeneration of the ammoniacal brine is accomplished in two stages. About 60% of the brine is evaporated at 150°C and thermally decomposed to yield SO_2, ammonia and water vapour:

$$(NH_4)_2SO_3 \rightarrow 2NH_3 + H_2O + SO_2 \quad (14.29)$$

$$NH_4HSO_3 \rightarrow NH_3 + H_2O + SO_2 \quad (14.30)$$

Because the sulphates in the brine do not decompose in the evaporators, a concentrated sulphate slurry is withdrawn from the evaporator bottom and sent to the second stage of regeneration. The slurry is reduced with recycled molten sulphur at a temperature of between 300 and 370°C to yield SO_2, ammonia, water vapour and a small amount of SO_3:

$$(NH_4)_2SO_4 \rightarrow NH_4HSO_4 + NH_3 \quad (14.31)$$

$$2NH_4HSO_4 + S \rightarrow 3SO_2 + 2NH_3 + 2H_2O \quad (14.32)$$

$$(NH_4)_2SO_4 \rightarrow SO_3 + 2NH_3 + H_2O \quad (14.33)$$

The combined gases from the two stages of regeneration are partly reacted with a reducing gas of CO and H_2 to form H_2S. The reaction is controlled to produce a gas having the stoichiometric ratio 2:1 H_2S/SO_2 which is required for the subsequent liquid phase Claus reaction.

The process has been demonstrated successfully in a pilot plant at the Champagne sur Oise power station near Paris, France. The plant treats the equivalent of 25 MW of flue gas from two 250 MW oil-fired boilers (Blanc, 1978). Greater than 90% removal of SO_2 at an inlet concentration of 1500 ppm has been demonstrated (Engdahl and Rosenberg, 1978). The liquid sulphur collected was greater than 99.9% pure. Similar SO_2 removal rates were achieved from a pilot plant at the Calvert City, Kentucky chemicals plant of Air Products and Chemicals, Inc. (Quackenbush *et al.*, 1978).

A recent process modification is reported to be the simultaneous removal of NO_x by reacting NO with NH_3 and SO_2 to form ammonium sulphate in a solution containing iron and EDTA (ethylenediaminetetracetic acid; Jahnig and Shaw, 1981a).

Chapter 15

Dry flue gas desulphurisation processes

Dry SO_2 removal refers to systems in which the product of the reaction between SO_2 and reagent is a solid. There are three methods of achieving this. The first involves injecting a dry sorbent into the flue gas where it absorbs SO_2. The resulting spent sorbent and fly ash are collected in an integrated particular removal step using baghouses or electrostatic precipitators. The sorbent is normally discarded after use. Such processes are often referred to as 'dry sorbent injection' processes. The second method uses an aqueous solution or suspension of reagent, dispersed as fine droplets, to absorb SO_2 from the flue gas. The droplets are dried by the heat of the flue gas and collected together with fly ash in a manner similar to that described above. The process is often referred to as 'dry scrubbing' or 'semi-dry scrubbing' but is more correctly termed 'spray drying'. There are both regenerable and non-regenerable spray dryer processes. The third method involves the adsorption of SO_2 onto a bed of reagent. The sorbent is normally regenerated. These processes, which can also incorporate the removal of NO*x* and fly ash, are often referred to as 'dry adsorption' processes. The classification of dry flue gas desulphurisation processes used in this chapter is similar to that used for wet scrubbing processes in Chapter 14 and is represented in Table 15.1.

Table 15.1 Classification of dry FGD processes

Non-regenerable		*Regenerable*	
Dry injection	*Spray dry*	*Spray dry*	*Dry adsorption*
Wheelabrator-Frye/ Rockwell Int.	Sodium carbonate-Wheelabrator-Frye/ Rockwell Int. lime	Aqueous sodium carbonate	Shell/UOP Activated carbon Cat-ox

The advantages of dry SO_2 removal systems over wet scrubbing systems include: a dry product; lower capital costs; lower energy and water requirements; and a more simple design which should be reflected in increased availability and reduced maintenance (Kelly and Dickerman, 1981). The lower energy and water requirements are a result of two main factors:

1. Reheat of flue gases is often not required.
2. Relatively small volumes of scrubbing solution or slurry are recirculated.

Dry injection processes

A pilot plant programme involving a dry injection process was initiated at the Lelands Olds station, ND, USA of the Basin Electric Power Cooperative in 1976. The plant was designed by Wheelabrator-Frye/Rockwell International to investigate SO_2 removal by the injection of powdered nahcolite, calcium oxide, and calcium hydroxide into the flue gas with subsequent collection in fabric filter bags (Devitt and Laseke, 1980). Nahcolite, a naturally occurring form of sodium bicarbonate, was found to be the most effective (Donovan, 1979). However, the availability of this mineral is limited: the only known deposit in the USA is in Colorado. As a result Basin Electric Power Cooperative ceased investigation of dry injection processes.

The Electric Power Research Institute (EPRI) is continuing research into dry injection processes. A demonstration programme is in progress at the 22 MW Cameo station of the Public Service Company of Colorado, USA. The SO_2 removal efficiency of the process has been evaluated as a function of the nahcolite feed rate and baghouse operating parameters. Results are presented in Figure 15.1. The SO_2 removal efficiency is also a function of nahcolite particle size; the finer the particles the greater the reactive surface area. The nahcolite did not distribute equally among the baghouse compartments which resulted in differing SO_2 removal efficiencies. The plant is to be modified to improve particle and flow distributions. However, the dry injection process was found to have minimal effect on the overall operation of the baghouse. The pressure drop across the baghouse did not increase significantly although this will have to be confirmed in long-term tests (Yeager, 1981). In view of the limited availability of nahcolite EPRI is also evaluating the

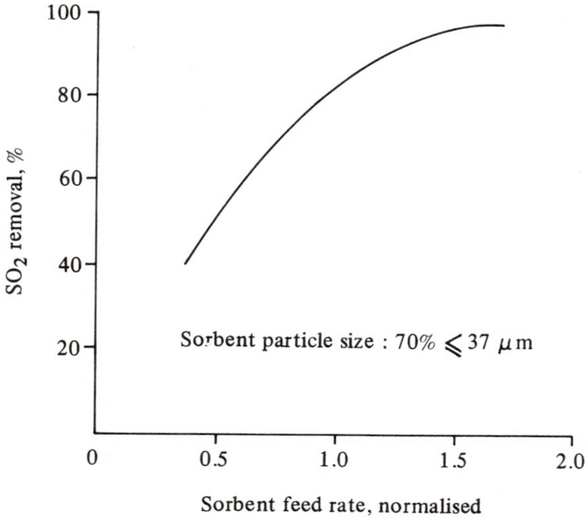

Figure 15.1 SO_2 removal efficiency as a function of sorbent feed rate for a dry injection process using nahcolite (after Yeager, 1981)

use of trona, a naturally occurring mixture of sodium bicarbonate and sodium carbonate.

Spray dryer processes

The contacting of flue gas with a solution or suspension widens the range of possible reagents for absorption of SO_2 compared with that for dry injection processes. Sodium carbonate (soda ash), trona, lime, limestone and slurries of fly ash have been investigated although sodium carbonate and lime have been shown to be the most effective. However, the processing equipment is more complex than that of dry injection processes. Meyler (1981) describes spray dryer processes in five steps:

1. reagent preparation;
2. droplet formation (atomisation);
3. droplet contact with flue gas;
4. evaporation/SO_2 absorption;
5. waste product removal.

Sodium carbonate is the reagent used in the first utility spray dryer application in the USA. Preparation of the reagent simply involves dissolving the dry material in water (Meyler, 1981). However, sodium carbonate is usually more costly than lime and has the disadvantage that the reaction products, sodium sulphate and sodium sulphite are very soluble and present significant waste disposal problems. Sodium ions will leach into water and present ground water contamination problems (Felsvang and Masters, 1979). Lime is the reagent used in all other planned utility installations of non-regenerable spray dryer processes in the USA.

Atomisation is normally achieved by the use of a nozzle or rotary atomiser. Although nozzles are less expensive than rotary atomisers they possess certain operating disadvantages: they have a tendency to plug; do not operate uniformly over a wide range of loads; and are susceptible to erosion by abrasive slurries. Rotary atomisers are the most commonly used in spray dryer processes. Details of the designs of atomisers are given by Meyler (1981) and by Andreasen (1980).

Absorption and evaporation occur simultaneously in the spray dryer chamber. Getler *et al.* (1979) describe two stages of evaporation. During the first stage moisture is driven off from the slurry droplet, the size of which decreases. Contact between individual particles occurs. The surface of the agglomerate begins to dry and the second stage of evaporation begins. Moisture diffuses from the interior of the agglomerate to the surface and the rate of evaporation decreases. Absorption of SO_2 occurs most rapidly in the liquid–gas phase. Pilot plant tests have shown that no further SO_2 absorption takes place when the reagent is completely dry (Meyler, 1981). Reisinger and Gehri (1979) refer to SO_2 absorption as a two-stage process in which SO_2 is removed initially in the spray dryer and subsequently, by a different mechanism, in the baghouse. However, Meyler (1981) claims that absorption of SO_2 occurs continuously throughout the equipment, provided that the particles are not completely dry. The majority of spray dryer FGD applications employ 'two-point product collection'. Part of the mixture of dry powder and fly ash settles by gravity and exits the bottom of the spray chambers via a hopper. The remainder of the mixture leaves through a separate duct to be collected in baghouses or by electrostatic precipitators. One disadvantage of spray dryer processes is that, generally, reagent utilisation is low because the scrubbing medium is not easily recirculated (Stern, 1981).

Non-regenerable spray dryer processes

The first utility application of a spray dryer process was a pilot plant installed at the Lelands Olds station of the Basin Electric Power Cooperative, USA. Following this test programme a contract was awarded to Wheelabrator–Frye/Rockwell International for a full scale plant on the 440 MW coal-fired boiler at the Coyte station of the Otter Tail Power Company. This plant was commissioned in 1981. As discussed above the FGD unit uses sodium carbonate (soda ash) as reagent.

Basin Electric Power Cooperative evaluated the use of other alkaline reagents because of the potential disposal problem and the relatively high cost of sodium reagents. Four pilot plants were built to demonstrate the use of lime as reagent (Davis *et al.*, 1979). Subsequent to the operation of these plants Basin Electric Power Cooperative awarded a contract to Joy/Niro for the installation of a spray dryer FGD system using lime on the 440 MW No. 1 boiler at the Antelope Valley Station. The unit was due to be in operation by April 1982 (Davis *et al.*, 1979). A spray dryer FGD facility of 130 MW capacity at the Riverside station of the Northern States Power Company was commissioned in 1981. Coals of low, medium and high sulphur contents will be tested (Andreason, 1981). Electrostatic precipitators will be used initially to be replaced later by a baghouse. A total of 1890 MW of FGD capacity in the USA is now committed to spray dryer units employing lime as reagent (Jahnig and Shaw, 1981a).

Regenerable spray dryer process

The only regenerable spray dryer process currently receiving significant attention is the aqueous sodium carbonate process being developed by Rockwell International (Katz *et al.*, 1978; Oldenkamp *et al.*, 1979). The spray dryer stage of this process is similar to that of non-regenerable processes. However, sodium carbonate is used as the reagent. The spent reagent, consisting of sodium sulphite, sodium sulphate and fly ash, is reacted with coal in a molten salt reactor at between 925 and 1040°C to reduce the sulphur compounds to Na_2S. The latter is quenched with and dissolved in water, filtered to remove unreacted coal ash and passed to a carbonation tower. Sulphur is stripped out as H_2S by reaction with CO_2 and water, thereby regenerating sodium carbonate solution. HCl in the flue gas is removed in the spray dryer and enters the molten salt reactor where it is vaporised as NaCl. The gases from the molten salt reactor are washed with water to separate a chloride stream which is purged and ponded (Jahnig and Shaw, 1981a).

The regeneration stage adds considerably to the complexity of the process compared with non-regenerable spray dryer processes. More than eighty steps are involved in the regeneration, the most significant of which is the molten salt reduction. This produces a very corrosive melt. Other potential problem areas are reported to be the quench and solids filtration steps (Pruce, 1981b).

A 100 MW demonstration plant is under construction at the Charles R. Huntly station of the Niagara Mohawk Power Company (Binns and Aldrich, 1978). Startup was scheduled for mid 1982. EPA, EPRI and the New York Energy Research and Development Authority are sponsoring the project.

Dry adsorption

Dry adsorption processes include those based on supported metal oxides, activated carbon and the Cat-ox process.

Supported metal oxides

Flue gas desulphurisation can be achieved by using supported metal oxides as sorbents. The oxides of copper, manganese, iron and cobalt are suitable for this purpose (van der Linde *et al.*, 1979). Shell is developing the UOP process which involves passing flue gas over a fixed bed of cupric oxide supported on alumina. The copper oxide reacts readily with SO_2 and oxygen at 400°C to yield $CuSO_4$ which can subsequently be reduced at the same temperature to release a concentrated SO_2 stream:

$$CuO + SO_2 + 1/2O_2 \rightarrow CuSO_4 \tag{15.1}$$

The beds of CuO operate in a cyclic manner. When the acceptor becomes saturated with SO_2 in the form of $CuSO_4$ to the extent that the limiting removal efficiency is reached, the flue gas stream is diverted to another reactor. The spent acceptor is then purged with a gas containing hydrogen to reduce the $CuSO_4$ to Cu. The copper later oxidises to CuO when the bed is put back on stream and exposed to flue gas containing oxygen. A cyclone may be used upstream of the beds to remove most of the fly ash although the process can operate with all the fly ash passing through the absorber. Blocking of the bed with fly ash is avoided by the use of a parallel passage reactor in which the flue gas passes over the bed rather than through it.

A feature of this process is that it provides the basis for the simultaneous reduction of NO_x and SO_x. Both CuO and $CuSO_4$ catalyse the reduction of NO_x with ammonia under conditions similar to those used for SO_2 removal; the sulphate being the more reactive. Hence, a FGD unit can provide simultaneous NO_x reduction if ammonia is injected into the flue gas. The mechanism of denitrification, catalysis with ammonia, is independent of the chemisorption mechanism for desulphurisation. Figure 15.2 shows the concentration of NO_x and SO_x in the reactor outlet gas as a function of time for different NH_3/NO ratios. The concentration of NO_x in the exit flue gas may initially exceed that in the inlet. Because elemental copper is present at the beginning of acceptance NO_x reduction is not catalysed and ammonia oxidation occurs. As CuO and $CuSO_4$ are formed the concentration of NO_x drops sharply, approaching asymptotically a value dependent on flow-rate, temperature and NH_4NO_x ratio. This initial 'slip' of NO_x may be reduced by: delaying the injection of ammonia into the flue gas until after the copper is oxidised; prior oxidation of the copper to copper oxide; or by leaving part of the acceptor/catalyst unregenerated in the form of the sulphate.

A pilot plant of 0.6 MW capacity was operated in the desulphurisation mode between 1974 and 1975 on a side stream of flue gas from a 400 MW coal-fired boiler at the Big Bend station, FL, USA. The plant operated with dust loadings of 22.9 g/m^3 and the acceptor retained its activity for desulphurisation after two years and 13 000 cycles. In 1980, under an EPA contract, the plant was used to demonstrate simultaneous NO_x and SO_x reduction. Both SO_2 and NO_x removal efficiencies were designed to be 90% for a 90 day demonstration run (Nooy and Pohlenz, 1979).

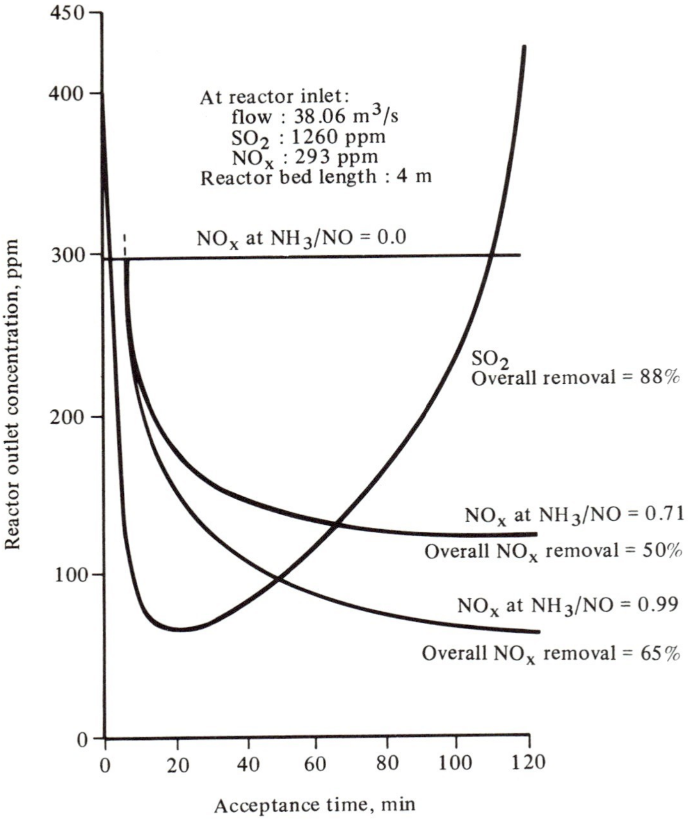

Figure 15.2 Removal of SO_2 and NObx versus time for different NH_3/NO ratios in the Shell UOP process (after Nooy and Pohlenz, 1979)

A similar process, but using a fluidised bed instead of a fixed bed, is being investigated at the Pittsburgh Energy Technology Center (PETC), USA (Strakey *et al.*, 1980). It is claimed that a fluidised bed has several advantages over a fixed bed for this application; the absorber and regenerator can operate at different temperatures; a constant flow of regenerator offgas and fly ash should pass more easily through the bed avoiding the plugging that can occur in fixed beds. However, a fluidised bed has a higher pressure drop across the bed and attrition of the sorbent may occur. Results of pilot plant work are shown in Figure 15.3. Greater than 90% removal of SO_2 is possible with a bed height of 1 m.

Activated carbon process

Bergbau–Forschung GmbH of the FRG is developing a dry bed process which uses activated carbon as the adsorbent (Knoblauch *et al.*, 1981). SO_2 adsorption takes place at between 120 and 150°C in a moving bed of 1.3 cm carbon pellets. Flue gas flows through the bed horizontally at a velocity of about 0.3 m/s (Jahnig and Shaw,

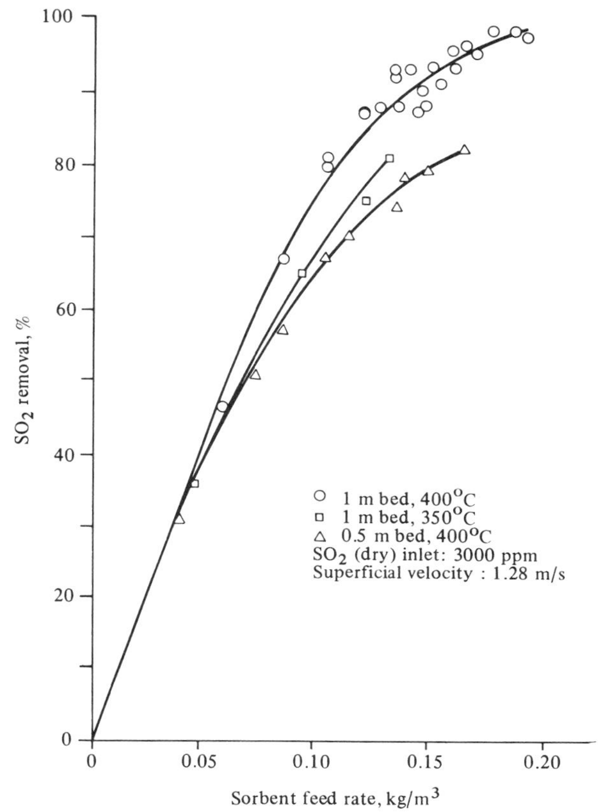

Figure 15.3 SO_2 removal efficiency versus sorbent feed rate in a fluidised bed of supported copper oxide (after Stakey *et al.*, 1980)

1981a). The adsorbed SO_2 reacts with oxygen and water to form sulphuric acid, filling the pores of the bed. Passage of the flue gas through the bed also affects removal of fly ash particulates from the flue gas as well as some removal of nitrogen oxides. The flue gas leaves the adsorber at a temperature of between 135 and 150°C which reflects a 15 to 20°C temperature rise due to the exothermic heat of reaction within the adsorber (Beychok, 1980). The saturated carbon from the bottom of the bed is screened to remove most of the fly ash collected in the absorber before passing to the regeneration section.

Regeneration is effected by heating the carbon to 650°C by contact with hot sand in an inert atmosphere. The adsorption reactions are reversed:

$$H_2SO_4 \rightarrow SO_3 + H_2O \tag{15.2}$$

$$2SO_3 + C \rightarrow CO_2 + 2SO_2 \tag{15.3}$$

The reduction of SO_3 to SO_2 during the regeneration utilises some of the carbon as a reductant. Theoretical carbon consumption is 6 wt% on sulphuric acid absorbed or 0.1 kg/kg SO_2. However, this consumption plus other oxidation weakens the carbon structure and increases the attrition losses so that in practice carbon makeup is 0.12 to 0.18 kg/kg SO_2 removed.

A 30 MW demonstration plant has been operated at a power station in Lunen, FRG (Engdahl and Resenberg, 1978; Dalton and Preston, 1979). The work was sponsored jointly by Bergbau–Forschung GmbH, Foster Wheeler, STEAG AG, Deutsche Babcock AG and the Umweltbundesamt (Environmental Agency of FRG). A removal efficiency of 80% was achieved for flue gas containing 1000 ppm of SO_2. A 20 MW pilot plant has been operated at the Scholz station of Gulf Power, FL, USA in combination with the Foster Wheeler Resox process which converts SO_2 gas to elemental sulphur (Rush and Edwards, 1978a,b,c; Steiner *et al.*, 1980).

A similar process using activated char and producing elemental sulphur is being developed in Japan by Electric Power Development Co., Sumito Heavy Industries and Ishikawajima Harima Heavy Industries (Takeuchi *et al.*, 1980; Takenouchi *et al.*, 1981).

Cat-ox process

In this process SO_2 is oxidised to SO_3 by passing the flue gas over a catalyst at 425°C. The SO_3 is absorbed in sulphuric acid to make additional acid. Prescrubbing of the flue gas is required to ensure that fly ash does not block the catalyst.

Illinois Power Company together with EPA support installed a demonstration plant of 110 MW capacity on a coal-fired boiler at Wood River, USA. The plant, which started up in 1972, used natural gas to achieve the relatively high temperature in the catalytic reactor. When natural gas became unavailable, the system was converted to No. 2 fuel oil and soot problems occurred in the in-line burners. Further problems were experienced with corrosion of equipment to the extent that the plant operated for only 27 days in 4 years (Engdahl and Rosenberg, 1978; Erskine and Schmitt, 1978). However during the period of operation particulate removal by the electrostatic precipitator was greater than 99% and the conversion of SO_2 to SO_3 was greater than 90%. The product was 78% sulphuric acid.

There appears to be little current interest in the Cat-ox process, although a similar process is being investigated at bench scale by CIT-ALCATEL at the Villarceaux Centre, France (Salaun and Trempu, 1978). The high temperature required in the catalytic reactor means that the process is unsuitable for retrofit installations.

Chapter 16

Disposal of flue gas desulphurisation waste

The production of a sludge of calcium sulphite and calcium sulphate is the major disadvantage of wet non-regenerable processes which represent the majority of FGD units in use today. The sludge has extremely poor dewatering characteristics (Wilhelm *et al.*, 1979). The volume of sludge produced is significant: a 500 MW station burning coal with a 3% sulphur content produces more than 450 000 t/year of sludge containing 50% solids (Pruce, 1981a). The sludge is usually stored in ponds or used as landfill material depending on the degree of dewatering that is practised and the geological/hydrological conditions of the site (Goodwin, 1977). Disposal in underground mines has been examined but is not widely practised (Duvel *et al.*, 1980). However, concern about the potential pollution of ground water is leading to the development of processes which first stabilise the sludge before disposal. Even when chemically stabilised, flue gas desulphurisation sludges may emit sulphurous gases into the atmosphere (Adams, 1979; Adams and Farwell, 1981).

Direct ponding

'Direct ponding' or the storage of sludge in ponds adjacent to the power plant is the most widely practised method of sludge disposal. Scrubber sludge generally contains about 15% solids as it exits the absorber. It is thickened to 30% solids before being pumped directly to a disposal pond. Excess water from the thickener and pond is recycled to the FGD system. It is normally uneconomic to pump the sludge further than 5 km (Goodwin and Gleason, 1978).

Even in its thickened state the sludge is thixotropic and may remain so indefinitely depending upon the conditions under which it is stored. Unless the base material of the pond is considered to be impermeable the pond must be lined with impermeable soil or clay. Even so, considerable seepage may occur which, depending on the hydrogeology of the disposal site may lead to pollution of the ground water. Table 16.1 is a comparison of the concentration of trace elements and major species in sludge liquors compared with those in the public water supply in the USA (Brown and Brown, 1978). Even if the storage area is designed to minimise seepage, a water imbalance in the scrubbing process can lead to overflow and the direct discharge to the surface waters. Possible flooding or other unusual events must be considered during the design of a pond. Some states in the USA require sludge ponds to be located above the 50 year or 100 year flood level (Ellison and Kanfmann, 1978).

Table 16.1 Trace metals in FGD sludge liquors (after Brown and Brown, 1978)

Scrubber type:	*Lime*	*Limestone*	*Limestone*	*Limestone (Western USA coal)*	*Lime (Eastern USA coal)*	*Water quality criteria (USA, 1973)*
Scrubber size, MW;	10	10	<1	>100	>100	
Solids in discharge, %	50–50	35–40	75–80	15	10–40	
Arsenic, ppm	0.02	0.20	0.30	<0.004	0.085	0.1
Beryllium, ppm	<0.002	0.01	0.001	0.18	0.012	—
Cadmium, ppm	0.10	0.005	0.05	0.01	0.023	0.01
Chromium (total), ppm	0.03	0.15	0.005	0.21	0.40	0.05
Copper	<0.002	0.02	0.08	0.20	0.048	1.0
Lead	0.05	0.1	0.01	0.02	0.18	0.05
Mercury	<0.001	0.06	0.0012	0.12	0.045	0.002
Selenium	<0.2	0.30	0.12	2.5	0.80	0.01
Zinc	0.8	0.30	0.12	0.12	0.09	5.0
pH	9	8	7	4	9	5.0 to 9.0

Most sludge liquors contain significant quantities of fly ash

Large sludge disposal ponds are expensive to construct, utilise large areas of land and, since the sludge may contain in excess of 70% water, represent wastage of large quantities of water. It is rarely possible to reclaim the land after the retirement of a sludge pond although this has been achieved in areas of low annual rainfall and high rates of evaporation.

Landfill of untreated sludge

Untreated sludge may sometimes be used as landfill material although its properties are not really suitable for this method of disposal. The sludge is thickened from 15 to 30% solids as described above. It is then dewatered in a vacuum filter to 65% solids for lime/limestone FGD processes or 55% solids for the double alkali FGD process (Haynes *et al.*, 1980). The sludge has the appearance of damp solids and may be thixotropic. It has a high coefficient of permeability and no unconfined compressive strength. Careful consideration must be given to the hydrogeology of a site before this method of disposal is used to avoid contamination of ground water from runoff or leachate.

Treatment of sludge

There are essentially three methods of stabilising FGD sludge:

1. chemical stabilisation;
2. oxidation of the calcium sulphite in the sludge to calcium sulphate;
3. blending sludge with fly ash.

Chemical stabilisation

FGD sludge may be stabilised chemically by the addition of lime and additives containing alumina and silica (Duvel *et al.*, 1978b). The chemistry is similar to that employed in the Portland cement industry (Goodwin and Gleason, 1978). Sulpho-alumina hydrates (ettringites) and calcium silica hydrates (tobermorite) are formed:

$$\left.\begin{array}{l}3CaO\cdot Al_2O_3\cdot 3CaSO_4\cdot 28\text{-}32H_2O\\ 3CaO\cdot Al_2O_3\cdot 3CaSO_3\cdot 28\text{-}32H_2O\end{array}\right\}\ \text{forms of ettringite}$$

$$CaO\cdot SiO_2\cdot nH_2O \quad \text{tobermorite}$$

These minerals assume crystalline habits exhibiting the density and particle interlocking indicative of materials having the relatively high compressive strength required for landfill material. A chemically stabilised sludge has a coefficient of permeability within the range of 10^{-5} to 10^{-6} cm/s, the minimum of which is an order of magnitude lower than that of untreated sludges (Elnagger *et al.*, 1977). Whilst the concentrations of salt constituents in the leachate from the sludge are reduced by approximately 50% those of trace elements are not reduced appreciably. Hence the leaching of chemically treated sludge should be avoided if possible (Duvel *et al.*, 1978c). A general procedure for managing rainfall runoff is to collect it in a peripheral ditch which directs the water to a settling pond. Depending on the quality of the water in this pond, it can be decanted to a watercourse or returned to the scrubber system (Rossoff *et al.*, 1978).

Although several companies in the USA offer chemical stabilisation processes only Dravo Corporation and IU Conversion Systems Inc. (IUCS) have full-scale utility operating experience with FGD sludge (Workman and Rothfuss, 1978; Briggs and Freas, 1980; Haynes *et al.*, 1980). The Dravo Corporation process uses a proprietary reagent named Calcilox to stabilise the sludge. This consists essentially of ground blast furnace slag. The slag, a waste product from the production of pig iron, consists of iron ore impurities and limestone. It is tapped off from the furnaces as a homogeneous mixture of lime, silica and alumina in the approximate, respective percentages: 35 to 45; 28 to 38; and 8 to 18. The molten slag is quenched to a glass form before being finely ground. An activator is added to accelerate the stabilisation of the sludge. Lime is used to control the pH of the sludge at between 10.5 and 11.0.

The IUCS process, also known as the Poz-O-Tec process, employs lime and fly ash. The composition of fly ash is generally in the following range (Haynes *et al.*, 1980):

SiO_2	30–50%
Al_2O_3	15–30%
CaO	1.5–4.5%
Fe_2O_3	10–30%

Fly ash is formed in a molten state. It cools to a glass state which has the form of mullite, $3Al_2O_3\cdot 2SiO_2$. The relative proportions of sludge, fly ash and lime vary depending upon the availability of material. A mixture of 0.5 to 1.0 part of fly ash to 1 part of sludge on a dry basis is typical. Lime is added at a rate of up to 4% of the dry weight of the mixture. The chemical reactions are similar to those of the Dravo process except that the rates are slower for the Poz-O-Tec process and the product has a lower compressive strength (Haynes *et al.*, 1980). In the first full scale application involving the landfill of sludge treated by the Poz-O-Tec process several problems were encountered. The permeability of the treated sludge was found to be an order of magnitude greater than that predicted from laboratory tests. Unacceptable concentrations of calcium, total dissolved solids, sulphate and magnesium were detected in the ash pond area as a result of leachate from either the treated sludge or from an emergency sludge pond (Hupe and Shoemaker, 1979; Golden, 1981). However, the material has been used for the construction of an impermeable evaporation pond (Brown and Brown, 1978) and, more recently, for the construction of an artificial reef in the Atlantic Ocean near Fire Island, NY, USA (Woodhead *et al.*, 1981). Whilst these projects have demonstrated the possible uses of the product their economic viability is less certain (Pruce, 1981a).

Oxidation of sludge

Oxidation of FGD sludge converts $CaSO_3 \cdot 1/2H_2O$ to $CaSO_4 \cdot 2H_2O$ or gypsum. The flat plate crystalline structure of the sulphite, which gives sulphite sludge its thixotropic property is modified to large needle-like sulphate formations (Goodwin, 1978). These large crystals exhibit improved settling rates and dewatering characteristics. Hence, gypsum sludge can be centrifuged or dewatered by vacuum filtration to a solids content of between 75 and 85%. The material is suitable as landfill or as a source of gypsum. However, in many countries the market for gypsum is limited. The stacking of gypsum produced from a FGD installation has been demonstrated at the Scholz station of Gulf Power Company in Florida, USA (Pruce, 1981a; Garlanger and Ingra, 1980; Golden, 1981). This minimises the amount of land required for waste disposal. However, as with the other disposal methods discussed so far there is the potential problem of contamination of ground water by the leachate from a gypsum disposal site. The leachate can contain concentrations of sulphate, chloride, calcium, and magnesium several orders of magnitude greater than natural background concentrations. Also, the addition of fly ash to gypsum has been found to decrease the coefficient of permeability, the sedimentation rate and the shear strength of gypsum (Pruce, 1981a). Forced oxidation can be accomplished in an additional processing step (Massey *et al.*, 1981) or within the absorber loop of a FGD process provided that a pH of between 4 and 5.5 can be maintained (Jackson, 1980). The Chiyoda Thoroughbred 121 process incorporates oxidation within the absorber.

Blending with fly ash

This is the simplest form of treating FGD sludge. Dewatered sludge and fly ash are mixed to form a physically stable material (Hagerty *et al.*, 1977). There are few chemical reactions unless the ash is particularly alkaline when partial ‘curing’ of the sludge may occur. The sludge may be dewatered in centrifuges and vacuum filters to a solids content of between 50 and 80%. This material is suitable as landfill although careful control of leaching is required (Groenewold *et al.*, 1981).

Chapter 17

Conclusions

Much of the sulphur in coal is emitted to the atmosphere as SO_2 during the combustion process unless specific control measures are practised. Measures to reduce SO_2 emissions may include the physical, chemical or microbiological pretreatment of coal to remove sulphur, the absorption of SO_2 by a sorbent during the combustion process and the removal of SO_2 from flue gases by a FGD process.

Physical coal cleaning techniques are applicable only to the removal of inorganic material which includes pyritic sulphur. Organic sulphur is not removed. Prior comminution of coal is required in order to first liberate the pyrite. The more closely the particle size of raw coal approaches that of finely disseminated pyrite the greater is the potential for the removal of sulphur. However, the degree of size reduction practised must be balanced against the cost of energy required and with consideration for the problems involved in subsequent handling and dewatering of fine coal. As with most physical separation processes product losses increase with increasing removal of impurities. As much as 50% of the heating value of a coal may be rejected with the pyrite material when attempting to maximise the removal of sulphur. In current practice there are restrictions on the degree of crushing which can be economically achieved and on the amount of rejection of combustible material that can be tolerated. These limit the removal of pyrite by physical processes to about 50% of the total pyritic sulphur.

Commercial physical coal cleaning techniques are based on differences in relative densities and in surface properties of coal and pyrite. Research into cleaning techniques based on differences in magnetic properties is in progress. Of those processes based on differences in relative densities only those which can operate with fine coal in a medium with a relative density close to that of coal (approx. 1.2) are capable of removing significant quantities of inorganic sulphur. Dense media vessels and dense media cyclones are in this category. Froth flotation and oil agglomeration are techniques which depend on differences in the surface properties, mainly affinity to water, of coal and inorganic material to effect a separation. Whilst these techniques are well suited to the cleaning of fine coal particles the similarities in the surface properties of fine coal particles and fine pyrite particles reduce their effectiveness as a means of sulphur removal. Chemical additives must be used to enhance the differences in surface properties. Reductions of up to 90% in the pyritic sulphur content of some coals have been obtained in laboratory scale equipment. Magnetic separation processes have only recently been applied experimentally to the removal of pyrite from coal. By using air as the transport medium it is possible to obtain a dry coal product. However, some degree of drying of run-of-mine coal may be required and the coal must be carefully classified and screened prior to magnetic separation. Wet magnetic separation processes can treat wider ranges of coal particle size but have the disadvantage of wetting the coal.

Chemical cleaning of coal has the potential advantage of removing both inorganic and organic sulphur. However, such processes are still at the research stage. Many rely on the oxidation or reduction of sulphur in coal. The problem is to treat selectively the sulphur. Conditions under which significant quantities of sulphur are removed also lead to degradation of the coal. Much more fundamental research is required to determine the occurrence and nature of sulphur forms in coal and their reaction under possible desulphurisation conditions.

Microbiological desulphurisation of coal is, at present, of only academic interest. Reaction times are of the order of several days. To process sufficient coal for a power plant would require a prohibitively large storage volume.

The potential for absorption of SO_2 by a sorbent during the fluidised bed combustion of coal is widely known as one of the advantages of this combustion technique. By removing the sulphur in the form of SO_2 the technique is not dependent on the form or chemical nature of sulphur in coal. Reductions in SO_2 emissions of greater than 90% can be achieved for most coals. However, at the particle sizes used in FBCs, only a small part (about 20%) of the active constituent in the sorbent is available for reaction with SO_2 and high sorbent to sulphur ratios are required. If limestone is used as sorbent it must be quarried, transported and crushed prior to use. Ultimately the spent sorbent must be disposed of. Much of the current research into sulphur removal during FBC is directed at reducing the quantities of sorbent rquired by increasing its utilisation. Pretreatment of sorbent, modification of the operating parameters of a FBC and post sulphation treatment of sorbent have been investigated. Hydration of spent sorbent has been found to increase the utilisation of the sorbent by a factor of four. However, a fully regenerable sorbent has yet to be found.

There is no commercial process for the absorption of SO_2 during the combustion of pulverised coal. Flue gas desulphurisation, a commercial and widely used technique, must be employed. As with sulphur removal during FBC the technique is not dependent on the form or chemical nature of sulphur in coal. Processes can consistently remove more than 90% of the SO_2 from flue gases. The majority of processes in use today are of the wet, non-regenerable type using calcium in the form of lime or limestone. The major disadvantage of these processes is the production of large quantities of calcium sulphate/sulphite sludge. The sludge is usually stored in large ponds or lagoons adjacent to the power plant although it may be rendered suitable as landfill by chemical treatment. The wet, regenerable processes eliminate this problem by regenerating the sorbent producing a concentrated stream of SO_2 gas for subsequent processing to sulphur or sulphuric acid. However, there is a penalty of increased energy requirements over non-regenerable processes. More recently, dry processes of both regenerable and non-regenerable types have been developed. The advantages of dry SO_2 FGD processes over wet scrubbing processes include: a dry product; lower capital costs; lower energy and water requirements; and a more simple design which should be reflected in increased availability and reduced maintenance. The technique of spray drying is particularly suited to dry FGD.

References

Adams, DF (1979) Sulfur gas emissions from flue gas desulfurization sludge ponds. *Journal of the Air Pollution Control Association*; **29** (9); 963–968

Adams, DF and Earwell, SO (1981) Sulfur gas emissions from stored flue gas desulfurization sludges. *Journal of the Air Pollution Control Association*; **31** (5); 557–604

Albanese, AS and Sethi, DS (1980) *Regenerative process for desulfurization of high temperature combustion and fuel gases. Progress report No. 14 October 1–December 31, 1979*; BNL-51223; Upton, NY, USA; Brookhaven National Laboratory; 24 pp

Ando, J. (1979) SObx and NO*bx* abatement in Japan. *American Chemical Society, Division of Petroleum Chemistry, Preprints*; **24** (2); 506–516

Ando, J. and Laseke, BA (1977) *SO_2 abatement for stationary sources in Japan. Final task report, March 1976 August 1977*; PB-272986, Research Triangle Park, NC, USA; Environmental Protection Agency; 206pp

Andreasen, J (1980) Flue gas desulphurisation by spray dryer absorption. In: *Energy for our world, Division 3, Energy and Environment. Proceedings of the Eleventh World Energy Conference, Munich, FRG, 8–12 Sep 1980*; Vol 3; London, UK; World Energy Conference; 285–304

Andreason, J (1981) Spray adsorption: a new concept within flue gas desulphurisation. In: *Third Seminar on Desulphurization of Fuels and Combustion Gases, Salzburg, Austria, 18–22 May 1981*; ENV/SEM 13–COM 2; United Nations; Economic Commission for Europe; 7pp

Andrews, RL (1977) Current assessment of flue gas desulfurization technology. *Combustion*; **49** (4); 20–25

Aplan, FF (1977) Use of the flotation process for desulfurization of coal. In: *Coal Desulfurization, Chemical and Physical Methods*, (ed. TD Wheelock) Washington, DC, USA; American Chemical Society; *ACS Symposium Series*; (64); 70–82

Atzger, J (1979) Rauchgasentschwefelungsanlage fuer 700 MW-Kohlekraftwerk (Flue gas desulphurisation plant for 700 MW coal-fired power station). *Brennstof-Warme-Kraft*; **31** (7); 303–304 (In German)

Baker, DC. and Attar, A (1981) Sulfur pollution from coal combustion. Effect of the mineral components of coal on the thermal stabilities of sulfated ash and calcium sulfate. *Environmental Science and Technology*; **15** (3); 288–293

Bakke, E (1980) Flue gas desulfurization for industrial coal fired boilers. In: *Second Annual Industrial Coal Utilization Symposium, Charleston, SC, USA, 17 Apr 1980*; CONF-8004129; Paper 17; 24 pp

Ban, TE (1980) Converting high-sulfur coal to clean pellet fuel. *Mining Congress Journal*; **66** (5); 25–28

Barrier, JW, Faucett, HL. and Henson, L J (1978a) *Economics of disposal of lime/limestone scrubbing wastes: untreated and chemically treated wastes*; PB–281391; Research Triangle Park, NC, USA; Environmental Protection Agency; 446 pp

Barrier, JW, Faucett, HL. and Henson, L J (1978b) Comparative economics of FGD sludge disposal. *Proceedings of the Air Pollution Control Association*; **5**; 1–22

Bechtel National, Inc. (1980) *Economic and design factors for flue gas desulfurization technology. Final report*; EPRI-CS–1428; Palo Alto, CA, USA; Electric Power Research Institute; 206 pp

Behrens, GP and Hargrove, OW (1980a) *Evaluation of Chiyoda Thoroughbred 121 FGD process and gypsum stacking. Volume 1. Chiyoda evaluation. Final report*; EPRI-CS–1579 (Vol. 1); Palo Alto, CA, USA; Electric Power Research Institute; 240 pp

Behrens, GP and Hargrove, OW (1980b) *Evaluation of Chiyoda Thoroughbred 121 FGD process and gypsum stacking. Volume 2. Chiyoda evaluation appendixes. Final report*; EPRI-CS–1579 (Vol. 2); Palo Alto, CA, USA; Electric Power Research Institute; 237 pp

Best, RJ and Yates, JG (1977) Removal of sulfur dioxide from a gas stream in a fluidized bed reactor. *Industrial and Engineering Chemistry, Process Design and Development*; **3** (16); 347–352

Bettelheim, J and Morris, GF (1979) Prospects of partial pyrites removal. In: *Proceedings of the International Symposium on Sulphur Emissions and the Environment, London, UK, 8–10 May 1979*; London, UK; Society of Chemical Industry; pp 347–355

Beychok, MR (1980) *Comparative economics of advanced regenerable flue gas desulfurization processes. Final report*; EPRI-CS–1381; Palo Alto, CA, USA; Electric Power Research Institute; 131 pp

Biedell, EL and Stevens, NJ (1979) Tests validate lime/limestone scrubber for SO_2 and particulate control. *Power (NY)*; **123** (5); 68–69

Binns, DR and Aldrich, RG (1978) Design of the 100 MW Atomics International Aqueous Carbonate process regenerative FGD demonstration plant. In: *Proceedings of Symposium on Flue Gas Desulfurization, Hollywood, FL, USA, November 1977*; PB–282091; Research Triangle Park, NC, USA; Environmental Protection Agency; pp. 665–695

Blanc, H (1978) Procedes industriels de desulfuration des fumees de centrales thermiques (Processes for desulfurization of waste gases tried out at 'Electricite de France'). *Revue Generale de Thermique*; **17** (195); 199–206

Bluhm, DD, Fanslow, GE and Beck-Montgomery, SR (1981) *Selective magnetic enhancement of pyrite in coal by dielectric heating. Final report, Oct 1, 1978–Dec 31, 1980*; IS-4766; Ames, IA, USA; Ames Laboratory, Iowa State University; 200 pp

Borgwardt, RH (1980) Combined flue gas desulfurization and water treatment in coal-fired power plants. *Environmental Science and Technology*; **14** (3); 294–298

Boron, DJ, Dietz, AG and Taylor, SR (1981) Sulphur removal from coal via hydrogen peroxide oxidation. *Fuel*; **60** (10); 991–992

Bottenbruch, H and Kaemmer, K (1978) Optimized stacks reduce SO_2 emission. *Erdoel und Kohle, Erdgas, Petrochemie, vereinigt mit Brennstoff-Chemie*; **31** (10); 479–481

Braden, HH (1978) Double-loop operating offers best of both worlds approach to sulfur dioxide scrubbing. *Public Utilities Fortnightly*; **102** (4); 54–55, 58–59

Briggs, RW and Freas, RC (1980) Sludge to fertile soil—research results. *Journal of Testing and Evaluation*; **8** (5); 265–269

Broeker, G (1978) Trockene Rauchgasentschwefelung bei Braunkohlenkraftwerken (Dry gas desulphurization in brown coal power plants). *VGB Kraftwerkstechnik*; **58** (4); 305–308 (In German)

Brown, WD and Brown, EC (1978) Removal of SO_2 from stack gases—a review of technology and waste byproduct disposal. *CIM Bulletin*; **71** (795); 82–89

Buder, MK, Clifford, KL, Huettenhain, H and McGowin, CR (1980) The effects of coal cleaning on power generation economics. *Combustion*; **51** (10); 19–25

Bulewicz, E, Wievzorek-Ciurowa, K, Kandafer, S, Janicka, E, Jurys, C and Klimek, J (1980) Laboratory investigation of the desulfurizing capacity of limestone during coal combustion in a fluidized bed. *Energetyka*; **34** (9); 367–370

Burbank, DA, Wangs, SC, McKinsey, RR and Williams, JE (1981) Test results of adipic acid-enhanced limestone scrubbing at the EPA Shawnee Test Facility—third report. In: *Proceedings of Symposium on Flue Gas Desulfurization, Houston, TX, USA, 28–31 Oct 1980*; EPA-600/9-81-091a; Washington, DC, USA; Environmental Protection Agency; pp 233–286

Burdett, NA, Gliddon, BJ, Hotchkiss, RC and Squires, RT (1981) Fluidised bed desulphurisation and the role of SO_3. In: *Fluidized Combustion Conference, University of Cape Town, South Africa, 28–30 Jan 1981*; Vol 2; Cape Town, South Africa; Energy Research Institute, University of Cape Town, pp. 424–442

Burdett, NA, Hotchkiss, RC and Fieldes, RB (1979) SO_2 formation and retention in coal-fired fluidized bed combustors. In: *The Control of Sulphur and other Gaseous Emission, Salford, UK, Apr 1979*; IChemE Symposium Series No. 57; pp M1–M9 (1979)

Burger, JR (1980) Froth flotation is on the rise. *Coal Age*; **85** (3); 99–108

Burnett, TA, Anderson, KD and Torstrick, RL (1981) Spray dryer FGD: technical review and economic assessment. In: *Proceedings of Symposium on Flue Gas Desulfurization, Houston, TX, USA, 28–31 Oct 1980*; EPA-600/9-81-091a; Washington, DC, USA; Environmental Protection Agency; pp. 713–730

Byron, RA. and Saleem, A (1978) Particulate and sulfur dioxide control options for conventional coal combustion. In: *Fifth Energy Technology Conference, Washington, DC, USA, 27 Feb–1 Mar 1978*; CONF-780222–; Washington, DC, USA; Government Institutes, Inc.; pp. 142–156

Calderbank, PH (1979) A new flue–gas SO_2 removal process using activated chalk in a fluidised bed at (460–550°C). In: *The Control of Sulphur and Other Gaseous Emissions, Salford, UK, Apr 1979*; IChem E Symposium Series No. 57; pp D1–D9

Capes, CE (1979) Agglomeration. In: *Coal Preparation*, (ed. JW Leonard) New York, NY, USA; American Institute of Mining, Metallurgical and Petroleum Engineers; **10** (4); pp. 10/105 — 10/116

Capes, CE, Coleman, RD, Germain, RJ and McIlhinney, AE (1978) Coal beneficiation by ash agglomeration. In: *Proceedings of the Coal and Coke Sessions, 28th Canadian Chemical Engineering Conference, Halifax, Nova Scotia, Canada, 22–25 Oct 1978*; Ottawa, Canada; Canadian Society for Chemical Engineering, pp. 118–131 (1978)

Capes, CE, Sproule, GI and Taylor, JB (1979) Effect of crystal structure on the removal of pyrite from coal by oil agglomeration. *Fuel Processing Technology*; **2** (4); 323–329 (1979)

Cavallaro, JA and Deurbrouck, AW (1977) An overview of coal preparation. In: *Coal Desulfurization, Chemical and Physical Methods*, (ed. TD Wheelock) Washington, DC, USA; American Chemical Society; *ACS Symposium Series*; (64); pp. 35–57

Chandra, D, Chakrabarti, JN and Swamy, YV (1982) Auto-desulphurization of coal. *Fuel* ; **61** (2); 204–205

Chandra, D, Roy, P, Mishra, AK, Chakrabarti, JN and Sengupta, B (1979) Microbial removal of organic sulphur from coal. *Fuel* **58** (7); 549–550

Chandra, D, Roy, P, Mishra, A.K, Chakrabarti, JN, Prasad, NK and Chaudhuri, SG (1980) Removal of sulphur from coal by *Thiobacillus ferrooxidans* and by mixed acidophilic bacteria present in coal. *Fuel*; **59** (4); 249–252

Chang, C-S. and Borgwardt, RH (1980) Effect of scrubbing operating conditions on adipic acid degradation. In: *Proceedings of Conference on Flue Gas Desulfurization, Morgantown, WV, USA, 6–7 Nov 1980*; CONF-801176; Berkeley, CA, USA; Lawrence Berkeley Laboratory; pp. 1–28

Chang, C-S and Dempsey, JH (1982) Effect of limestone type and grind on SO_2 scrubber performance. *Environmental Progress*; **1** (1); 59–64

Chemical Engineering (1981) Full-scale tests confirm adipic acid's promise for flue-gas desulfurization. *Chemical Engineering*; **88** (12); 17

Chemical Week (1980) Scrubber users want a home for waste. *Chemical Week*; **126** (20); 44–45

Chemie Ingenieur Technik (1978) Davy Powergas-Technologie fuer die Rauchgasreninigung (Davy Powergas Technology for flue gas cleaning). *Chemie Ingenieur Technik*; **50** (9); A484-A490 (In German)

Cleve, U (1979) Filternde Abscheider fuer Kohlenkraftwerke. Teil II Einsatz von Schaunchfilteranlagen bei der Schwefeldioxidentfernung (Fabric filters for coal-fired power stations. Part II Use of bag filter installations for sulphur dioxide elimination). *Staub-Reinhaltung der Luft*; **39** (10); 367–373 (In German)

Cole, R (1979) Coal cleaning plus scrubbing proves economic desulfurization combination. *Electric Light and Power (Boston)*; **57** (4); 25–26

Cole, RM, Kelso, TM, Moore, ND and Robards, RF (1978) *Evaluation of 1-MW horizontal scrubber. Final Report*; EPRI-FP–752; Palo Alto, CA, USA; Electric Power Research Institute; 245 pp

Cole, RM, Kelso, TM and Robards, RF (1979) *Reheat study and the corrosion-erosion tests at TVA's Colbert Pilot Plant*; EPRI-FP–940; Palo Alto, CA, USA; Electric Power Research Institute; 68 pp

Cooper, DW (1980) Energy consumption reduction using two-stage return scrubbers. *Journal of the Air Pollution Control Association*; **30** (10); 1130–1132

Cooper, JC, Ostermeier, RM and Donnelly, RJ (1978) Feasibility study of condensation of flue gas cleaning (CFGC) system. *Environment Science and Technology*; **12** (10); 1183–1188

Corbett, WE, Hardgrove, OW and Merrill, RS (1977) *Summary of the effects of important chemical variables upon the performance of lime/limestone wet scrubbing systems. Interim report*; EPRI-FP–639; Palo Alto, CA, USA; Electric Power Research Institute; 55 pp

Crocker, L, Martin, DA and Nissen, WI (1979) *Citrate-process pilot-plant operation at the Bunker Hill Company. Report of Investigations*; PB-80-122708; Washington, DC, USA; US Bureau of Mines; 83pp

Crowe, RB., Lane, JF and Petti, VJ (1981) Early operation of the Celanese Fibers Company coal-fired boiler using the dry flue gas cleaning system. *Combustion*; **52** (8); 34–37

Cullis, CF and Hirschler, MM (1980) Atmospheric sulphur: natural and man-made sources. *Atmospheric Environment*; **14**; 1263–1278

Dalton, SM and Preston, GT (1979) Demonstration of the Resox process in Germany. *American Chemical Society, Division of Petroleum Chemistry, Preprints*; **24** (2); 587–592

Davids, P (1979) Optimization of wet and dry processes for simultaneous removal of particulates and gaseous air pollutants from coal fired power stations. In: *Proceedings of the first workshop of particulate control, Juelich, FRG, 16–17 Mar 1978*; INKA Conf-78-409-009; pp. 252–267

Davis, RA, Meyler, JA and Gude, KE (1979) Dry SO_2 scrubbing at Antelope Valley station. *Combustion*; **51** (4); 21–27

Davis, WT. and Fielder, MA (1982) The retention of sulfur in fly ash from coal-fired boilers. *Journal of the Air Pollution Control Association*; **32** (4); 395–397

Delleney, RD, Dickerman, JC and Menzies, WR (1978) Application of flue gas desulfurization to industrial boilers. *Proceedings of the Air Pollution Control Association*; **5**; 1–22

Detz, CM and Barvinchak, G. (1979) Microbial desulfurization of coal. In: *American Mining Congress Coal Conference, St Louis, MO, USA, 20 May 1979*; CONF-7905148–; Washington, DC, USA; American Mining Congress; pp. 75–82, 86

Devitt, TW, Laseke, BA and Kaplan, N (1980) Utility flue gas desulfurization in the USA *Chemical Engineering Progress*; **76** (5); 45–57

Dijkhuis, JP and Kerkdijk, CBW (1981) Upgrading of coal using cryogenic HGMS. *Institute of Electrical and Electronics Engineers Transactions on Magnetics*; **MAG-17** (4); 1503–1505

Dimitry, P, Gambhir, SP and Heil, TJ (1980) Emission control for industrial coal fired boilers. *Combustion*; **51** (7); 32–41

Dobby, GS and Kelland, DR (1982) Magnetite-coal separation by continuous HGMS Presented at *3rd Joint Intermag—Magnetism and Magnetic Materials Conference*, Montreal, Quebec, Canada, 20–23 Jul 1982. *Institute of Electrical and Electronics Engineers Transactions on Magnetics*; **18** (6); 1698–1700

Domahidy, G (1979) Integration of power plant water treatment system with flue gas desulfurization. *Combustion*; **51** (4); 28–30

Donovan, RP (1979) SO_2 removal by a fabric filter using nahcolite injection. In: *Proceedings of the first workshop of particulate control, Juelich, FRG, 16–17 Mar 1978*; INKA Conf-78-409-009; pp. 189–209

Dugan, PR and Apel, WA (1977) Microbiological removal of sulfur from a pulverized coal blend. In: *Third Symposium on Coal Preparation, Louisville, KY, USA, 18 Oct 1977*; CONF-7710113; pp. 10–21

Dunne, PG and Gasner, LL (1980) Agglomeration methods of improving FBC sorbent utilization and combustion. In: *Proceedings of the Sixth International Conference on Fluidized Bed Combustion, Atlanta, GA, USA, 9–11 Apr 1980;* CONF-800428-Vol. 3; Washington, DC, USA; US DOE; pp. 1004–1014

Durkin, TH, Van Meter, JA and Legatski, LK (1981) Operating experience with the FMC double alkali process. In: *Proceedings of Symposium on Flue Gas Desulphurization, Houston, TX, USA, 28–31 Oct 1980*; EPA-600/9-81-901a; Washington, DC, USA; Environmental Protection Agency; pp. 473–496

Dutkiewicz, RK. and Naude, DP (1981) Sulphur retention in shallow fluidized beds. In: *Fluidized Combustion Conference, Cape Town, South Africa, 28–30 Jan 1981*; Vol 2; Cape Town, South Africa; Energy Research Institute University of Cape Town, pp. 409–423

Duvel, WA, Gallagher, WR, Knight, RG, Kolarz, CR and McLaren, RJ (1978a) *State-of-the-art of FGD sludge fixation. Final report*; EPRI-FP–671; Palo Alto, CA, USA; Electric Power Research Institute; 270 pp.

Duvel, WA, McLaren, RJ, Knight, RG, Baker, M and Morasky, TM (1978b) State-of-the-art of FGD sludge disposal. *Proceedings of the Air Pollution Control Association*; **5**; 13 pp.

Duvel, WA, Rapp, JR, Atwood, RA and Merritt, GL (1978c) Leachate from disposal of FGD sludge in deep mines. *Proceedings of the Air Pollution Control Association*; **5**; 16 pp.

Duvel, WA Jr.; Call, P, Buckley, B and Moore, R (1980) *Coal mine disposal of flue gas cleaning wastes. Final report*; EPRI-CS-1376; Palo Alto, CA, USA; Electric Power Research Institute; 166 pp.

Eissenberg, DM, Hise, EC and Silverman, MD (1979) ORNL program for development of magnetic beneficiation of dry pulverized coal. In: *Proceedings of International Conference on Industrial Applications of Magnetic Separation, Rindge, NH, USA, 30 Jul–4 Aug 1978*; New York, NY, USA; Institute of Electrical and Electronic Engineers; pp. 91–94

Eldring, AK (1979) The Stauffer SO_2 abatement system. In: *The Control of Sulphur and other Gaseous Emissions, Salford, UK, Apr 1979*; IChem E Symposium Series **57**; pp. J1–J20

Elk, FA (1978) Common sense coal preparation. *Coal Mining and Processing*; **15** (3); 58–62

Ellison, W. and Kanfmann, RS (1978) Towards safe scrubber-sludge disposal. *Power (NY)*; **122** (7); 54–57

Elnaggar, HA, Rahim, AS. and Selmeczi, JG (1977) Chemical stabilization of SO_2 scrubber sludges. In: *Conference on Geotechnical Practices for Disposal of Solid Waste Materials, Ann Arbor, MI, USA, 13 Jun 1977*; CONF-7706158–; New York, NY, USA; American Society of Civil Engineers; pp. 645–660

Energy Recycling Corporation (1980) Coal-cleaning process holds promise for reducing pollutants. *Power (NY)*; **124** (1); 103

Engdahl, RB and Rosenberg, HS (1978) Status of flue gas desulfurization. *Chemtech*; **8** (2); 118–128

Environmental Protection Agency (1979) *Sulfur emission; control technology and water management*; EPA/600/9–79-019; Washington, DC, USA; Environmental Protection Agency; 33 pp.

EPRI Journal (1979) Shifting SO_2 from the stack. *EPRI Journal*; **4** (6); 15–18

Erskine, G and Schmitt, JC (1978) Status of the catalytic oxidation (Cat-ox) flue gas desulfurization system. In: *Proceedings of Symposium on Flue Gas Desulfurization, Hollywood, FL, USA, Nov 1977*; PB–282091; Research Triangle Park, NC, USA; Environmental Protection Agency; pp. 695–706

Esche, M (1975) Results of testing a flue gas desulphurization plant using the Hoelter process. *Brennstof-Waerme-Kraft*; **27** (12); 457–459

Evans, JM (1977) Alternatives to stack-gas scrubbing. *Coal Processing Technology*; **3**; 47–52

Federal Power Commission (1977a) *Status of flue gas desulfurization applications in the United States. A technological assessment. Highlights*; DOE/TIC–11370; Washington, DC, USA; US DOE; 80 pp.

Federal Power Commission (1977b) *Status of flue gas desulfurization applications in the United States. A technological assessment. Report in full*; DOE/TIC–11369; Washington, DC, USA; US DOE; 541 pp.

Fedorova, SK, Alekseeva, LA and Gladkii, AV (1978) Kinetics of oxidation of magnesium sulfite suspensions. *Journal Applied Chemistry of the USSR (English Translation)*; **51** (1); 20–23

Fee, DC, Wilson, WI, Shearer, JA, Smith, GW, Lenc, JF, Fan, LS, Myles, KM and Johnson, I (1980) *Sorbent utilization prediction methodology: sulfur control in fluidized-bed combustors*; ANL/CEN/FE-80-10; Research Triangle Park, NC, USA; Environmental Protection Agency; 443 pp.

Felsvang, K and Masters, K (1979) Flue gas desulphurization of coal fired power plant by dry absorption in spray dryers. In: *The Control of Sulphur and Other Gaseous Emissions, Salford, UK, Apr 1979*; IChem E Symposium Series No. 57; pp. I1–I15

Flament, G (1982) Simultaneous reduction of NO_x and SO_2 in pulverized-coal flames by the combined application of staged combustion and the direct injection of lime. *Rev. Gen. Therm.*; **21** (248/249); 649–663

Fraser Ross, F (1978) British approach to sulphur dioxide emissions. *Mechanical Engineering*; **100** (8); 42–45

Garlanger, JE and Ingra, TS (1980) *Evaluation of Chiyoda Throughbred 121 FGD process and gypsum stacking. Volume 3: testing the feasibility of stacking FGD gypsum*; EPRI-CS-1579 (Vol 3); Palo Alto, CA, USA; Electric Power Research Institute; 162 pp.

Gasner, LL. and Setesak, SE (1978) Limestone utilization optimized in fluidized bed boilers. In: *Proceedings of the Fifth International Conference on Fluidized Bed Combustion. II Near-term implementation, Washington, DC, USA, 12 Dec 1977*; Vol 2; M–78-68, CONF- 771272–P2; McLean, VA, USA; The MITRE Corporation; pp. 763–774

Gautney, J, Kim, YK and Hatfield, JD (1982) Melamine: a regenerative SO_2 absorbent. *Journal of the Air Pollution Control Association*; **32**; 260–265

Getler, JL, Shelton, ML and Furlong, DA (1979) Modelling the spray absorption process for SO_2 removal. *Journal of the Air Pollution Control Association*; **29** (12)

Gibbs, LL, Rorste, DS and Shah, YM (1978) *Particulate and sulfur dioxide emission control cost for large coal-fired boilers. Final report*; PB-281 271; Research Triangle Park, NC, USA; Environmental Protection Agency; 168 pp.

Gibson, J (1981) Present status and future prospects for fluidised combustion. In: *Fluidized Combustion Conference, Cape Town, South Africa, 28–30 Jan 1981*; Vol 3; Cape Town, South Africa, Energy Research Institute, University of Cape Town, pp. 2.1–2.35

Gille, JA and MacKenzie, JS (1978) Philadelphia Electric's experience with magnesium oxide scrubbing. In: *Proceedings of Symposium on Flue Gas Desulfurization, Hollywood, FL, USA, Nov 1977*; PB-282 091; Research Triangle Park, NC, USA; Environmental Protection Agency; pp. 737–751

Glamser, JH, Rock, KL and Esche, M (1979) A lime based FGD process utilizing forced oxidation to produce stable gypsum by-product. *Combustion*; **51** (1); 22–26

Goar, BG (1979) Processes sited for SO_2 removal from furnace stacks. *Oil and Gas Journal*; **77** (4); 58–61

Golden, DM (1981) EPRI FGD sludge disposal demonstration and site monitoring projects. In: *Proceedings of Symposium on Flue Gas Desulfurization, Houston, TX, USA, 28–31 Oct 1980*; EPA-600/9-81-091a; Washington, DC, USA; Environmental Protection Agency; pp. 625–656

Goldschmidt, K (1978) Zusammenfassung der Ergebruisse mit bisher in der Bundersrepublik Deutschland betriebenen Rauchgas (Summary of the results achieved by flue-gas desulphurization plants hitherto operated in the Federal Republic of Germany). *Bergbau*; **29** (4); 150–155 (In German)

Goodwin, RW (1977) Site specific burial of unfixed flue gas sludge. *Journal of the Environmental Engineering Division, American Society of Civil Engineers*; **103** (6); 1105–1114

Goodwin, RW (1978) Oxidation of flue gas desulfurization waste and the effect on treatment modes. *Journal of the Air Pollution Control Association*; **28** (1); 35–39

Goodwin, RW and Gleason, RJ (1978) Options for treating and disposing of scrubber sludge. *Combustion*; **50** (4); 37–41

Granat, L, Rodhe, H. and Hallberg, RO (1976) The global sulphur cycle. *Ecological Bulletins*; (22); 89–134

Granger, L, Scotti, LJ and Yarze, JC (1979) Economics of Pullman Kellog's magnesium promoted FGD system. In: Sixth Energy Technology Conference and Exposition, Washington, DC, USA, 26 Feb 1979. *Energy Technology (Washington, DC)*; **6**; 602–616

Grimm, C, Abrams, JZ, Leffmann, WW, Raben, IA and LaMantia, C (1978) The Colstrip flue gas cleaning system. *Chemical Engineering Progress*; **74** (2); 51–57

Grimm, U (1981) *Evaluation of the sulfur sorption properties of chemically improved limestones during fluidized-bed combustion.* US/DOE/METC/RI-128; Morgantown, WV, USA; Morgantown Energy Technology Center; 84 pp.

Groenewold, GH, Cherry, JA, Manz, OE, Gullicks, HA, Hassett, DJ and Bernd, WR (1981) Potential effects on groundwater of fly ash and FGD waste disposal in lignite surface mine pits in North Dakota. In: *Proceedings of Symposium on Flue Gas Desulfurization, Houston, TX, USA, 28–31 Oct 1980*; EPA-600/9-81-091a; Washington, DC, USA; Environmental Protection Agency; pp. 657–694

Gruenberg, NR (1979) Instrumentation and control for double-loop limestone scrubbers. *Power Engineering*; **83** (6); 72–75

Hagen, RI and Kolderu, H (1979) *Flue gas desulfurization process, study. Phase I, survey of major installations. Appendix 95-D, sea water scrubbing flue gas desulfurization process*; PB–295005; Washington, DC, USA; Environmental Protection Agency; 65 pp.

Hagerty, DJ, Ullrich, CR and Thacker, BK (1977) Engineering properties of FGD sludges. In: *Conference on Geotechnical Practices for Disposal of Solid Waste Materials, Ann Arbor, MI, USA, 13 Jun 1977*; CONF-7706158–; New York, NY, USA; American Society of Civil Engineers; pp. 23–40

Hall, EH, Hoffman, L, Hoffman, J and Schilling, RA (1978) *Physical coal cleaning for utility boiler SO_2 emission control*; PB-277408; Ohio, CO, USA; Battelle Memorial Institute; 112 pp.

Hancock, BL (1978) *Sulfur removal from flue gas. Final report*; FE–2240-51; Alhambra, CA, USA; Braun (CF) and Co.; 52 pp.

Haque, R, Duttam, ML and Chakrabarti, RK (1979) Fluidized bed combustion of high sulphur coals. *Journal of the Institute of Energy*; **52** (413); 173–177

Hartman, M, Pata, J and Coughlin, RW (1978) Influence of porosity of calcium carbonates on their reactivity with sulfur dioxide. *Industrial and Engineering Chemistry, Process Design and Development*; **17** (4); 411–419

Haynes, LH, Ansari, AH and Ovens, JE (1980) Ash/FGD waste disposal options. A comparative study for CILCO Duck Creek site. *Combustion*; **51** (7); 21–27

Head, HN, Wang, SC, Rabb, DT, Borgwardt, RH, Williams, JE and Maxwell, MA (1979) Recent results from EPA's lime/limestone scrubbing program—adipic acid as a scrubber additive. Presented at: *Fifth Flue Gas Desulfurization Symposium, Las Vegas, NV, USA, 5 Mar 1979*; PB-80-133168; Research Triangle Park, NC, USA; Environmental Protection Agency; pp. 342–385.

Hewitt, RA and Saleem, A (1981) Operation and maintenance experience of the world's largest spray tower SO_2 scrubbers. In: *Proceedings of Symposium on Flue Gas Desulfurization, Houston, TX,*

USA, 28–31 Oct 1980; EPA-600/9-81-091a; Washington, DC, USA; Environmental Protection Agency; pp. 433–452

Highley, J (1975) Fluidised combustion as a means of firing gas turbines and reducing atmospheric pollution. In: *Institute of Fuel Symposium Series No. 1: Fluidised Combustion, London, UK, 16–17 Sep 1975*; London: UK; Institute of Fuel; paper RAP/D-1, 8 pp.

Highton, NH. and Webb, MG (1980) Sulphur dioxide from electricity generation. Policy options for pollution control. *Energy Policy*; **8** (1); 61–76

Hise, EC (1979) *Correlation of physical coal separations—part 1; ORNL-5570; Springfield, VA, USA; National Technical Information Service; 47 pp.*

Hise, EC (1980) Sulfur and ash reduction in coal by magnetic separation. In: Proceedings of the Second US Department of Energy Environmental Control Symposium, Reston, VA, USA, 17 Mar 1980; Vol 1; CONF-800334, pp. 67–79

Hise, EC (1982) Oak Ridge National Laboratory, DOE, USA. Private communication

Hise, EC, Wechsler, I. and Doulin, JM (1979) *Separation of dry crushed coals by high-gradient magnetic separation*; ORNL-5571; Springfield, VA, USA; National Technical Information Services; 80 pp.

Hise, EC, Wechsler, I. and Doulin, JM (1981) *The continuous separation of dry crushed coal at one ton per hour by high-gradient magnetic separation*; ORNL-5763; Springfield, VA, USA; National Technical Information Service; 63 pp.

Hoffmann, MR, Faust, BC, Panda, FA, Koo, HH and Tsuchiya, HM (1981) Kinetics of the removal of iron pyrite from coal by microbial catalysis. *Applied and Environmental Microbiology*; **42** (2); 259–271

Hollinden, GA and Massey, CL (1979) *Preheat study and the corrosion-erosion tests at TVA's Colbert pilot plant*; EPRI-FP-940; Palo Alto, CA, USA; Electric Power Research Institute; 68 pp.

Hollinden, GA and Massey, CL (1979) TVA compliance programs for SO_2 emission. Presented at: *Fifth Flue Gas Desulphurization Symposium, Las Vegas, NV, USA, 5 Marh 1979*; PB-80-133168; Research Triangle Park, NC, USA; Environmental Protection Agency; pp. 386–417

Hollinden, GA, Robards, RF, Moore, ND, Kelson, TM and Cole, RM (1979b) *Concurrent scrubber evaluation: TVA's Colbert lime-limestone wet-scrubbing pilot plant*; EPRI-FP-941; Palo Alto, CA, USA; Electric Power Research Institute; 271 pp.

Holman, AS, Hise, EC and Jones, JE (1982) *Initial investigation of open-gradient magnetic separation*; ORNL-5764; Oak Ridge, TN, USA; Oak Ridge National Laboratory; 37 pp.

Hucko, RE (1979) DOE research in high gradient magnetic separation applied to coal beneficiation. In: *Proceedings of International Conference on Industrial Applications of Magnetic Separation, Rindge, NH, USA, 30 Jul–4 Aug 1978*; New York, NY, USA; Institute of Electrical and Electronic Engineers; p 7782

Hucko, RE. and Miller, KJ (1980) *A technical performance comparison of coal-pyrite flotation and high-gradient magnetic separation*; RI-PMTC-10(80); Pittsburgh, PA, USA; Pittsburgh Mining Technology Center, US DOE; 46 pp.

Huettenhain, H (1978) Coal preparation and cleaning for the power industry. In: *Proceedings of the American Power Conference, Chicago, IL, USA, Apr 1978*; Vol. 40; CONF-780440–; pp. 515–521

Hupe, DW. and Shoemaker, SH (1979) *Monitoring the fixed FGD sludge landfill, Conesville, OH, Phase 1. Final report*; EPRI-FP-1172; Palo Alto, CA, USA; Electric Power Research Institute; 159 pp.

Hurst, TB and Bielawski, GT (1981) Dry scrubber demonstration plant—operating results. In: *Proceedings of Symposium on Flue Gas Desulfurization, Houston, TX, USA, 28–31 Oct 1980*; EPA-600/9-81-091a; Washington, DC, USA; Environmental Protection Agency; pp. 853–860

Hutcheson, R and Johnson, CA (1981) Alabama Electric Cooperative flue gas desulfurization operating and maintenance experience. *Combustion*; **52** (8); 29–33

Idemura, H, Kanai, T and Yanagioka, (1978) Jet bubbling flue gas desulfurization. *Chemical Engineering Progress*; **74** (2); 46–50

Idemura, H, Kanai, T. and Yanagioka, (1978) Jet bubbling flue gas desulfurization. *Chemical Engineering Progress*; **74** (2); 46–50

Inculet, II, Quigley, RM, Bergougnou, MA, Brown, JD and Faurschou, DK (1980) Electrostatic beneficiation of Hat Creek coal in the fluidised state. Presented at *82nd Annual General Meeting of The Canadian Institute of Mining and Metallurgy, Toronto, Ontario, Canada, 20–23 Apr 1980*; Paper

13; Montreal, Quebec, Canada; The Canadian Institute of Mining and Metallurgy; 40 pp.

International Energy Agency (1978) *Steam coal prospects to 2000*. Paris, France; OECD; 157 pp.

Jackson, SB (1980) Limestone scrubbing with forced oxidation. In: *Workshop on sulfur chemistry in flue gas desulfurization, Morgantown, WV, USA, 7 Jun 1979*; DOE/METC/8333/11; Morgantown, WV, USA; Morgantown, Energy Technology Center; pp 134–168

Jackson, SB (1981) Cocurrent scrubber test, Shawnee test facility. In: *Proceedings of Symposium on Flue Gas Desulfurization, Houston, TX, USA, 28–31 Oct 1980*; EPA-600/9-81-091a; Washington, DC, USA; Environmental Protection Agency; pp. 287–310

Jackson, SB, Dene, CE and Smith, DB (1981) DOWA process tests, Shawnee test facility. In: *Proceedings of Symposium on Flue Gas Desulfurization, Houston, TX, USA, 28–31 Oct 1980*; EPA-600/9-81-091a; Washington, DC, USA; Environmental Protection Agency; pp. 311–326

Jahnig, CE and Shaw, H (1981a) A comparative assessment of flue gas treatment processes. Part 1—Status and design basis. *Journal of the Air Pollution Control Association*; **31** (4); 421–428

Jahnig, CE and Shaw, H (1981b) A comparative assessment of flue gas treatment processes. Part II—Environmental and cost comparison. *Journal of the Air Pollution Control Association*; **31** (5); 596–604

Jenkinson, DE and Cammack, P (1981) Coal preparation and the reduction of pyritic sulphur. In: *Third Seminar on Desulfurization of Fuels and Combustion Gases, Salzburg, Austria, 18–22 May 1981*; ENV/SEM 13–COM 17; United Nations: Economic Commission for Europe; 22 pp.

Johnson, CA (1978) Minimizing FGD operating costs. *Power Engineering*; **82** (2); 62–65

Johnson, CA (1979) Economic benefits of using flyash as an alkali for SO_2 removal systems. In: *The Control of Sulphur and Other Gaseous Emissions, Salford, UK, Apr 1979*; IChemE Symposium Series No. 57; pp. H1–H9

Johnson, CA (1981) Minnesota Power's operating experience with integrated particulate and SO_2 scrubbing. *Journal of the Air Pollution Control Association*; **31** (6); 701–705

Johnson, CA and Hutcheson, R (1980) Alabama Electric Cooperative flue gas desulfurization operating and maintenance experience. *Journal of the Air Pollution Control Association*; **30** (7); 744–748

Johnson, I, Vogel, GJ, Lee, SHD, Shearer, JA, Snyder, RB, Smith, GW, Swift, WM, Teats, FG, Turner, CB, Wilson, WI and Jonke, AA (1978) *Support studies in fluidized-bed combustion. Quarterly report: Jan–Mar 1978*; ANL/CEN/FE–78-3; Argonne, IL, USA; Argonne National Laboratory

Jones, DG, Slack, AV and Campbell, KS (1978) *Lime/limestone scrubber operation and control study. Final report*; EPRI–FP–627; Palo Alto, CA, USA; Electric Power Research Institute; 244 pp.

Karlsson, HT and Rosenberg, HS (1980a) Technical aspects of lime/limestone scrubbers for coal-fired power plants. Part 1. Process chemistry and scrubber systems. *Journal of the Air Pollution Control Association*; **30** (6); 710–714

Karlsson, HT and Rosenberg, HS (1980b) Technical aspects of lime/limestone scrubbers for coal-fired power plants. Part 2. Instrumentation and technology. *Journal of the Air Pollution Control Association*; **30** (7); 822–826

Katz, B, Gehri, DC and Oldenkamp, RD (1978) Atomics International open loop and closed loop aqueous carbonate processes: a current assessment. *American Society of Mechanical Engineers—Papers*; (78-JPGC-PWR-15); pp. 1–8

Kelly, ME and Dickerman, JC (1981) Current status of dry flue gas desulfurization systems. In: *Proceedings of Symposium on Flue Gas Desulfurization, Houston, TX, USA, 28–31 Oct 1980*; EPA-600/9-81-091a; Washington, DC, USA; Environmental Protection Agency; pp. 761–776

Kempton, AG, Moneib, N, McCready, RGL and Capes, CE (1980) Removal of pyrite from coal by conditioning with *Thiobacillus ferrooxidans* followed by oil agglomeration. *Hydrometallurgy*; **5**; 117–125

Killmeyer, RP (1980) *Performance characteristics of coal-washing equipment: Baum and Batac jigs*; RI-PMTC-9(80); Pittsburgh, PA, USA; Pittsburgh Mining Technology Center; 42 pp.

Kirkby, DR (1979) Flue gas desulfurization produces saleable gypsum. *Chemical Engineering (NY)*; **86** (1); 56–57

Knoblauch, K, Richter, E and Juntgen, H (1981) Application of active coke in processes of SO_2- and NO_x-removal from flue gases. *Fuel*; **60** (9); 832–838

Kragh, OT, Brna, TG and Ostop, RL (1981) SO_2 removal by dry FGD In: *Proceedings of Symposium on Flue Gas Desulfurization, Houston, TX, USA, 28–31 Oct 1980*; EPA-600/9-81-091a; Washington, DC, USA; Environmental Protection Agency; pp. 801–852

Kucera, KM (1978) The case for coal with scrubbers: economics and technology. *Combustion*; **50** (4); 30–33

Lallai, A, Mura, G, Viola, A and Gioia, F (1979) Removal of sulfur during the combustion of coal by adding limestone. *International Chemical Engineering*; **19** (3); 445–453

Lamantia, CR, Lunt, RR, Jashanani, IL, Donnelly, RG, Interess, E, Woodland, LR and Adams, ME (1977) *Application of scrubbing systems to low sulfur alkali ash coals. Final report*; EPRI-FP-595; Palo Alto, CA, USA; Electric Power Research Institute; 393 pp.

Land, GW (1979) Problems of removing sulfur from coal. In: 82nd National Western Mining Conference, Denver, CO, USA, 31 Jan 1979. *Mining Year Book 1979*; pp. 158–170

Laros, TJ (1977) *Physical desulfurization of Iowa coal by flotation*; IS-ICP-47; Ames, IA, USA; Iowa State University

Laseke, BA (1979a) *Survey of flue gas desulfurization systems: Duck Creek Station, Central Illinois Light Co. Final Report Jul–Dec 1978*; PB-80-126279; Research Triangle Park, NC, USA; Environment Protection Agency; 93 pp.

Laseke, BA (1979b) *Survey of flue gas desulfurization systems: Sherburne Country generating plant, Northern States Power Co. Final Report Jul–Dec 1978*; PB-80-126287; Research Triangle Park, NC, USA; Environmental Protection Agency; 121 pp.

Le, HV (1977) *Flotability of coal and pyrite*; IS-T-779; Ames, IA, USA; Iowa State University

Lee, SS (1978) *Evaluation of Radian's report, 'Water pollution impact of controlling sulfur dioxide emissions from coal-fired steam electric generators'*; CRESS–49; Menlo Park, CA, USA, SRI International; 14 pp.

Leivo, CC (1978a) *Flue gas desulfurization systems: design and operating considerations. Volume 1. Executive summary. Final report April–December 1977*; PB-280253; San Francisco, CA, USA; Bechtel Corp.; 29 pp.

Leivo, CC (1978b) *Flue gas desulfurization systems: design and operating considerations. Volume II Technical report. Final report April–December 1977*; PB-280254; San Francisco, CA, USA; Bechtel Corp.; 221 pp.

Lihach, N (1979a) More coal per ton. *EPRI Journal* **4** (5); 6–13

Lihach, N (1979b) Steps to sludge disposal. *EPRI Journal*; **4** (6); 19

Liu, YA (1982) *Physical cleaning of coal*. New York, NY, USA; Marcel Dekker, Inc.; 568 pp.

Luborsky, FE (1978) *High-gradient magnetic separation for removal of sulfur from coal*; FE-8969-1; Pittsburgh, PA, USA; US DOE; 82 pp.

Lutz, SJ and Chatlynne, CJ (1979) Dry FGD systems for the electric utility industry. Presented at: *Fifth Flue Gas Desulfurization Symposium, Las Vegas, NV, USA, 5 Mar 1979*; PB-80-133168; Research Triangle Park, NC, USA; Environmental Protection Agency; pp. 508–525

Madenburg, RS and Seesee, TA (1980) H_2S reduces SO_2 to desulfurize flue gas. *Chemical Engineering (NY)*; **87** (14); 88–89

Maiva, PS (1980) *State-of-the-art review of materials-related problems in flue gas desulfurization systems*; ANL-80-59; Argonne, IL, USA; Argonne National Laboratory; 45 pp.

Mann, EL and Adams, RC (1981) Status report on the Wellman-Lord/Allied Chemical flue gas desulfurization plant at Northern Indiana Public Service Company's Dean H Mitchell station. In: *Proceedings of Symposium on Flue Gas Desulfurization, Houston, TX, USA, 28–31 Oct 1980*; EPA-600/9-81-091a; Washington, DC, USA; Environmental Protection Agency; pp. 497–542

Marcus, EG, Wright, TL. and Wells, WL (1981) Magnesium FGD at TVA: pilot and full-scale designs. In: *Proceedings of Symposium on Flue Gas Desulfurization, Houston, TX, USA, 28–31 Oct 1980*; EPA-600/9-81-091a; Washington, DC, USA; Environmental Protection Agency; pp. 543–558

Martin, JR, Malki, KW. and Graves, N (1979) The results of a two-stage scrubber/charged particulate separator pilot program. *Combustion*; **51** (4); 12–20

Massey, CL, Moore, ND, Munson, GT, Runyan, RA and Wells, WL (1981) Forced oxidation of limestone scrubber sludge at TVA's Widow's Creek Unit 8 steam plant. In: *Proceedings of Symposium on Flue Gas Desulfurization, Houston, TX, USA, 28–31 Oct 1980*; EPA-600/9-81-091a; Washington, DC, USA; Environmental Protection Agency; pp. 371–390

Master, WA and Zellmer, SD (1979) *Effects of fertilizer made from flue gas and fly ash on selected crops and soils. Summary report*; ANL/EES-TM-43; Argonne, IL, USA; Argonne National Laboratory; 28 pp.

Maxwell, MA (1979) Recent developments in flue gas desulfurization. *American Chemical Society, Division of Petroleum Chemistry, Preprints*; **24** (2); 517–526

Maxwell, JD, Faucett, HL and Burnett, TA (1977) *Environmental control implications of generating electric power from coal: 1977 technology status report. Appendix G State-of-the-art review for simultaneous removal of nitrogen oxides and sulfur oxides from flue gas*; ANL/ECT-3 (App. G); Argonne, IL, USA; Argonne National Laboratory; 228 pp.

McCain, JD (1979) Performance tests of the Montana Power Company Colstrip station flue gas cleanup system. In: *Proceedings of the First Workshop of Particulate Control, Juelich, FRG, 16–17 Mar, 1978*; INKA Conf-78-409-009; pp. 210–238

McCarthy, JE (1977) Start-up and operation of the Research-Cotterell/BAHCO scrubber for removal of SO_2 and particulates from flue gas at Rickenbacker Airforce Base, Columbus, OH, USA In: *Energy Technology '77 23rd IES Annual Technical Meeting and Exposition, Los Angeles, CA, USA, 24 Apr 1977*; CONF0770415–MT; Prospect, IL, USA; Institute of Environmental Sciences; pp. 90–94

McConnell, JF (1979) Multi-stream coal cleaning strategy. In: *Proceedings of the Coal Processing and Conversion Symposium, Morgantown, WV, USA, 1 Jun 1976*; CONF-7606193; Morgantown, WV, USA; West Virginia Geological and Economic Survey; pp. 31–47

McKaveney, JP. and Stivers, D (1978) Capacity of some California alkaline brines for sulfur dioxide. *American Chemical Society, Division of Fuel Chemistry Preprints*; **23** (2); 214–220

McKisson, RL, Baverle, GL, Bodine, JE, Rennick, RD, Stewart, AE and Tsang, S (1980) *Aqueous carbonate process design study. Final report*; EPRI-CS-1574; Canoga Park, CA, USA; Rockwell International Corp.; 466 pp.

Megonnell, WH (1978) Efficiency and reliability of sulfur dioxide scrubbers. *Journal of the Air Pollution Control Association*; **28** (7); 725–731

Meserole, FB (1980) Recent data concerning fate of adipic acid in FGD systems. In: *Proceedings of Conference on Flue Gas Desulfurization, Morgantown, WV, USA, 6–7 Nov 1980*; CONF-801176; Berkeley, CA, USA; Lawrence Berkeley Laboratory; pp. 29–34

Meserole, FB, Lowell, PS and Hargrove, OW (1979) A test for limestone reactivity and magnesium availability in SO_2 wet scrubbing. *American Chemical Society, Division of Petroleum Chemistry, Preprints*; **24** (2); 553–560

Mesich, FG and Jones, DG (1979) Process control for environmental systems: case studies in desulfurization. In: *Proceedings of the 1979 symposium on instrumentation and control for fossil energy processes, Denver, CO, USA, 20 Aug 1979*; ANL-79-62, CONF-790855; Argonne, IL, USA; Argonne National Laboratory; pp. 376–398

Meyler, J (1981) Dry flue gas scrubbing—a technique for the 1980s. *Combustion*; **52** (8); 21–28

Midkiff, LA (1979a) Lime system scrubs SO_2 yields gypsum. *Power (NY)*; **123** (5); 65–67

Midkiff, LA (1979b) New trends update FGD systems. *Power (NY)*; **123** (6); 103–105

Miles, AJ, Holmes, J and Draffin, CW (1979) Cost comparison of pre-combustion and post-combustion cleanup technologies *Energy and the Environment (NY)*; 448–455

Miller, I (1979) Dry scrubbing looms large in SO_2 cleanup plans. *Chemical Engineering (NY)*; **86** (18); 52, 54

Miller, KJ (1978) *Desulfurization of various midwestern coals by flotation*; RI8262; Washington, DC, USA; US Bureau of Mines; 14 pp.

Min, S (1977) *Physical desulfurization of Iowa coal*; IS-ICP-35; Ames, IA, USA; Iowa State University

Mobley, JD and Chang, JCS (1981) The adipic acid enhanced limestone flue gas desulfurization process. *Journal of the Air Pollution Control Association*; **31** (12); 1249–1253

Mohn, U (1979) *Process for removing sulfur oxides from gases with direct production of a usable finished reaction product*; BMFT-FB-T-79-90; Bonn-Bad Godesberg, FRG; Bundesministerium fuer Forschung und Technologie; 25 pp.

Monostory, FP (1979) Reducing the sulphur content of power-station coal. *Glueckauf*; **115** (18); 420–423

Montagna, JC, Lenc, JF, Vogel, GJ and Jonke, AA (1977a) Regeneration of sulfated dolomite from a coal-fired FBC process by reductive decomposition of calcium sulfate in a fluidized bed. *Industrial and Engineering Chemistry, Process Design and Development*; **16** (2); 230–236

Montagna, JC, Vogel, GJ, Smith, GW and Jonke, AA (1977b) *Fluidized-bed regeneration of sulfated dolomite from a coal-fired FBC process by reductive decomposition*; ANL-77-6; Argonne, IL, USA; Argonne National Laboratory

Montagna, JC, Nunes, FF, Smith, GW, Smyk, EB, Teats, FG, Vogel, GJ. and Jonke, AA (1978a) Development of a process for regenerating partially sulfated limestone from FBC boilers. In: *Proceedings of the Fifth International Conference on Fluidized Bed Combustion, Washington, DC, USA, 12–14 Dec 1977*; Vol 2; M-78-68 CONF-771272; McLean, VA, USA; The MITRE Corporation; pp. 776–794

Montagna, JC, Swift, WM, Smith, GW, Vogel, GJ and Jonke, AA (1978b) Regeneration of sulfated dolomite by reductive decomposition with coal in a fluidized bed reactor. *American Institute of Chemical Engineers, Symposium Series*; **74** (176); 203–211

Morasky, TM, Burford, DP and Hargrove, OW (1981) Results of the Chiyoda Thoroughbred-121 prototype evaluation. In: *Proceedings of Symposium on Flue Gas Desulfurization, Houston, TX, USA, 28–31 Oct 1980*; EPA-600/9-81-091a; Washington, DC, USA; Environmental Protection Agency; pp. 347–370

Mori, T, Matsuda, S, Nakajima, F, Nishimura, T and Arikawa, Y (1981) Effect of Al^{3+} and F^{-} on desulfurization reaction in the limestone slurry scrubbing process. *Industrial and Engineering Chemistry, Process Design and Development*; **20** (1); 144–147

Morrison, GF (1978) *Combustion of low grade coal*; ICTIS/TR02; London, UK; IEA Coal Research; 90 pp.

Morrison, GF (1980) *Nitrogen oxides from coal combustion—abatement and control*; ICTIS/TR11; London, UK; IEA Coal Research; 86 pp.

Morrison, GF (1981) *Chemical desulphurisation of coal*; ICTIS/TR15; London, UK; IEA Coal Research; 72 pp

Mozes, MS (1978) Regenerative limestone slurry process for flue gas desulfurization. *Environmental Science and Technology*; **12** (2); 163–169

Murray, HH (1977) Magnetic desulfurization of some Illinois Basin coals. Prestned at *173 National Meeting of the American Chemical Society, New Orleans, LA, USA, 20 Mar 1977*; CONF-770301-P4 American Chemical Society, Division of Fuel Chemistry, *Preprints*; **22** (2); 106–112

Munzner, H. and Bonn, B (1980) Sulfur capturing effectivity of limestone and dolomites in fluidized bed combustion. In: *Proceedings of the Sixth International Conference on Fluidized Bed Combustion, Atlanta, GA, USA, 9–11 Apr 1980*; Vol 3; CONF-800428; Washington, DC, USA; US DOE; pp. 997–1003

Nack, H, Weller, AE and Liu, KT (1977) Battelle's Multisolid Fluidized-Bed Combustion (MS-FBC) process. In: *Proceedings of the Fluidized Bed Combustion Technology Exchange Workshop, Reston, VA, USA, 13 Apr 1977*; CONF-7700447–P1; pp. 221–235

National Coal Board (1980) *Fluidised bed combustion of coal*. London, UK, National Coal Board; 44 pp.

Newby, RA, Bachovchin, DW, Peterson, CH, Rohatgi, ND, Ulerich, NH and Keairns, DL (1980) An assessment of advanced sulfur removal system for electric utility AFBC In: *Proceedings of the Sixth International Conference on Fluidized Bed Combustion, Atlanta, GA, USA, 9–11 Apr 1980*; CONF-800428-Vol. 3; Washington, DC, USA; US DOE; pp. 1044–1059

Newby, RA and Keairns, DL (1978) Initial assessment of alternative SO_2 sorbents for fluidized-bed combustion power plants. In: *Proceedings of the Fifth International Conference on Fluidized Bed Combustion, Volume II Near-term implementation, Washington, DC, USA, 12 Dec 1977*; Vol. 2; M–78-68, CONF-771272-P2; pp. 680–699

Newby, RA, Ulerich, NH, O'Neill, EP, Ciliberti, DF and Keairns, DL (1978) *Effect of SO_2 emission requirements on fluidized-bed combustion systems: preliminary technical/economic assessment*; PB-286971; Pittsburgh, PA, USA; Westinghouse Research and Development Center; 192 pp.

Nissen, WI and Madenburg, RS (1979) *Citrate process demonstration plant design*; BM-IC-8806; Washington, DC, USA; US Dept. of the Interior; 18 pp.

Nooy, FM and Pohlenz, JB (1979) The Shell flue gas treating process for the reduction of nitrogen and sulfur oxides. In: *Proceedings of the Second NO_x* Control Technology Seminar, Denver, CO, USA, 8–9 Nov 1978; FP-1109-SR; Palo Alto, CA, USA; Electric Power Research Institute; pp. 23/1-23/13

Norwest Resource Consultants Ltd (1981) *A review of dry cleaning processes*; Calgary, Alberta, Canada; Norwest Resource Consultants Ltd; 57 pp.

Oberteuffer, JA (1979) High gradient magnetic separation: basic principles, devices and applications. In: *Proceedings of International Conference on Industrial Applications of Magnetic Separation, Rindge, NH, USA, 30 Jul–4 Aug 1978*; New York, NY, USA; Institute of Electrical and Electronic Engineers; pp. 3–7

O'Brien, WE. and Anders, WL (1979) *Potential production and marketing of FGD byproduct sulfur and sulfuric acid in the US (1983 projection). Final report, June 1976–December 1978*; PB–299205; Research Triangle Park, NC, USA; Environmental Protection Agency; 59 pp.

Ohtsuka, T. and Ishihara, Y (1977) Emission control technology for sulphur oxides and nitrogen oxides from flue gases in Japan. *Journal of the Institute of Fuel*; **82**; 82–90

Oldenkamp, RD, Gehri, DC and Katz, B (1979) *Aqueous carbonate process for flue gas desulfurization: a status report*; American Chemical Society, Division of Petroleum Chemistry, *Preprints*; **24** (2); 613–619

Patterson, EC, Le, HV, Ho, TK and Wheelock, TD (1979) Better separation by froth flotation and oil agglomeration. *Coal Processing Technology*; **51** 171–177

Paul, G (1978) Corrosion resistant materials for SO_2 scrubbers. *Power Engineering*; **82** (5); 54–57

Pedco-Environmental, Inc. (1979) *Lime FGD systems data book. Final report*; EPRI-FP-1030; Palo Alto, CA, USA; Electric Power Research Institute; 714 pp.

Pedroso, RI and Press, KM (1979) Sulphur recovered from flue gas at large coal fired power plants. In: *The Control of Sulphur and Other Gaseous Emissions, Salford, UK, Apr 1979*; IChemE Symposium Series No. 57; pp F1-F20

Ponder, WH, Stern, RD and McGlamery, GG (1978) SO_2 control technologies: commercial availabilities and economics. *Energy Communications*; **4** (2); 175–212

Potts, JM and Jordan, JE (1978) *Advanced concepts: SO_2 removal process improvements. Final report*; PB–294471; Research Triangle Park, NC, USA; US EPA; 168 pp.

Potts, JM and Jordan, JE (1978) *Advanced concepts: SO_2 removal process improvements. Final report*; PB–294471; Research Triangle Park, NC, USA; Environmental Protection Agency; 168 pp.

Princiotta, FT (1978) Advances in SO_2 stack gas scrubbing. *Chemical Engineering Progress*; **74** (2); 58–64

Princiotta, F (1979) *Flue gas desulfurization pilot study. Phase I, survey of major installations. Summary of survey reports on flue gas desulfurization processes. Report No. 95 and appendices*; PB-295001 onward; Washington, DC, USA; Environmental Protection Agency

Prior, M (1977) *The control of sulphur oxides emitted in coal combustion*; EAS-B1/77, London, UK; IEA Coal Research, 108 pp.

Pruce, LM (1978) Coal cleaning at Homer City: an alternative to scrubbers. *Power (NY)*; **122** (11); 213–216

Pruce, LM (1981a) Evaluating the newest disposal options for scrubber sludge. *Power (NY)*; **125** (5); 63–65

Pruce, LM (1981b) Why so few regenerative scrubbers? *Power (NY);* **125** (6); 73–76

Quackenbush, VC, Polek, JR and Agarwal, D (1978) Ammonia scrubbing pilot activity at Calvert City. In: *Proceedings of Symposium on Flue Gas Desulfurization, Hollywood, FL, USA, Nov 1977*; PB-282091; Research Triangle Park, NC, USA; Environmental Protection Agency; pp. 794–818

Ratcliffe, CT and Pap, G (1979) Chemical reduction of SO_2 with lignite or coal: kinetics and proposed mechanism. *American Chemical Society, Division of Petroleum Chemistry, Preprints*; **24** (2); 577–586

Raymond, WJ, Hackensack, NJ and Sliger, AG (1978) The Kellog-Weir scrubbing system. *Chemical Engineering Progress*; **74** (2); 75–80

Reeves, AM (1979) New scrubber process is sludge-free. *Coal Mining and Processing*; **16** (6); 50–54

Reisinger, AA and Gehri, DC (1979) Two stage dry FGD and particulate removal. *Journal of the Air Pollution Control Association*; **29** (4); 419–421

Rittenhouse, RC (1981) Air pollution control for power plants. *Power Engineering*; **85** (9); 62–70

Robards, RF, Moore, ND, Kelso, TM and Cole, RM (1979) *Concurrent scrubber evaluation TVA's Colbert lime–limestone wet-scrubbing pilot plant. Final report*; EPRI-FP–941; Palo Alto, CA, USA; Electric Power Research Institute; 307 pp.

Rochelle, GT and King, CJ (1977) The effect of additives on mass transfer in $CaCO_3$ or CaO slurry scrubbing of SO_2 from waste gases. *Industrial Engineering Chemistry, Fundamentals*; **16** (1); 67–75

Rochelle, GT and King, CJ (1978) Alternatives for stack gas desulfurization by throwaway scrubbing. *Chemical Engineering Progress*; **74** (2); 65–70

Rosenberg, HS (1978) How good is flue gas desulphurization? *Hydrocarbon Processing*; **57** (5); 132–135

Rosenberg, HS and Grotta, HM (1979) Scale suppression in lime/limestone FGD systems. *American Chemical Society, Division of Petroleum Chemistry, Preprints*; **24** (2); 561–568

Ross, FF (1978) A British approach to sulfur oxide emissions. *Mechanical Engineering*; **100** (8); 42–45

Ross, FF (1979) The sizing of tall-enough stacks. In: *The Control of Sulphur and Other Gaseous Emissions, Salford, UK, Apr 1979*; IChemE Symposium Series No. 57; pp. B1–B11

Rossoff, J, Leo, PP and Fling, RB (1978) Landfill and ponding concepts for FGD sludge disposal. *Proceedings of the Air Pollution Control Association*; **5**; 1–15

Rowland, CH and Abdulsattar, AH (1978) Equilibrium for magnesium wet scrubbing of gases containing sulfur dioxide. *Environmental Science and Technology*; **12** (10); 1158–1162

Rubin, ES and Nguyen, DG (1978) Energy requirements of a limestone FGD system. *Journal of the Air Pollution Control Association*; **28** (12); 1207–1212

Rubin, ES (1981) Air pollution constraints on increased coal use by industry: an international perspective. *Journal of the Air Pollution Control Association*; **31** (4); 349–360

Rush, RE and Edwards, RA (1978a) *Evaluation of three 20 MW prototype flue gas desulfurization processes. Final report*; EPRI-FP–12-SY (Vol. 1); Palo Alto, CA, USA; Electric Power Research Institute; 77 pp.

Rush, RE and Edwards, RA (1978b) *Evaluation of three 20 MW prototype flue gas desulfurization processes. Final report*; Vol. 2; EPRI-FP-713; Palo Alto, CA, USA; Electric Power Research Institute; 455 pp.

Rush, RE and Edwards, RA (1978c) *Evaluation of three 20 MW prototype flue gas desulfurization processes. Final report*; Vol. 3; EPRI-FP–713; Palo Alto, CA, USA; Electric Power Research Institute; 177 pp.

Ruth, LA and Keairns, DL (1976) Combustion and desulfurization of coal in a fluidized bed of limestone. In: *Fluidization Technology, International Fluidization Conference, Pacific Grove, CA, USA, 15 Jun 1975, Washington, DC, USA*; Vol. 2; Hemisphere Publishing Corp.; pp. 321–327

Salaun, A and Trempu, R (1978) Le procede ST (CIT-ALCATEL) de desulfuration des fumees. (The ST (CIT-ALCATEL) process for desulphurization of waste gases). *Revue Generale de Thermique*; **17** (195); 207–210

Sayre, WG (1980) Selenium: a water pollutant from flue gas desulfurization. *Journal of Air Pollution Control Association*; **30** (10); 1134

Schaeffer, SC (1978) Coal preparation versus stack-gas scrubbing to meet SO_2 emission regulations. In: *Proceedings of the Gulf Coast Lignite Symposium: geology, utilization and environment aspects, Austin, TX, USA, 2 Jun 1976*; CONF-7606131; pp. 266–272

Schulte, H (1977) Sulfur gained from flue gas, a demonstration unit of the Wellman-Lord process annexed to a black coal power plant. *VDI (Verein Deutscher Ingenieure) Nachrichten*; **31** (50); 2–5

Science Applications, Inc. (1980) *Report on subtask 1: data gathering, analysis and reporting of scrubber instrumentation survey*; SAI-442-810-01; McLean, VA, USA; Science Applications, Inc.; 64 pp.

Selmeczi, JG and Stewart, DA (1978) The thiosorbic flue gas desulfurization process. *Chemical Engineering Progress*; **74** (2); 41–45

Sessler, G (1980) *Sammis generating station: meeting SO_2 and particulate standards with cleaned Ohio coals. Final report Mar–Jul 1979*; PB-80-147077; Berkeley, CA, USA; Teknekron Research, Inc.; 103 pp.

Shearer, JA *et al.* (1979) Effects of sodium chloride on limestone calcination and sulfation in fluidized-bed combustion. *Environmental Science and Technology*; **13** (9); 1113–1118

Shearer, JA, Johnson, I and Turner, CB (1980a) Interaction of NaCl with limestones during calcination. *American Ceramic Society Bulletin*; **59** (5); 521–528

Shearer, J, Johnson, I and Turner, C (1980b) The effects of $CaCl_2$ on limestone sulfation in fluidized-bed coal combustion. *Thermochimica Acta*; **35**; 105–109

Shearer, JA, Smith, GW Moulton, DS, Smyk, EB, Myles, KM, Swift, WM and Johnson, I (1980c) Hydration process for reactivating spent limestone and dolomite sorbents for reuse in fluidized-bed coal combustion. In: *Proceedings of the Sixth International Conference on Fluidized Bed Combustion, Atlanta, GA, USA, 9–11 Apr 1980*; CONF-800428-Vol. 3; Washington, DC, USA; US DOE; pp. 1015–1027

Shearer, JA, Smith, GW, Myles, KM and Johnson, I (1980d) Hydration enhanced sulfation of limestone and dolomite in the fluidized bed combustion of coal. *Journal of the Air Pollution Control Association*; **30** (6); 684–688

Siddiqi, AA and Tenini, JW (1977) FGD: A viable alternative. *Hydrocarbon Processing*; **56** (10); 104–110

Slack, AV (1978) Lime-limestone scrubbing: design considerations. *Chemical Engineering Progress*; **74** (2); 71–74

Slater, SM and Rizzone, MS (1980) Simultaneous oxidation of SO_2 and NO in flue gas by ozone injection. *Fuel*; **59** (12); 897–899

Slaughter, WW and Karlson, FV (1979) Process design considerations for the cleaning of wet pulverized coal by high gradient magnetic separation. In: *Proceedings of International Conference on Industrial Applications of Magnetic Separation, Rindge, NH, USA, 30 Jul–4 Aug 1978*; New York, NY, USA; Institute of Electrical and Electronic Engineers; pp. 83–90

Smith, EO, Morgan, WE, Noland, JW, Quinlan, RT, Stresewski, JI and Swenson, DD (1980) *Lime FGD system and sludge disposal case study. Final Report*; EPRI-CS-1631; Palo Alto CA, USA; Electric Power Research Institute; 169 pp.

Smith, GW, Hajicek, DR, Myles, KM, Goblirsch, GM, Mowry, RW and Teats, FG (1981) *Demonstration of a hydration process for reactivating partially sulfated limestone sorbents*; ANL/CEN/FE-80-23; Argonne, IL, USA; Argonne National Laboratory; 29 pp.

Smock, R (1979) Are double-alkali scrubbers the answer to the sulfur problem? Electric Light and Power (Boston); **57** (3); 21, 23

Smyk, EB, Swift, WM, Podolski, WF, Mules, KM and Johnson, I (1980) Methods of improving limestone utilization in fluidized-bed combustion. In: *Proceedings of the 15th Intersociety Energy Conversion Engineering Conference, Seattle, WA, USA, 18–22 Aug 1980*; New York, NY, USA; American Institute of Aeronautics and Astronautics; pp. 62–66

Snyder, RB, Wilson, WI and Johnson, I (1978) Prediction of limestone requirements for SO_2 emission control in atmospheric pressure fluidized-bed combustion. In: *Proceedings of the Fifth International Conference on Fluidized Bed Combustion Volume II. Near-term implementation, Washington, DC, USA, 12 Dec 1977*; Vol. 2; M–78-68; CONF-771272–P2; pp. 748–761

Snyder, RB, Wilson, WI, Johnson, I and Jonke, AA (1977) Synthetic SO_2 sorbents for fluidized-bed coal combustors. *Journal of the Air Pollution Control Association*; **27** (10); 975–981

Spring, RA (1981) La Cygne station unit No. 1 wet scrubber operating experience. In: *Proceedings of Symposium on Flue Gas Desulfurization, Houston, TX, USA, 28–31 Oct 1980*; EPA-600/9-81-091a; Washington, DC, USA; Environmental Protection Agency; pp. 391–414

Steiner, P, Dalton, SM, Knoblauch, K (1980) Capture and conversion of SO_2 RESOX prototype demonstration in Germany. *Combustion*; **51** (7); 28–31

Steiner, P and Gutterman, C (1980) *Laboratory testing of Resox ETEM process. Final report*; Vol. 1; EPRI-CS-1602; Palo Alto, CA, USA; Electric Power Research Institute; 190 pp.

Stern, JL (1981) Dry scrubbing for flue gas desulfurization. *Chemical Engineering Progress*; **77** (4); 37–42

Stevens, NJ (1981) Dry SO_2 scrubbing pilot test results. In: *Proceedings of Symposium on Flue Gas Desulfurization, Houston, TX, USA, 28–31 Oct 1980*; EPA-600/9-81-091a; Washington, DC, USA; Environmental Protection Agency; pp. 777–800

Stone, RC (1978) Relative environmental impact of two 570 MW atmospheric fluidized-bed electric power generating plants compared to a pulverized coal fired plant equipped with a wet limestone flue gas desulfurization system. In: *Proceedings of the Fifth International Conference on Fluidized Bed Combustion Vol II Near-term implementation, Washington, DC, USA, 12 Dec 1977*; M–78-68; CONF-771272–P2; pp. 892–986

Strakey, JP, Joubert, JI, Ruether, JR and Peters, WC (1980) Flue gas clean-up research at the Pittsburgh Energy Techology Center. In: *Proceedings of the Second US DOE Environmental Control Symposium, Reston, VA, USA, 17 Mar 1980*; Vol. 1; CONF-800334;, pp. 175–186

Sugarek, RL and Sipes, TG (1978a) *Controlling SO_2 emissions from coal-fired steam-electric generators: water pollution impact. Vol. 1. Executive Summary. Final Task report April-December 1977;* PB–

279635; Austin, TX, USA; Radian Corp.; 34 pp.

Sugarek, RL and Sipes, TG (1978b) *Controlling SO_2 emissions from coal-fired steam-electric generators: water pollution impact. Volume II Technical discussion. Final task report April-December 1977*; PB-279636; Austin, TX, USA; Radian Corp.; 268 pp.

Swift, WM, Montagna, JC, Smith, GW and Smyk, EB (1980) *Process costs and flowsheets, bed defluidization characteristics, stone reactivity changes and attrition losses for a regenerative fluidized-bed combustion process*; ANL-CEN/FE-78-14; Springfield, VA, USA; National Technical Information Service; 103 pp.

Takahashi, Y, Fujima, S, Takamoku, H and Fujioka, Y (1980) Reclamation/regeneration of desulfurization agent (limestone) in fluidized bed coal combustion. In: *Symposium on Research on Coal Utilization Technology, Tokyo, Japan, Aug 1980*; Tokyo, Japan; Coal Mining Research Centre; pp. 63–81 (In Japanese)

Takenouchi, S, Fijii, J and Atsumi, T (1981) Application of dry desulfurization processes to total flue gas cleaning systems. In: *Proceedings of the Third Symposium on Coal Utilization Technology Research, Tokyo, Japan, 15 Sep 1981*; Tokyo, Japan; Coal Mining Research Centre; pp. 99–113 (In Japanese)

Takeuchi, S, Fujii, J and Atsumi, T (1980) Dry-type desulfurization facilities using carbon catalysts in coal fired power plant. In: *Symposium on Research on Coal Utilization Technology, Tokyo, Japan, Aug 1980*; Tokyo, Japan; Coal Mining Research Centre; pp. 34–47 (In Japanese)

Tarkington, TW, Kennedy, FM and Patterson, JG (1979) *Evaluation of physical/chemical coal cleaning and flue gas desulfurization. Final report June 78–Oct 79*; PB-80-147622; Muscle Shoals, AL, USA; Tennessee Valley Authority; 378 pp.

Thomas, WC (1978a) *Energy requirements for controlling SO_2 emissions from coal-fired steam/electric generators. Final report*; PB–281331; Research Triangle Park, NC, USA; Environmental Protection Agency; 136 pp.

Thomas, WC (1978b) *Energy requirements for controlling SO_2 emissions from coal-fired steam/electric generators – executive summary. Final report*; PB–281332; Research Triangle park, NC, USA; Environmental Protection Agency; 18 pp.

Toprac, AJ and Rochelle, GT (1982) Limestone dissolution in stack gas desulfurization. *Environmental Progress*; **1** (1); 52–58

Troupe, JS (1978) Handwriting on the power plant wall: flue gas treatment. *Combustion*; **50** (4); 42–47

Ulerich, NH, O'Neill, EP and Keairns, DL (1977) *The influence of limestone calcinations on the utilization of the sulfur-sorbent in atmospheric pressure fluid-bed combustors*; EPRI-FP-426; Palo Alto, CA, USA; Electric Power Research Institute; 158 pp.

Ulerich, NH, O'Neill, EP, Alvin, MA and Keairns, DL (1979) *Criteria for the selection of SO_2 sorbents for atmospheric pressure fluidized-bed combustors. Volume 2. Tasks II and III report: experimental work, sorbent selection technique. Final report, November 1, 1976–July 31, 1977*; EPRI-FP–1307 (Vol. 2); Palo Alto, CA, USA; Electric Power Research Institute; 125 pp.

Ulerich, NH, Vaux, WG, Newby, RA and Keairns, DL (1980) *Experimental/engineering support for environmental protection agencies fluidized bed combustion (FBC) program: final report. Volume 1. Sulfur oxide control*; PB80-188402; Research Triangle Park, NC, USA; Environmental Protection Agency; 230 pp.

Ulerich, NH, Newby, RA and Keairns, DL (1981) *Calcium-based sorbent desulfurization in pressurized fluidized-bed combustion power plants*; EPRI-CS-1847; Pittsburgh, PA, USA; Westinghouse Electric 140 pp.

Van der Linde, BJ, Visser, L. and de Jong, WA (1979) An experimental study of MnO_x/α-alumina accepts for flue gas desulphurization. In: *The Control of Sulphur and Other Gaseous Emissions, Salford, UK, Apr 1979*; IChemE Symposium Series No. 57; E1–E13

Van Houste, G, Rodrique, L, Genet, M and Delmon, B (1981) Kinetics of the reaction of calcium sulfite and calcium carbonate with sulfur dioxide and oxygen in the presence of calcium chloride. *Environmental Science and Technology*; **15** (3); 327–332

Van Ness, RP, Somers, RC, Weeks, RC, Frank, T, Ramans, GJ, Lamantia, CR, Lunt, RR and Valencia, JA (1979) *Full-scale dual-alkali demonstration system at Louisville Gas and Electric Co. Final design and system cost*; EPA-600/7-79-221b; Washington, DC, USA; Environmental Protection Agency; 104 pp.

Van Ness, RP, Kingston, WH and Borsare, DC (1980) Operation of C-E flue gas desulfurization system for high sulfur coal at Louisville Gas and Electric Co., Cane Run 5. *Combustion*; **51** (8); 10–16

Van Ness, RP, Kaplan, N and Watson, DA (1981) Dual alkali demonstration project-interim report. In: *Proceedings of Symposium on Flue Gas Desulfurization, Houston, TX, USA, 28–31 Oct 1980*; EPA-600/9/081-091a; Washington, DC, USA; Environmental Protection Agency; pp. 453–472

Verhoff, FH and Choi, M.-K (1979) The effect of sulfuric acid condensation on flue gas processing equipment. *American Chemical Society, Division of Petroleum Chemistry, Preprints*; **24** (2); 601–612

Wallace, WE Jr. (1980) *Workshop on sulfur chemistry in flue gas desulfurization*; DOE/METC/8333-11, CONF-7906165; Morgantown, WV, USA; Morgantown Energy Technology Center; 253 pp.

Weeter, DW (1981) Utilization of dry calcium based flue gas desulfurization waste as a hazardous waste fixation agent. *Journal of the Air Pollution Control Association*; **31** (7); 751–760

Wells, WL *et al.* (1978) TVA's experiences with limestone scrubbers at the 550 MW Widows Creek Unit 8. *Combustion*; **50** (4); 19–26

Wen, CY and Chang, CS (1978) Absorption of SO_2 in lime and limestone slurry: pressure drop effect on turbulent contact absorber (TCA) performance. *Environmental Science and Technology*; **12** (6); 703–707

Wheelock, TD (1982) Development and demonstration of selected fine coal beneficiation methods. In: *Physical cleaning of coal.* (ed. YA Liu) New York, NY, USA; Marcel Dekker, Inc.; 568 pp.

Wheelock, TD and Ho, TK (1979) Modification of the floatability of coal pyrites. Presented at *American Institute of Mining, Metallurgical and Petroleum Engineers Annual Meeting, New Orleans, LA, USA, 18 Feb 1979;* 1S-M–169, CONF-790219–7; 11 pp.

Wilhelm, JH, Kobler, RW, Naide, Y and Redfield, G (1979) *Sludge dewatering for FGD products. Final report*; EPRI-FP-937; Palo Alto, CA, USA; Electric Power Research Institute; 280 pp.

Willmes, O and Schiffers, A (1979) Einfluss des Wasserdampfgehaltes in Rauchgasen auf die Prozessfuhrung von Nassentschwefelungs-verfahren bei Braun- und Steinkohlenfuerungen (The influence of the water vapour content of flue gases on the wet desulfurization process for brown coal and hard coal combustion). *Braunkohle*; **31** (11); 356–359 (In German)

Wilson, CL (1980) *Coal — Bridge to the Future*, Cambridge, MA, USA; Ballinger; 272 pp.

Wilson, WI, Snyder, RB and Johnson, I (1980) The use of oil shale for SO_2 emission control in atmospheric pressure fluidized-bed coal combustion. *Industrial and Engineering Chemistry, Process Design and Development*; **19** (1); 47–51

Wilzbach, KE, Livengood, CD and Farber, PS (1980) Environmental control implications of coal use. In: *Proceedings of the Second US DOE Environmental Control Symposium, Reston, VA, USA, 17 Mar 1980*; CONF-800334-18; 20 pp.

Wood, R (1978) FGD demonstration points to regenerate future. *Process Engineering*; 146–147, 149

Woodhead, PMJ, Parker, JH and Duedall, IW (1981) Environmental compatibility and engineering feasibility for utilization of FGD waste in artificial fishing reef construction. In: *Proceedings of Symposium on Flue Gas Desulfurization, Houston, TX, USA, 28–31 Oct 1980*; EPA-600/9-81-091a; Washington, DC, USA; Environmental Protection Agency; pp. 695–700

Workman, KH and Rothfuss, EH (1978) FGD waste disposal effective despite surprises. *Power Engineering*; **82** (11); 60–63

Yarze, JC and Beiersdorf, W (1979) Sulfur dioxide removal process using magnesium promoted limestone slurry. In: *The Control of Sulphur and Other Gaseous Emissions, Salford, UK, Apr 1979*; IChemE Symposium Series No. 57; G1–G25

Yeager, K (1981) Coal combustion systems division. R&D status report. *Electric Power Research Institute Journal*; **6** (4); 36–39

Zabel, SA (1979) Comparative controls for sulfur oxides in four countries. In: *Proceedings of the International Symposium on Sulphur emissions and the Environment, 8–10 May 1979, London, UK*; London, UK; Society of Chemical Industry; pp. 314–324

Ziegler, EN. and Meyers, RE (1979) Control technology for coal-fired combustion in northeastern US. Part A. Overview and sulfur emissions control. *Water, Air and Soil Pollution*; **12** (3); 355–369

Part 3

Sulphates in the atmosphere

Summary

Recent concern about acidic secondary pollutants formed by atmospheric reactions of sulphur dioxide from coal combustion waste gases stems from the discovery of adverse effects on environmental systems in areas remote from the main sulphur dioxide emission sources. This review covers the atmospheric processes involving sulphates, from emission of sulphur dioxide to deposition of sulphates, and the direct effects of atmospheric sulphates. Possible ecological pathways and mechanisms of acidification of the environment by pullutants from coal combustion will be treated in a forthcoming review. The global sulphur cycle is examined to determine the magnitude of natural and man-made sulphur emissions and to outline the pathways of sulphur species through the environment. The atmospheric reactions of sulphur from the reduced to the fully oxidised states are treated with emphasis on the possible mechanisms of sulphur dioxide oxidation to the sulphate ion and sulphuric acid, including aerosol particle formation. The emission of primary sulphates and the oxidation of sulphur dioxide in plumes from coal-fired power plants are reviewed and compared to sulphate production taking place in plumes from other sulphur dioxide sources, including area sources. Concentrations of sulphate aerosol in the atmosphere and aerosol transport processes in the medium- and long-range are outlined. Dry, occult and wet deposition are investigated, including historical trends in precipitation chemistry. The results of mathematical modelling of transport and deposition processes are briefly examined. Possible atmospheric effects of sulphates are outlined, covering effects on visibility, climate, materials and direct health effects on humans, with special reference to respiratory effects. Neither traditional scientific methods nor mathematical models have been able to pinpoint relationships between source emissions and deposition in areas where emissions sources are dense and widespread. However they have established the long-term source and receptor areas of man-made pollutants over regions the size of Europe or eastern North America, although they cannot account for deposition on a day-to-day basis. What is clear is that transport of sulphur pollution is taking place over political boundaries in both Europe and eastern North America. Although mathematical models have not been shown to predict accurately the effects of alternative control strategies on sulphur deposition patterns it is quite clear that reducing sulphur dioxide emissions would reduce the total deposition of sulphur. The only direct effect which can be attributed to atmospheric sulphates at present day ambient concentrations is reduction of visibility in some areas. The association of secondary sulphates with visibility reduction could possibly widen the geographical spread of combustion plants affected by the pressure for reduction of sulphur dioxide emissions.

Acronyms and abbreviations

BaP	benzo(a)pyrene
CANSAP	Canadian Network for Sampling Precipitation
CCN	cloud condensation nuclei
d	day(s)
EACN	European Air Chemistry Network
EMEP	European Monitoring and Evaluation Programme (Co-operative Programme for Monitoring and Evaluation of Long-range Transport of Air Pollutants in Europe)
EPA	Environmental Protection Agency (USA)
EOFA	empirical orthogonal function analysis
ESP	electrostatic precipitator
FGD	flue gas desulphurisation
GAMETAG	Global Atmospheric Measurements on Tropospheric Aerosols and Gases Programme
h	hour(s)
IERE	International Electric Research Exchange
ILWAS	Integrated Lake Watershed Acidification Study
IN	ice nuclei
LC50	lethal concentration causing 50% mortality
LRTAP	Co-operative Technical Programme to Measure the Long-Range Transport of Air Pollutants
MAP3S	Multistate Atmospheric Power Production Pollution Study
MISTT	Midwest Interstate Sulfur Transformation and Transport
mmd	mass median diameter
NOAA	National Oceanic and Atmospheric Administration (USA)
NADP	National Atmospheric Deposition Programme (USA)
NMHC	non-methane hydrocarbons
NRC	National Research Council (USA)
OECD	Organisation for Economic Co-operation and Development
PCA	principal component analysis
PAN	peroxyacetyl nitrate
r.h.	relative humidity
SMSA	standard metropolitan statistical area
SURE	Sulfate Regional Experiment
TSP	total suspended particles
UNECE	United Nations Economic Commission for Europe
WMO	World Meteorological Organization
y	year(s)

Chapter 18

Introduction

Sulphur is an abundant element which is essential to life. It is transported through the environment while undergoing transformation from one oxidation state to another. Sulphates, in the fully oxidised state, are the most stable form. On a global scale, the main natural emission sources of sulphur to the troposphere are biogenic emissions (50 to 100 Tg/y) and sea spray (175 Tg/y ± 50%), although only about 10% of marine emissions affect the land. Volcanic sources contribute 2 to 28 Tg/y of sulphur. Human activities, mainly the combustion of fossil fuels, produce 90–113 Tg/y. Because of the relatively short residence time of sulphur dioxide and sulphate aerosols in the atmosphere, of the order of a few days, they are deposited on a regional scale. Since the bulk of the man-made sulphur emissions are produced over a small area of the planet's surface, human influence on the sulphur cycle varies widely from one region to another. On a global scale human activities produce about 60% of total sulphur emissions but over north-western Europe and eastern North America they contribute about 90% and 99% of total sulphur emissions respectively.

Gaseous sulphur dioxide emissions have increased by almost a factor of thirty since 1860. On a global scale, the present level of combustion of fossil fuels accounts for 82% of sulphur dioxide emissions, with 56% coming from coal. In industrialised areas, about 60% of sulphur dioxide is emitted from coal combustion. In the developed countries the majority of coal is burned for electricity generation. In eastern USA for example 57% of the total sulphur dioxide emissions are released by electricity generation, with 60 to 70% of that injected in the 100–700 m height range. About 1–3% of total sulphur is released as directly emitted primary sulphates by coal-fired boilers as opposed to 5–9% from oil-fired boilers burning fuels of a similar sulphur content. The use of particle and sulphur dioxide control systems often do not remove more than a small fraction of the primary sulphate in the flue gas. This is not of great importance since model estimates suggest that about 85% of surface sulphate concentrations originate from atmospheric sulphur dioxide oxidation.

Gaseous reduced sulphur compounds are either oxidised to sulphur dioxide in the troposphere, by a mechanism involving the hydroxyl radical, or pass into the stratosphere where they undergo photolytic reactions to form sulphuric acid particles. The probable average residence time of sulphur dioxide in the lower troposphere is about 1.5 days. Removal takes place by dry deposition or by a variety of chemical reactions all of which involve the oxidation of sulphur dioxide to sulphuric acid and the incorporation of the acid into cloud droplets and aerosols.

Although there are many uncertainties in the transformation process, the dominant mechanism in gas phase oxidation is undoubtedly reaction with hydroxyl radicals which are generated in the atmosphere by photochemical reactions. In the aqueous phase, the reaction with hydrogen peroxide is the single most important reaction over much of the pH range. However ozone oxidation rates may be competitive with hydrogen peroxide rates above about pH 4.5 and dominate at pH levels above 5.5. Other mechanisms also contribute above a pH of about 5.2 and, in the atmosphere, if one path is not available the reaction will go by another route. Gas phase hydroperoxyl and hydroxyl radicals and ozone are formed by photochemical reactions and may then dissolve in cloudwater, regenerating the aqueous phase reactants. Aqueous phase reactions are not entirely independent of gas phase photochemistry. Dependence on solar radiation to produce oxidant radicals leads to variations in sulphate formation rate with season and latitude.

Although these are the dominant reactions in the background atmosphere other mechanisms can contribute in more polluted areas. Catalytic oxidation may occur in urban fog conditions where the concentration of metal ions is likely to be high. The heavy metals manganese and iron are thought to promote sulphur dioxide oxidation. The presence of ammonia can affect aqueous phase reactions by increasing the pH of the solution. Nitrogen oxides could be significant oxidants in power plant flue gases and plumes, where the concentrations reach high levels. Reactions with carbonaceous and siliceous particles may make significant contributions to sulphate production but only under special circumstances which have not been well defined. It has not yet been conclusively determined whether the level of oxidants in the atmosphere is a limiting factor in the oxidation of sulphur dioxide.

Sulphates are active as cloud condensation nuclei in the atmosphere and so can influence cloud formation and precipitation processes. After gas phase oxidation combustion-generated sulphates form transient nuclei which quickly grow into accumulation mode particles (diameter $<2\,\mu m$). These are removed only slowly from the atmosphere and so may be transported long distances. Sulphates dominate the accumulation mode and the most common compound found is ammonium sulphate.

In the atmosphere the fate of a plume is dependent on meteorological and chemical conditions. The depth of the mixing layer is governed by input from solar radiation causing atmospheric turbulence and creating a diurnal cycle varying from stable to unstable, well-mixed conditions. Plumes rapidly expand to fill the vertical dimension of the mixing layer as they experience dilution and dry deposition of sulphur dioxide starts as soon as they intersect the ground. Conversion to sulphate takes place during plume transport and all species are subject to removal during transport either directly or by precipitation. Tall stacks can inject a plume above the mixing layer, during both night and daytime hours, where it experiences more rapid transport, less dilution and very much less deposition of sulphur species than is the case in the mixing layer.

The average sulphate formation rate seen in point source plumes is usually less than 4%/h with the extent of sulphur dioxide oxidation typically about 25 ± 10% in conditions equivalent to those in the eastern USA. Variations are due to external conditions rather than type of fuel burned. No single factor is dominant under all atmospheric conditions but some of the most important influences are insolation, relative humidity, oxidant and nitrogen oxides levels. Little photochemical oxidation occurs from late evening to early morning but the rate rises considerably around mid-day. There is a large seasonal variation, with the average conversion rate higher in the summer than in winter. Both gas and liquid phase mechanisms probably contribute to observed rates. Most aircraft studies take place in the dry, stable weather conditions which are most favourable to a gas phase mechanism, but possible liquid phase reactions are seen where a plume intersects clouds, fog or a cooling tower plume and here rates are up to an order of magnitude higher. The theoretical basis of predicting oxidation rates consistent with measurements for the more complex liquid phase mechanisms is not as far advanced as for gas phase mechanisms, which can also lead modellers to pay more attention to the latter. The generally higher rates seen in urban plumes are probably due to the higher levels of pollutants present to provide the necessary free radicals or catalysts.

Secondary sulphate particles occur predominantly in the fine particle range ($<2\,\mu m$). They are the dominant species in this range although the percentage present varies with location. The background concentration is still uncertain but there is evidence that it has increased in recent years. In the areas most affected by man-made sulphur pollutants there is a wide variability in atmospheric sulphate concentration of a highly episodic nature, involving sequential periods of two to six days of high concentration amidst periods of lower concentration. The highest sulphate concentrations are seen generally in the warmer months of the year. Statistical studies have shown that the relative importance of meteorological and pollutant loading factors in determining sulphate concentration varies with the locality.

Transport of sulphate aerosols over distances between ten and a few hundred kilometres can be seen in plumes and mesoscale weather systems such as sea–land breezes and urban heat island systems. Under certain conditions these weather systems, especially those involving wind reversal patterns, can concentrate sulphates and amplify the effects of local pollution sources. Long-range trans-boundary transport of sulphates and episodes of high sulphate concentration are associated with synoptic scale weather systems. Regional air mass stagnation and migrating high pressure systems are both related to elevated sulphate episodes. The sulphates form hazy air masses which can be tracked by satellite. The formation of sulphates takes place during long-range transport. The majority of regional atmospheric sulphate aerosol accumulation episodes occur in the summer months.

Man-made sulphate aerosols are transported to the Arctic during the winter months where they cause an increase in sulphate concentrations of more than an order of magnitude over summer levels. Aerosols are more persistent in the Arctic than at lower latitudes since the normal deposition processes do not operate as effectively. Sulphur is also the most common element in the Antarctic aerosol, but this is of natural origin.

Combustion-generated sulphate aerosols are predominantly found in the size range 0.1 to 1 μm where atmospheric dry deposition is not therefore considered a significant removal process in most areas. However, recent field measurements suggest that dry deposition may be more important than theory suggests. Wet deposition is the most important removal process for sulphates, with nucleation contributing an estimated 65% to wet-deposited sulphur and the solution and oxidation of sulphur dioxide contributing 20% in regions remote from sources. Once the sulphate ion is attached to cloud water elements it is subject to the storm behaviour of the associated water. In the past, deposition of sulphates in fog, dew and frost was overlooked in measurements of wet deposition. This can lead to a significant underestimation of total sulphate deposition, especially since the ion concentrations in these forms of deposition are relatively high. A large proportion of annual sulphate deposition can come from fog and cloud droplet capture even where wet deposition rates are high. Precipitation scavenging occurs in plumes from coal-fired power plants, removing mainly primary sulphates and dissolved sulphur dioxide, relatively close to the plants.

On a regional basis the relative importance of wet and dry sulphur deposition processes to total sulphate and hydrogen ion concentrations is dependent on factors such as distance from sulphur emission sources and meteorological conditions. Wet deposition of sulphate aerosol is of most relative significance in areas of high rainfall far from emission sources.

Precipitation contains a complex mixture of acidic and alkaline ions with the overall acidity of the solution governed by the ion balance. Unpolluted rainwater, in equilibrium with dissolved atmospheric carbon dioxide, would theoretically have a pH of 5.6 but natural variability in the sulphur cycle alone could cause pH values to range from 4.5 to 5.6 in areas unaffected by human activity. Only precipitation with a pH less than 5.6 is termed acidic. In areas receiving acidic precipitation the hydrogen ions are largely associated with sulphate and nitrate ions. Although the proportions vary from place to place, in remote areas of industrialised regions commonly about 70% of the acidity is due to sulphuric acid and 30% to nitric acid. In areas of Europe, up to 15% can be associated with hydrochloric acid. There is evidence of a recent increase in the contribution of nitric acid to precipitation acidity in some areas, possibly due to mobile nitrogen oxides sources. Basic species, especially calcium, magnesium and ammonium ions, have a great influence on precipitation acidity. The quality of precipitation chemistry is dominated by local and regional land use patterns and proximity to man-made emissions. In general, acidity in precipitation is highest where emissions of precursor gases derived from human activity are highest and concentrations of neutralising components are lowest.

The rate of sulphate and hydrogen ion deposition with precipitation is episodic in nature. In some remote areas up to 30% of the annual sulphate deposition can fall in one rainstorm. These sulphate episodes cannot be explained by the episodicity of precipitation alone, but by episodicity of sulphate concentrations in the ambient air associated with the precipitation.

In Europe there is evidence that there have been changes in sulphate ion deposition and rainfall acidity similar to trends in man-made emissions. Over the European area as a whole, it is not possible to conclude that there is a non-linear relationship between deposition and emissions, given the limitations of the data. In Scandinavia, where a longer data record exists, changes in precipitation sulphate were in fair agreement with changes in man-made sulphur dioxide emissions in countries known to contribute to deposition in the area.

In eastern North America, although the evidence is more fragmentary, an increase in precipitation acidity from the 1950s to the 1970s is seen. This has been ascribed to a decrease in basic cations rather than an increase in man-made anions. Trends in precipitation sulphate concentration over a 20-year period are consistent with a general proportionality between sulphur dioxide emissions and sulphate deposition. The National Research Council (1983) found 'no evidence for a strong non-linearity in the relationship between long term average emission and deposition' in this area. A linear relationship between sulphur dioxide emissions and sulphate concentration in wet deposition has been seen in western USA.

Sulphate concentration has increased by up to a factor of 3.7 in Greenland ice over the last 200 years, probably due to emissions from fossil fuel combustion, whereas no change is seen in Antarctic ice where sulphate concentrations are of natural origin.

The use of mathematical models has established the long-term source and receptor areas of man-made pollutants on a large scale but it is still not possible to account for deposition on a day-to-day basis. Model results suggest that considerable transport of sulphur pollution is taking place over national boundaries in both Europe and North America. In most European countries sulphur deposition due to foreign emissions is an important contributor to total deposition and in 13 countries it greatly outweighs indigenous sources. In eastern North America, US emissions provide about half of the total sulphur deposited in Canada while less than five per cent of total sulphur deposition in the US can be attributed to Canadian emissions. In the long term, up to about 20% of emitted sulphur is deposited within 50 to 100 km of its source, leaving most of the emitted sulphur available for long-range transport. However, some investigators find a significant influence of local sources on precipitation chemistry (up to 50%) while others can detect no contribution.

Understanding of the complex processes governing the transport, transformation and deposition of pollutants is incomplete and there is a lack of field data for both input and validation purposes. Current models have not been demonstrated to predict accurately the locations where high deposition levels would occur under given circumstances. They have not been shown to predict accurately the effects of alternative control strategies on patterns of sulphur deposition, although it is clear that reducing present sulphur dioxide emissions would reduce total deposition of sulphur.

Atmospheric sulphates are suspected of having direct detrimental effects on human health. Toxicological studies in animals suggest that there is a causal link between high concentrations of acidic airborne sulphates and health damage. Human exposure studies have shown comparable pulmonary reactions to submicron sulphuric acid as to fresh cigarette smoke. Sulphuric acid is more reactive than neutral sulphates so it may be acidity, or hydrogen ion concentration, which is the active component. Epidemiological studies can only give an indication of the pollution levels at which exposure effects are likely to be seen, under some circumstances, among certain human populations. Effects from acute exposure are seen in the elderly and those with cardio-respiratory diseases. Increased risk from chronic exposure has been convincingly demonstrated only in children. It must be stressed that no respiratory effects have been demonstrated at the current ambient sulphate levels. Modelling studies give a fairly high upper boundary to the level of mortality connected with exposure to sulphates. However, there is no agreement on the likely level of mortality that would result below this upper limit. Whilst eye and skin irritation from atmospheric sulphates has been seen it is thought to occur only in specific meteorological conditions and in the presence of organic air contaminants. A study of occupational exposure showed a positive association between upper respiratory tract cancers and exposures to high concentrations of sulphuric acid. At the much lower concentrations found in the atmosphere the combination of acidic air pollution and smoking, or exposure to other carcinogens, may enhance the risk of lung cancer. Statistical associations of airborne sulphates with lung cancer have been proposed, but there is no experimental evidence for a connection at present-day atmospheric sulphate concentrations.

On the regional scale, sulphates are the largest single contributors to observed light extinction in polluted atmospheres but in urban areas the contribution of sulphates is more variable. In some urban areas carbonaceous aerosol is the most important species in visibility reduction because of its ability both to absorb and to scatter light while in others sulphate aerosol is the predominant species but a seasonal change in light scattering efficiency may be seen. The visibility-reducing effect of sulphate aerosols is at its greatest in the warmer months of the year. Some correspondence has been seen in eastern USA between patterns of coal use and meteorologically-adjusted visibility trends, though this does not prove that a cause and effect relationship exists.

Although theory suggests that tropospheric atmospheric aerosols would have an influence on climatic change this has not yet been proved. Climate modellers are not yet able to agree on whether there will be an increase or decrease in temperature as a result of changes in aerosol concentrations on a global scale. While there is some evidence for sulphates as cloud condensation nuclei affecting cloud dynamics and the probability and amount of rainfall this is not significant on the global scale. The stratospheric aerosol has a greater potential for climatic change on the global scale but at the present only volcanic eruptions are known to affect aerosol concentrations. Fossil fuel consumption would affect stratospheric aerosol levels only if massive increases in uncontrolled carbonyl sulphide emissions were to take place and this seems unlikely at present.

Materials damage is caused largely by locally emitted, dry deposited, sulphur dioxide rather than airborne sulphates although the effects of nitrogen oxides, oxidants and hydrocarbons have not yet been fully evaluated. Large carbonaceous particles, originating from the combustion of oil, may catalyse the oxidation of deposited sulphur dioxide on the surface of stone. Ambient pollution levels have decreased over the past few years in most developed countries and sulphur dioxide has been one of the main targets for reduction. It therefore seems possible that much of the damage to building materials seen at present was largely initiated during periods of higher pollution earlier in the century.

The only direct effect which can be attributed to atmospheric sulphates at present day ambient concentrations is reduction of visibility in some areas. The association of secondary sulphates with visibility reduction could possibly widen the geographical spread of combustion plants affected by the pressure for reduction of SO_2 emissions.

Estimates of the biogeochemical sulphur cycle indicate what proportion of the sulphur in the environment is emitted as waste products of human activities on the global and regional scales, compared with natural emissions. These are discussed in Chapter 19 with special reference to coal combustion, together with regional sulphur budgets for industrialised regions. The possible mechanisms of conversion of reduced sulphur compounds to sulphur dioxide and sulphur dioxide to sulphates under atmospheric conditions are covered in Chapter 20. Chapter 21 considers the formation of sulphate as it is seen in the field, comparing the plumes of coal-fired power plants with other point source and urban plumes to determine if the type of fuel burnt makes a significant difference to the course of events once sulphur dioxide reaches the atmosphere. Environmental influences on sulphate formation are also outlined. The variability in sulphate concentration patterns is discussed in Chapter 22 to determine the relative importance of meteorological and emission factors in producing these patterns. In Chapter 23 the scale of sulphate transportation in the atmosphere is considered together with the role of meteorological processes in transport and the build-up of high ambient concentrations. How the various deposition processes contribute to the deposition of sulphates is discussed in Chapter 24, together with an outline of the deposition patterns of sulphates in precipitation. The possible relationship between trends in sulphate deposition and sulphur dioxide emissions is considered. After looking at the traditional scientific methods in earlier sections, Chapter 25 looks at results from mathematical models used to simulate the relationship between source emissions and deposition regions.

Later chapters deal with the direct atmospheric effects attributed to atmospheric sulphates. These are the effects on human health (mainly respiratory), visibility, climate and materials. Indirect health effects, mediated through water or food consumption, and ecological effects will be covered in a later review.

Sulphur dioxide emissions were perceived as a local problem earlier this century and efforts to reduce local pollution started over twenty years ago. This led to the use of tall smokestacks to disperse and dilute the pollutant plumes of large-scale emitters. This reduced the local problem by transporting the pollutants further from their origins before they were deposited and by depositing them over a wider area. In 1968 the Swedish scientist Oden published a scientific paper in which he presented evidence for the increase in acidity in air and precipitation over the Scandinavian countries and its effects on soils, vegetation and surface waters. This change in the chemistry of air and precipitation was attributed to large quantities of acidic sulphur compounds being transported long distances from the industrial areas of Central Europe and the United Kingdom. Sweden presented a report on this at the UN Conference on the Human Environment in Stockholm in 1972 and this led to a variety of research projects being set up to deal with the issue of transport of air pollutants over national boundaries (Cowling, 1982). More recently, 'acid rain' has become of concern to the general public, with reports of the depletion of freshwater fish stocks in vulnerable areas, most notably Scandinavia and Canada, although damage to similar ecosystems has been seen to a smaller extent in many other areas such as the USA and the UK. A threat to forests has now appeared especially in the Federal Republic of Germany (FRG) where damage is now widespread. Only 8% of trees were estimated to be danaged in 1983. By 1984 the estimate had risen to 34%, affecting 1.8 m hectares of forest covering 10% of the FRG In 1984, 50% of the forests were affected. In southern Sweden 10% of the trees are reported damaged. In Czechoslovakia a similar proportion have died. In Vermont, USA, half of the red spruce trees may be dying, but no direct damage to trees has yet been claimed in Canada (Leone and Brennan, 1985; Economist, 1984). The popular press presents this information to the general public in an alarmist way. The damage is uncritically attributed to 'acid rain' caused by deposition of sulphur emanating from coal-fired power stations often hundreds of kilometres away, since the weather systems carrying the pollution do not respect national boundaries. A very simplistic view is given of the processes which lead to incorporation of sulphur dioxide into rain and entry of pollution-derived ions into the ecosystem. It is assumed that the whole process has been thoroughly investigated and explained and that simple, if costly, action will remedy the situation given the political will. This view is opposed by those who claim that there is no evidence of a direct connection between sulphur dioxide emissions from power stations and ecosystem damage.

In reality the acid deposition problem stems from a series of complex and varied chemical, meteorological and physical interactions. There is considerable uncertainty about many of the processes. Even the level at which precipitation is considered acidic is more a matter of convention than a reality. Sulphur dioxide is not the only pollutant which contributes to acid deposition. Nitrogen oxides formed during combustion of fuel can be oxidised to nitrates in the atmosphere and also contribute to the formation of ozone. Volatile organic hydrocarbons and fine particulate matter are released into the atmosphere as by-products of human activities and also have an effect on the chemical composition of precipitation. However, in the past emphasis has been on the contribution of sulphur compounds possibly because in industrialised regions the largest proportion of sulphur in the atmosphere arises from human activities and the bulk of man-made sulphur emissions originate from large point sources which would be relatively easy to regulate.

Pollution from fossil fuel combustion is not a new problem. The UK was experiencing local pollution from coal smoke at least as early as the thirteenth century. In an attempt to combat pollution by legislation Edward I in 1306 issued an apparently ineffective decree 'compelling all but smiths to eschew the obnoxious material [coal] and return to the fuel they used of old'. The use of tall smokestacks to control pollution was suggested in the late fourteenth century and coal cleaning was apparently tried as long ago as the fifteenth century. International transboundary air pollution between England and France was noted in 1661 by John Evelyn (National Research Council, 1979; Meyer, 1977).

The first scientific study of the acid deposition phenomenon, although dealing mainly with local effects, was by R.A. Smith, the General Inspector of Alkali Works for the British government. He was the first to use the term 'acid rain' and he noted several important patterns which are still being investigated today. Smith (1872) stated 'we may therefore find easily three kinds of air – that with carbonate of ammonia in the fields at a distance; that with sulphate of ammonia in the suburbs; and that with sulphuric acid, or acid sulphate, in the town' and noted the deposition of sulphate and ammonia together in rain. Values measured for sulphates in rain varied from 2.06 ppm for rural Scotland, 5.52 ppm for rural England, 16.50 ppm for Scottish towns, 34.27 ppm for English towns, 47.99 ppm for Manchester, up to 70.19 ppm for Glasgow which is little different from the value of 73.30 ppm found in rain near an alkali works. Smith remarked 'how much less [the sulphate concentration] is in places where house-fires only are burnt than where high chimneys consume great amounts of coal in a small space, and frequently also an inferior quality of coal, such as would be too sulphurous for domestic fires'. He also noted that where rain contained more than 40 ppm of sulphuric acid vegetation could not grow and blamed sulphate pollution for damage to buildings, rusting of metals, rotting of blinds and fading of colours in prints and dyed goods. He concluded 'acidity is caused almost entirely by sulphuric acid, which may come from coal or the oxidation of sulphur compounds from decomposition, but it may also be caused in manufacturing towns by other acids, and in country places to a small extent by nitric acid and by acids from combustion of wood, peat, turf, etc.' The historical development of research on the effect of human activities on the composition of precipitation and its influence on the chemical climate of the Earth is outlined by Cowling (1982).

Today the combustion of sulphur-emitting fossil fuels for electricity generation industry and transport is the main source of tropospheric sulphur dioxide in industrialised regions. The environmental and health effects of gaseous sulphur dioxide have been extensively studied over many years but it has recently been considered that many of these effects are more strongly related to the aerosols formed by oxidation of sulphur dioxide in the atmosphere. Aerosols are dispersions or suspensions of solid particles, liquid droplets or a mixture of both in air. The components of an aerosol are too small to have an appreciable deposition velocity and thus can remain in the atmosphere for a considerable time. Rodhe and Isakson (1980) estimate that the global average residence time of sulphur dioxide and sulphate ions in the atmosphere is about 1.5 and 5 days respectively and about 30% of sulphur dioxide is transformed to sulphate (Rodhe, 1978). The transport time required to carry sulphur molecules several hundred kilometres is longer than the average atmospheric residence time of sulphur dioxide so the sulphur species concerned in long-range transport is predominantly sulphate aerosol.

These secondary sulphur aerosols are often referred to as 'sulphates' although there is evidence that a wide variety of sulphur compounds exists in the atmosphere. This may be due both to the relative ease of sulphate ion analyses on collected samples and to the stability of sulphate as the fully oxidised state. Some earlier sulphate measurements are thought to contain errors due to sulphate compounds formed on the collecting device from some other atmospheric form (Charlson *et al.*, 1978).

Human health risks are thought to be related to sulphur compounds in the form of particles small enough to penetrate the lungs while the main environmental problems are visibility reduction and the possible increase in the acidity of precipitation which may cause adverse ecological effects in vulnerable areas.

Sulphate aerosols are always present in the atmosphere and are the most important sulphur compounds involved in the long-range transport and acidification issues. Where applicable in this review the term atmospheric sulphates will also be taken to include sulphites. Unless otherwise noted the term sulphate aerosol will be used to include all chemical combinations of the sulphate ion, including sulphuric acid.

Chapter 19

The sulphur cycle

The natural and man-made sources of sulphur are outlined in this chapter, together with estimates of the proportion of total atmospheric sulphur which arises from human activity, on both the global and regional scales. The amount of sulphur released from coal utilisation is compared with that released by other human activities to determine the contribution of coal. The proportion of sulphur released as acidic, primary sulphates is discussed, together with the likely range of primary sulphates. The effect of sulphur dioxide (SO_2) pollution control measures on primary sulphate emissions is considered.

Sulphur is a relatively abundant element essential to all life on Earth, and as such is constantly being transformed and transported through the environment. The physical environment contains a number of well-defined pathways which link up the atmosphere, the hydrosphere (all waters on the surface of the Earth), the lithosphere (the Earth's outer crust), and the pedosphere (the soil system of the lithosphere) together with the living matter of the biosphere. The biosphere exerts a considerable influence on the geochemistry of sulphur by transforming it from one oxidation state to another, and thus changing the chemical and physical properties of the sulphur compounds and their distribution in the environment. Sulphur exists in most oxidation states between the fully reduced (−II) and the fully oxidised (+VI), forming a large variety of organic and inorganic compounds and radicals. Sulphur has one of the most complex biogeochemical cycles in nature. Hydrogen sulphide (H_2S) in the atmosphere is oxidised to SO_2 and then to sulphate so that the oxidation state makes the complete transition. Naturally occurring oxidation states which have been identified in the atmosphere are S(−II), for example (H_2S); S(O), for example elemental sulphur; S(IV), for example SO_2; and S(VI) for example sulphuric acid (H_2SO_4). In addition S(V) in the form of the dithionate ion, $S_2O_6^{2-}$, may have been mistakenly identified as sulphate in ambient air samples.

The precise molecular and crystalline forms of the sulphate compounds studied and of other oxidation states of sulphur are of importance in determining some of the properties of the compound. The sulphate ion is almost inert chemically at normal temperatures but the associated cation may be strongly reactive. Although some sulphate compounds are toxic the ion itself is non-toxic. The molecular form of the aerosol particle determines its shape, state, toxicity, acidity, deliquescent/hygroscopic properties (which determine the change in size distribution when exposed to varying relative humidity), and optical properties, such as refractive index, among other properties. At high humidity the aerosol droplet is likely to consist of a large number of anions and cations in a highly hydrated state where the ionic composition will determine the behaviour of the system. The same aerosol at low humidity will exhibit properties related to the molecular characteristics (Charlson *et al.*, 1978).

The following sulphate compounds are found in the atmosphere, often in a relatively pure state, so at low humidity some of the physical properties can be close to those expected from the molecular or crystalline composition.

Sulphuric acid is very hygroscopic at normal tropospheric temperatures and humidities and so is usually found as a liquid. Above about 30% relative humidity the aerosol becomes highly dissociated, each molecule producing one sulphate ion and two hydrogen ions or protons, exhibiting strong acid characteristics and forming a variety of neutralisation products. The bisulphate ion (HSO_4^-) can be formed by dissociation of both H_2SO_4 and its neutralisation products with ammonia (NH_3) and is normally found in acid sulphate aerosols. It is a moderately strong acid and highly reactive with metal oxides. Acidic sulphate aerosols can only exist where the NH_3 gas concentration is very low. Lau and Charlson (1977) estimate that NH_3 gas concentration varies widely both in time and space. The midwest of the USA is considered a dominant source for North America, whereas areas with acidic soils or maritime air exhibit much lower concentrations, typically between 0.01 to 0.1 ppb.

Ammonium bisulphate (NH_4HSO_4) is a common form of acid sulphate which is usually found in the atmosphere as a hydrated, aqueous liquid aerosol. Letovicite ($(NH_4)_3H(SO_4)_2$) occurs less frequently than other ammonia neutralisation products. As it is deliquescent at about 68% relative humidity it is often found in the atmosphere as solution droplets. Ammonium sulphate ($(NH_4)_2SO_4$) is the fully neutralised ammonium salt of H_2SO_4. It is soluble in water, deliquescent at 80% relative humidity, only weakly acidic in aqueous solutions and is almost unreactive. A well-aged air mass will normally contain $(NH_4)_2SO_4$, provided that NH_3 is available in abundance. However from the point of view of soil and ecosystems, $(NH_4)_2SO_4$ is as much an acidifying agent as H_2SO_4 because hydrogen ions are released when the ammonium ion enters into the natural soil nitrification process. Binary mixtures of ammonium sulphate and nitrate, in two stable mixed phases ($(NH_4)_2SO_4 \cdot 3NH_4NO_3$; $(NH_4)_2SO_4 \cdot 2NH_4NO_3$), have been found in atmospheric aerosols (Harrison and Sturges, 1984; NRC, 1981; Charlson *et al.*, 1978; Oden, 1976).

Other common sulphates have been identified in the atmosphere but these in themselves are generally not acidic. Magnesium sulphate ($MgSO_4$) is found in the aerosol in many maritime areas where sea salt has a $MgSO_4$ content of 5.7% by weight. The comparatively insoluble calcium sulphate ($CaSO_4$) enters the atmosphere in the form of relatively large particles by erosion of gypsum rocks. Sodium sulphate (Na_2SO_4) is emitted by the paper manufacturing industry. It is not acidic but is deliquescent at about 85% relative humidity and so aerosols are active as cloud condensation nuclei. Zinc sulphate ($ZnSO_4$) and ammonium zinc sulphate ($(NH_4)_2SO_4 \cdot ZnSO_4$) have been found, probably associated with local metal processing industries. $ZnSO_4$ is almost as soluble as $(NH_4)_2SO_4$ but $(NH_4)_2SO_4 \cdot ZnSO_4$ is much less soluble (Charlson *et al.*, 1978).

The dominant size range of secondary sulphate particles is 0.1 to 1 μm diameter. Sulphates have the greatest influence on the physicochemical nature of sub-micron particles. These sulphates contain impurities ranging from water, the most common, to trace metals such as iron, lead, manganese and vanadium. Many organic compounds have been found associated with sub-micron sulphates, some of them also secondary particulate matter formed from gas phase reactants. Other gaseous pollutants such as ozone (O_3) and nitrogen dioxide (NO_2) are commonly associated with and react with sulphates. However mixed sulphate-nitrate salts are distributed into larger particle sizes, consonant with the behaviour of the nitrate component. Some of the properties ascribed to sulphates may be due not to sulphur but to one or many of the trace substances present (Harrison and Sturges, 1984; NRC, 1979; Charlson *et al.*, 1978).

Concentrations of many sulphur species in the atmosphere are so small that accurate concentration and rate transfer measurements have only recently become possible. In sediments the inhomogeneity of strata and deposits, which exhibit erratic changes in total sulphur content and in the ratio of isotopic forms, cause considerable problems of measurement. In the oceans, difficulty is still encountered in tracking the intermediates in reactions involving sulphur compounds, which occur at very low concentrations, due to the relatively high concentrations of background sulphate. In addition, some steps of the sulphur cycle involve geological time periods which are difficult to extrapolate from laboratory experiments, for example the residence time of sulphate in the oceans has been estimated at 40 million years; that of pyrite in sedimentary rocks at over 250 million years. The cycle is believed to have undergone drastic changes and alterations over the last 200 million years, leaving the present cycle not fully balanced. Many of the most important steps are known only qualitatively and at a relatively superficial level (Meyer, 1977; Moss, 1978; Charlson *et al.*, 1978; Zehnder and Zinder, 1980).

The global sulphur cycle has been described in a series of detailed models which are summarised in Table 19.1. It would lie outside the scope of this review to list at length the basic data from which these conclusions were drawn but the assumptions and justifications behind these figures can be found in the original references.

The global budget produced by Granat *et al.* (1976) was a first attempt to formulate a pre-industrial sulphur budget, balanced in the pedosphere. They showed that earlier estimates had been distorted by man-made emissions, and that the atmospheric component of the sulphur cycle was smaller than had previously been assumed. More recently, Moeller (1984b) estimated global biogenic sulphur emissions to the atmosphere to be in the range 50–100 Tg/y, sea spray emissions to be 175 ($\pm$ 50%) Tg/y, and volcanic sulphur emissions to be about 2 Tg/y.

After the Granat *et al.* (1976) study appeared, the international Scientific Committee on Problems of the Environment (SCOPE), the sponsors, commissioned a more detailed treatment of the sulphur cycle. This provided more reliable estimates of the sulphur reservoirs and fluxes of both man-made and natural origin. Ivanov (1983) compared previous estimates of sulphur fluxes with more recent information, including previously inaccessible work from the Soviet Union. He separated, wherever feasible, the natural and man-made contributions to these fluxes, as shown in Table 19.1. The differences between the different estimates are due, to a certain extent, to the better data available in recent years and the large uncertainties and changes with time associated with some of the fluxes. The pathways through the sources and sinks of the sulphur cycle are illustrated in Figure 19.1, after Ivanov (1983). Just considering the atmospheric fluxes, the total amount of sulphur emitted into the atmosphere is 164.5 Tg/y over land and 77.5 Tg/y over ocean. The amount removed from the atmosphere is 84 Tg/y to the land and 258 Tg/y to the oceans. Since the atmosphere as a whole should be a balanced

Table 19.1 Major fluxes of the global biogeochemical sulphur cycle in Tg/y of sulphur (after Ivanov, 1983)

Flux	*Nature of flux*	*Eriksson (1960)*	*Robinson & Robbins (1972)*	*Kellog* et al. *(1972)*	*Friend (1973)*	*Granat* et al. *(1976)*	*Ivanov (1983)*		*Total*
							Natural	*Man-made*	
Continental part of the cycle									
P1	Emission to the atmosphere from fuel combustion and metal smelting	39	70	50	65	65	–	113	113
P2	Volcanic emission	–	–	1.5	2	3	14	–	14
P3	Aeolian emission	–	–	–	–	0.2	20	–	20
P4	Biogenic emission	110	68	–	58	5	17.5	–	17.5
P5	Atmospheric transport of oceanic sulphate	5	4	4	4	17	20	–	20
P6	Deposition of large particles from the atmosphere	–	–	–	–	–	12	–	12
P7	Washout from the atmosphere, surface uptake and dry deposition	240	116	111	121	71	25	47	72
P8	Transport to the oceanic atmosphere	–	26	5	8	18	34.5	66	100.5
P9	Weathering	15	14	–	42	66	114.1	–	114.1
P10	River runoff to the world oceans	80	73	–	136	122	104.1*	104	208.1
P11	Underground runoff to world oceans	–	–	–	–	–	9.2	–	9.2
P12	River runoff to continental waterbodies	–	–	–	–	–	35	–	35
P13	Marine abrasion of shores and exaration	–	–	–	–	–	6.8	–	6.8
P14	Pollution of rivers with fertilisers	10	11	–	26	–	–	28	28
P15	Effluents from chemical industry	–	–	–	–	–	–	28	28
P16	Acid mine waters	–	–	–	–	–	–	1	1
Oceanic part of the cycle (see above, P2)									
P17	Volcanic emission	–	–	–	–	–	14	–	14
P18	Biogenic emission	170	30	–	48	27	23	–	23
P19	Marine sulphate	45	44	47	44	44	140	–	140
P20	Washout, surface uptake, and dry deposition	200	96	72	96	73	258	–	258
P21	Burial of reduced sulphur in sediments	–	–	–	–	–	111.4	–	111.4
P22	Burial of sulphate in sediments	–	–	–	–	–	27.8	–	27.8

*Another 4.8 Tg of sulphur should be added to the ionic runoff for particulates in river runoff
†About 5.6 Tg of sulphur are removed from farmlands annually in harvested crops; after use as food this sulphur is discharged in river runoff mainly as domestic sewage and effluents from livestock farms.

system the sulphur flux from the oceanic to the continental atmosphere is 20 Tg/y while the reverse flux from the continental to the oceanic atmosphere is 100.5 Tg/y of sulphur (Ryaboshapko, 1983). The following discussion of natural mechanisms is arranged by input into each sphere.

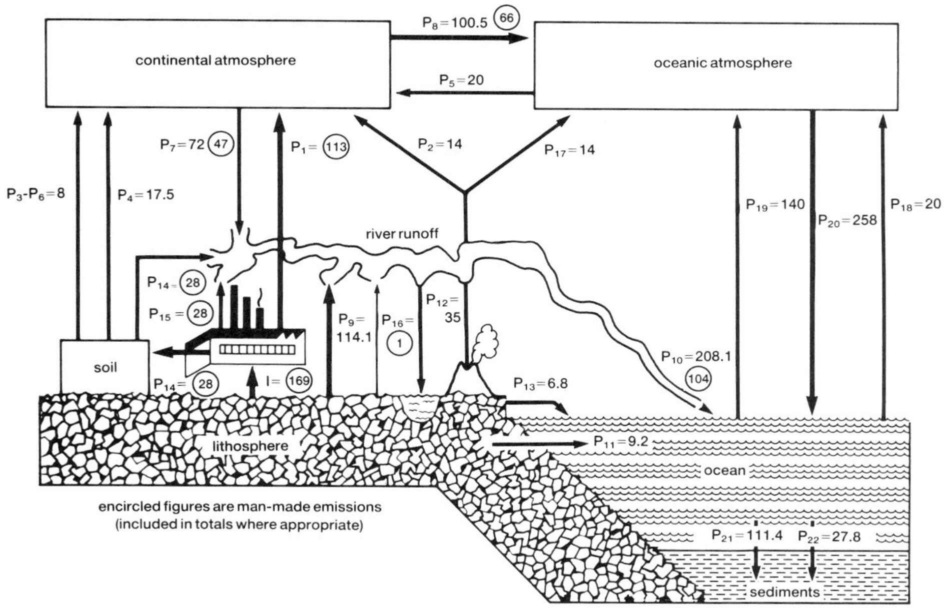

Figure 19.1 Major fluxes of the global biogeochemical sulphur cycle (flux designations are given in Table 1 of Ivanov, 1983)

Atmosphere

Volcanoes emit SO_2 and H_2S, which is mainly in the gaseous phase, with smaller amounts of sulphites, various sulphates and elemental sulphur. In the majority of major eruptions, particles are formed which enter the stratosphere to form the Junge sulphate layer which may affect global climate. A much smaller proportion enters the troposphere (the lowest 10–20 km of the atmosphere) mainly as easily-precipitated dust. Other sources of geothermal activity, such as sulphur springs, also provide a contribution which, while small in global terms, may be of regional importance. In New Zealand for instance geothermal boreholes produce H_2S at a concentration of 40 000 μg/m compared with an average range of 0.1 to 0.38 μg/m produced from biogenic emissions (Granat *et al.*, 1976; Cullis and Hirschler, 1980; Cope and Spedding, 1982).

Sea spray, formed by the bursting of bubbles in the oceans, with individual droplets evaporating to leave solid particles, is an important source of atmospheric sulphur over the oceans. Much of the sulphate is returned to the oceans by precipitation and sedimentation but it is generally accepted that about 10% is deposited on land surfaces (Eriksson, 1960). Garland (1981) showed that sulphate is enriched relative to sodium by 10–30% in spray generated artificially. He concluded that the enrichment ratio of sulphate in sea spray was in the range of 1.1–1.3. Several mechanisms have been advanced to account for this enrichment including chemical fractionation across the air–sea interface, soil erosion or pollution by man-made sulphate. Bonsang *et al.* (1980) concluded that sulphate enrichment in aerosols in marine regions far from continents was due to oxidation of SO_2 of biogenic origin. The SO_2 was possibly produced by oxidation of simple organic sulphur compounds released from seawater. This sulphate excess was found mainly on submicron particles. This enrichment was comparable to the flux of sea-salt sulphate estimated by Eriksson (1960) and used since in most global sulphur cycle models. This would indicate an underestimation of the sea spray flux in these models.

Volatile reduced sulphur compounds resulting from biological processes, mainly the decay of organic matter, may form from the decomposition of plant tissue in the soil by micro-organisms or through reduction of sulphates by sulphate-reducing bacteria, chiefly of the genera *Desulfovibrio* and *Desulfotomaculum*, in anoxic aquatic environments, including marshy and estuarine areas as well as oceans (Zinder and Brock, 1978). Direct measurements of these reduced sulphur species have only recently been made. Details of data found are given in Adams *et al.* (1980a, 1980b, 1981a, 1981b); Aneja (1980); Aneja *et al.* (1981); Delmas *et al.* (1980) and Henry and Hidy (1980). The great variation found in emission rate measurements for reduced sulphur compounds from land areas was associated primarily with wide variations in the surface and climatic environments of the study sites. While the maximum emission rates from cycle, the mean average values would not. Carbon disulphide (CS_2) and carbonyl sulphide (COS) diffuse through the tropopause and are converted to sulphate aerosols in the stratosphere. Using the mean maximum emission rates for these compounds (142 Tg/y of sulphur), their contribution to the stratospheric sulphur layer can be computed to be greater than 100%. However, using the mean average emission rates (4.1 Tg/y) their contribution is about 1% (Aneja *et al.*, 1982). Steudler and Peterson (1984) estimate that, globally, salt marshes may release more than double the quantity of COS and CS_2 needed to sustain the stratospheric sulphate layer.

Bandy and Maroulis (1980), Ko and Sze (1980) and Sze and Ko (1980a, 1980b) consider these reduced compounds to have some importance for the troposphere; the first proposing that up to 33% of the SO_2 measured in a marine atmosphere may have been converted, directly or indirectly, from CS_2. Panter and Penzhorn (1980) and Penzhorn and Panter (1980) postulate that alkyl sulphonic acids recently detected in the atmosphere are oxidation products of these reduced sulphur compounds.

Some higher plants are also known to emit H_2S, dimethyl sulphide ($(CH_3)_2S$) CS_2 as by-products of photosynthesis (Aneja, 1980; Winner *et al.*, 1981). Dry soils with crops receiving sulphur fertilisers emit considerable amounts of reduced sulphur compounds. This emission rate increases if other fertilisers are also used to encourage microbial growth (Bache and Scott, 1979; Cullis and Hirschler, 1980). In addition acid mine drainage is suggested by Rice *et al.* (1981) as a source of exchangeable sulphur which may be emitted to the atmosphere. In this process, bacteria of the genera *Ferrobacillus* and *Thiobacillus* catalyse the oxidation of pyrite (FeS_2) minerals to sulphuric acid, adding a large source of sulphur to the natural reservoir.

Relatively high concentrations of reduced sulphur species, in particular dimethyl sulphide, are present in oceanic surface waters and this is related to the productivity of marine algae and coral organisms. The flux of sulphur by this process is estimated to range between 17 and 72 Tg/y of sulphur (Andrae *et al.*, 1983; Nguyen *et al.*, 1983).

Since there has been, until very recently, a lack of data on the input from biological processes to the sulphur cycle this factor has been used to balance the flow of sulphur in some budgets. Sze and Ko (1980b) put forward a suggested atmospheric sulphur cycle with reduced sulphur compounds considered as the sole source for atmospheric sulphur based on the low sulphur budget of Granat *et al.* (1976) and recent atmospheric data from remote regions, where man-made SO_2 emissions have minimal effect. They conclude that a flux of reduced sulphur compounds of about 28 Tg/y of sulphur may account for much of the atmospheric SO_2 and sulphate, with a tropospheric burden of 2.2 Tg of sulphur as COS representing the major form of atmospheric sulphur on a global scale.

Moeller (1984b) concluded that, on the global scale, the amount of sulphur emitted from natural sources (excluding sea spray) into the atmosphere is currently about equal to man-made sulphur emissions. Since natural inland sulphur emissions amount to only about a quarter of global emissions, sulphur emissions from human activities over land are about four times as great as natural emissions. The importance of dust in the sulphur cycle is largely regional since the composition of the dust depends on the minerals occurring in the soil and weathered rocks. The large size of the majority of particles results in a short residence time in the atmosphere (Granat *et al.*, 1976). Mineral dust of high sulphate content is picked up by wind to produce a high natural background sulphate concentration in Central Asia (Meszaros, 1978). Mamane *et al.* (1980) report that 20% of particles of Saharan desert origin in the 0.3 to 2.0 μm size range, measured in Israel on the day of a dust storm, were pure, volatile sulphate with another 63% mixed sulphates. These desert particles took part in the conversion of SO_2 to sulphate during transport.

Petrenchuck (1980) estimates that whereas dust particles from weathering form 65% of the total aerosol input to the atmosphere, because of their larger size, they constitute only 26% of the total aerosol consumed in cloud formation and involved in rainout processes. The long distance transport of dust between continents by high altitude winds and the global dispersion of ash injected into the stratosphere by volcanic eruptions are only minor inputs to the global sulphur cycle (Granat *et al.*, 1976).

Pedosphere

Sulphur exists in unweathered sedimentary rocks mainly as sulphides of iron, nickel and copper and as gypsum ($CaSO_4 \cdot 2H_2O$). During weathering sulphides normally oxidise to sulphates and gypsum dissolves to enter streams as Ca^{2+} and SO_4^{2-} ions. It is assumed that all the sulphur finally enters streams as sulphates, but some is retained in the pedosphere for a period of time and enters the biosphere. Plants form reduced sulphur compounds from sulphates, animals transform these reduced compounds back into sulphates. Various microorganisms can either reduce sulphates or oxidise reduced sulphur compounds (Moss, 1978; Zinder and Brock, 1978; Zehnder and Zinder, 1980).

Wet and dry deposition processes, including rain, return much of the sulphur in the atmosphere to the pedosphere. By comparison with nuclear fallout data Robinson and Robbins (1972) assume that dry deposition of sulphate sulphur is about 20–25% of total sulphur deposition. These data are thought to underestimate the dry component for near-surface sources (Hales, 1978). Granat *et al.* (1976) consider that wet and dry deposition contribute roughly equal amounts of sulphur to land and oceans, based on Scandinavian data. Plants can take up SO_2, H_2S and dimethyl sulphide directly through their leaves, and soils absorb SO_2. These are then oxidised to sulphates (Granat *et al.*, 1976; Moss, 1978; Zehnder and Zinder, 1980).

Volcanic activity causes sulphur compounds and elemental sulphur to be emitted in the gas phase which later partially precipitate close to the source (Granat *et al.*, 1976; Zehnder and Zinder, 1980).

Hydrosphere

River runoff is the main sulphur sink of the pedosphere. Sulphur entering salt marshes would make a significant contribution to the global sulphur rivers as the result of erosion or weathering, from overland runoff of precipitation and from leaching of rocks and soils (Granat *et al.*, 1976). Ocean water absorbs SO_2 particularly well around pH 8 but as the pH decreases this absorbing ability also declines. The normal high pH and buffering capacity of the oceans makes them an efficient sink for atmospheric SO_2. Sulphur enters the hydrosphere in an oxidised form as sulphates, which can lose oxygen faster than it can be replaced in stagnant or deep sea waters and be reduced to sulphide by anaerobic microorganisms. These sulphides influence metal sulphide precipitation, carbonate formation and water pH and enhance methane formation (Meyer, 1977; Zehnder and Zinder, 1980).

Underwater disturbances of the lithosphere, for example submarine volcanoes and earthquakes, introduce sulphur species to the hydrosphere, but this is an unquantified input at present.

Lithosphere

Marine sedimentation is the main mechanism for sulphur entry into the lithosphere in the form of gypsum (sulphate), pyrite (sulphide) or organically bound sulphide in dead biogenic material, much of which is transformed subsequently, by biological means, into pyrite. On a geological time scale, marine evaporites, extracting dissolved salts from seawater by evaporation, were one of the major sinks for sulphate, although the rate of formation today is negligible. Another sulphate sink is ridge crest hydrothermal activity from vents in mid-ocean ridges, in which a thermally-driven process causes sulphate to be reduced by ferrous iron and precipitated as pyrite (Garrels *et al.*, 1975; Meyer, 1977; Zehnder and Zinder, 1980).

Man-made component

Human activity has influenced the sulphur cycle in a number of ways, the most important being direct emissions to the environment by (in order of importance): the combustion of fossil fuel products; smelting of non-ferrous ores (mainly copper, lead, nickel and zinc); manufacture of sulphuric acid; conversion of wood pulp to paper; incineration of refuse and the production of elemental sulphur. The sulphur flux through the hydrosphere by river runoff has been increased by the input of sulphur from chemical fertilisers and by an increase in the rate of rock weathering due to human agricultural and industrial activities. This has led to an increase in the supply of easily decomposable organic matter in coastal and marshy areas, thus increasing the bacterial production of H_2S (Garrels *et al.*, 1975; Granat *et al.*, 1976; Moss, 1978; Cullis and Hirschler, 1980).

Reduced sulphur compounds are also emitted by the chemical industry, in paper manufacture and during fuel refining and combustion processes. The emissions of coal combustion desulphurisation plants could generate up to 5 Tg/y of COS, which amounts to 6% of total sulphur emissions. Coal processing plants also generate COS but after treatment, emissions should typically amount to $<1\%$ of the sulphur in the coal. Certain catalysts used to reduce nitrogen oxide (NO*x*) in stack gases can also convert up to 6% of the accompanying SO_2 into COS (Turco *et al.*, 1980). Extra sources of reduced sulphur compounds are believed to come from acid mine drainage (Rice *et al.*, 1981) and from stored flue gas desulphurisation (FGD) sludges. Adams and Farwell (1981) measured total sulphur gas emissions ranging from <0.01 to $>0.26\,g/m^2 y^{-1}$ of sulphur or 0.01–0.3 kg of sulphur per day for a 40.5 hectare FGD impoundment surface.

A number of variables affected the composition and concentration of emissions including the type of sludge, the water content, age and sulphate/sulphite ratio. COS may comprise 1–10% of the sulphur released to the atmosphere as a result of organic fuel combustion. Turco *et al.* (1980) consider that more than half of the annual global COS emissions could result from human activity. Khalil and Rasmussen (1984) however suggest that the figure is less than 25%.

Over 90% of man-made sulphur released to the atmosphere is emitted as SO_2, the remainder being either H_2SO_4, neutralised sulphates (e.g. $CaSO_4$), SO_3 or sulphites. SO_3 combines rapidly with water vapour to form H_2SO_4 and has not been found separately in the atmosphere. Eatough *et al.* (1978) have demonstrated that both organic and inorganic sulphite S(IV) species are stable constituents of aerosols produced by fossil fuel combustion, smelters and urban areas, forming particles in the respirable size range. Moeller (1984a) estimated the historical increase in SO_2 emissions since 1860, shown in Table 19.2. Global emissions of SO_2 in 1977 were almost a factor of 30 greater than emissions in 1860. They are projected to rise still further, to 40 times as much (a maximum of 100 Tg/y) by the end of the century. A rough estimate over the whole 'era of fossil fuel combustion' indicates total man-made SO_2 emissions of about 5000 Tg of sulphur. Of this, 50% will be emitted between 1970 and 2030. Projections for OECD European countries (OECD, 1981) show a 25% increase in SO_2 emissions in 1985 over 1974 levels if there is no change in current control practices. The relative global importance of the primary man-made sources of SO_2 are shown in Table 19.3 (adapted from Varhelyi, 1985). The southern hemisphere contributed less than 10% of man-made SO_2 emissions, and this ratio did not change appreciably with time. On a global scale, coal combustion produced 56%, petroleum combustion 26%, wood combustion $<1\%$ and industry 17% of total SO_2 emissions. Coal combustion contributed 58–61% in North America and 61% in Europe.

In the US, most of the coal consumed was used by utilities for electricity generation, 85% in 1975 with a further 12% going to industrial use (Nader, 1980). In the eastern USA approximately 57% of the total SO_2 emitted was released during electricity generation while combined heavy industry and electricity generation produced about 94% of the SO_2 emissions burden. Up to 70% of SO_2 emissions were concentrated in the 100–700 m height range (Mueller and Hidy, 1983). In the UK, over 60% of total SO_2 emissions have been emitted from tall chimneys since 1976 (McInnes, 1980). Emissions are geographically widely

Table 19.2 Global production of sulphur dioxide in Tg/y of sulphur (after Moeller, 1984a)

	Coal	*Lignite*	*Oil*	*Copper smelting*	*Others**	*Total*
1860	2.4	0.0	0.0	0.1	0.0	2.5
1870	3.6	0.1	0.0	0.1	0.1	3.9
1880	5.6	0.2	0.0	0.2	0.1	6.1
1890	8.4	0.3	0.1	0.4	0.1	9.3
1900	12.6	0.6	0.2	0.5	0.2	14.1
1910	18.8	0.9	0.4	0.9	0.3	21.3
1915	19.0	1.0	0.5	1.2	0.5	22.2
1920	21.2	1.3	0.7	1.5	0.6	25.0
1925	21.1	1.6	1.2	1.5	0.7	25.1
1930	24.2	1.7	1.6	1.6	0.6	27.1
1935	19.4	1.6	1.8	1.4	0.8	24.6
1940	24.2	2.6	2.3	2.4	1.2	32.7
1945	21.2	1.6	2.8	2.2	1.0	27.8
1950	25.8	3.0	4.2	2.4	1.4	36.8
1955	24.9	4.5	6.1	2.9	1.5	39.9
1957	27.2	5.0	7.0	3.2	1.5	43.9
1959	30.2	5.2	7.8	3.3	1.5	48.0
1961	30.6	5.6	8.8	3.7	1.5	49.2
1963	31.2	6.0	10.3	3.7	1.4	52.6
1965	31.8	6.2	12.0	3.9	1.4	55.3
1967	31.5	6.1	13.9	3.6	1.4	56.5
1969	31.3	6.4	16.3	4.2	1.2	59.5
1971	33.5	6.8	18.9	4.1	1.2	64.5
1973	33.6	6.9	21.9	4.7	1.1	68.2
1975	35.0	7.2	21.0	4.8	1.0	69.0
1977	37.2	7.7	24.0	4.9	1.1	74.9
1985	48	9.5	25	6.5	1	90
2000	55	12	23	9	1	100

*Lead and zinc smelting, sulphuric acid production

distributed with most of the major sources outside urban areas. Areas of very high SO_2 emission density exist which coincide with areas of maximum sulphate concentration (Lavery *et al.*, 1981; Husar and Patterson, 1980). A variety of emission inventories have been compiled for the USA with no general agreement as to accepted emission values. Figure 19.2, after Wilson and Mohnen (1982), shows an emission distribution for the north-eastern USA which is essentially proportional to actual source strengths. In Europe, there is a large uncertainty in the SO_2 emission inventories for most countries, especially the Eastern European countries.

Global figures are of interest for assessing possible impacts on the radiation balance via aerosol formation, but when assessing the impact of man-made emissions on ecological systems regional budgets are of more significance. The relatively short residence time of sulphur compounds in the troposphere (of the order of a few days) and the very uneven geographical distribution of both man-made and natural emissions causes human influence upon the sulphur cycle, especially the atmospheric component, to vary widely from one region to another. Since the bulk of man-made emissions and their environmental impacts are concentrated in two regions, north-western Europe and north-eastern USA with the adjacent parts of Canada, most studies have concentrated on these areas.

Table 19.3 Sulphur dioxide emissions from man-made sources 1979 in Tg/y of sulphur (adapted from Varhelyi, 1985)

	Hard coal	*Brown coal*	*Coal coke*	*Total coal*	*Combustion of petroleum products*	*Ore smelting*	*H_2SO_4 production*	*Petroleum refineries*	*Wood burning (1976)*	*Total*	*Region as % of total*
Europe	6.5	7.9	0.5	14.9	6.4	1.76	0.19	1.30	0.024	24.57	26.3
North America	10.7	0.6	0.34	11.64	3.7	2.23	0.22	1.26	0.013	19.06	20.4
USSR	5.9	2.2	0.36	8.46	2.9	1.38	0.11	0.64	0.073*	13.56	14.5
Africa	0.56	–	0.02	0.58	0.26	1.24	0.01	0.09	0.182	2.36	2.5
South Africa	0.51	–	0.02	0.53	0.06	0.19	0.01	0.02	–	0.81	0.9
Other America	0.24	–	0.04	0.28	2.0	1.04	–	0.62	0.164	4.1	4.4
Asia	7.0	0.19	0.28	7.47	4.4	1.68	0.10	0.92	0.112	14.68	15.7
Middle East	0.05	0.06	0.01	0.12	0.6	–	–	0.20	0.29$^+$	0.95	1.0
Far East	1.9	0.03	0.18	2.11	3.2	–	–	0.59	0.253	6.15	6.6
Centrally planned	5.0	0.1	0.1	5.2	0.6	–	–	0.13	–	5.93	6.4
Oceania	0.17	0.14	0.01	0.32	0.07	0.26	0.01	0.05	0.002	0.71	0.8
Australia	0.16	0.14	0.01	0.31	0.06	–	–	0.04	–	0.41	0.4
Total	38.69	11.36	1.87	51.92	24.25	9.78	0.65	5.86	0.852	93.31	
Fuel as % of total				55.6	26.0	10.5	0.7	6.3	0.9		

*Includes the USSR and Eastern European countries
$^+$Includes Near Eastern countries

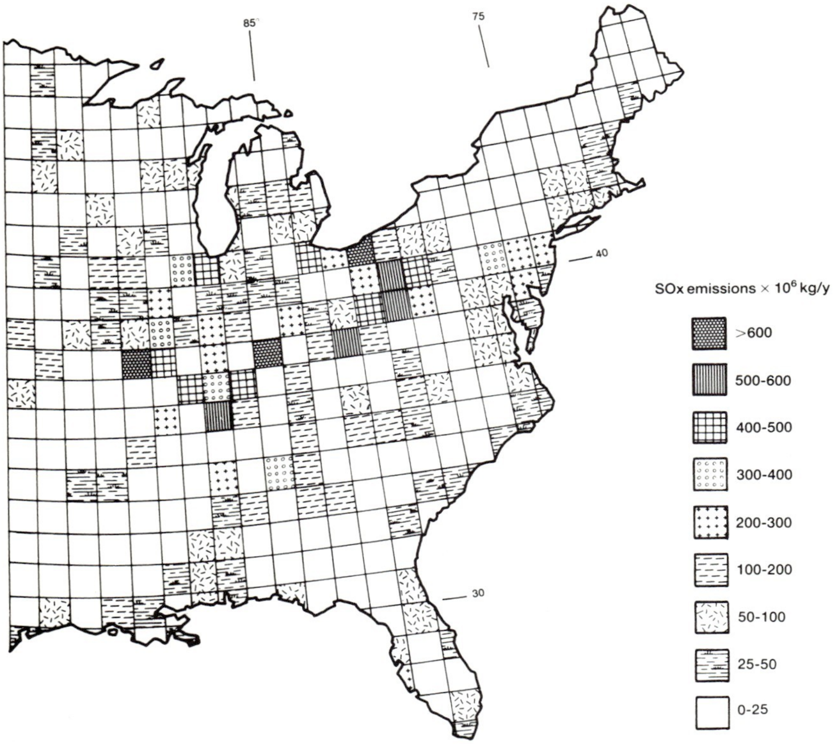

Figure 19.2 Total annual SO*x* emissions for the eastern USA (after Wilson and Mohnen, 1982)

Regional budgets for north-western Europe are proposed by Rodhe (1978, 1971) and Granat *et al.*, 1976). Oden and Ahl (1980) produced a budget for Sweden. The Sulfate Regional Experiment (SURE) study was designed to identify the influence of pollutants emitted by the electric power industry on ambient sulphate levels in the greater north-eastern USA, defined as an area extending in the east–west direction from eastern Kansas to the Atlantic seaboard and from mid-Alabama to south-eastern Canada in the north–south direction. The results suggest that biogenic emissions are a small fraction of total sulphur oxides (SO*x*) emissions and the contribution of sea salts to background airborne sulphates is negligible. The natural contribution is estimated to be a maximum of 0.5–1% of the ambient sulphate burden in this region (Mueller and Hidy, 1983). Biogenic emissions are at a maximum in south-eastern USA while emissions from human activities are at a maximum in the upper Ohio Valley. ASTRAP model results show that the biogenic contribution to sulphur deposition during summer is as high as 5% in Florida or 10% in Louisiana but only about 1% in New York State. Since the ratio of biogenic to man-made emissions in summer is about twice that for the year as a whole, the relative contribution of biogenic sources to annual sulphur deposition would be about half of that in summer simulations (Shannon, 1984).

Figure 19.3 is a schematic representation of an atmospheric sulphur budget for eastern North America formulated by Galloway and Whelpdale (1980). Man-made

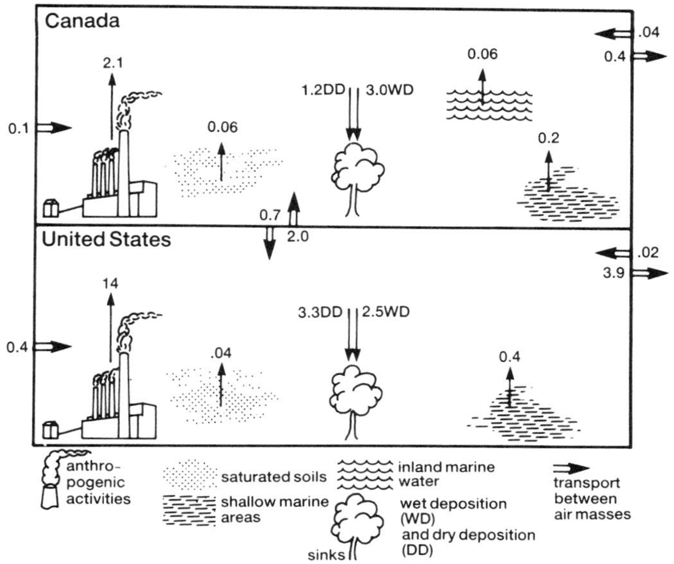

Figure 19.3 Schematic representation of an atmospheric sulphur budget for eastern North America in Tg/y of sulphur (after Galloway and Whelpdale, 1980)

emissions accounted for over 90% of total emissions, natural sources for about 4% and inflow from outside the region contributed the remainder. Of the 17.4 Tg/y of sulphur (16 Tg/y emitted by man) entering the atmosphere in eastern North America, 57% (62% of man-made emissions) was deposited in the region by wet and dry deposition processes. The remainder was transported out of the region in the atmosphere, primarily to the east over the Atlantic Ocean. Galloway *et al.* (1984a) estimated that about 34% (4.3 Tg/y) of the sulphur emitted to the atmosphere in this region was advected over the Atlantic. As a function of height approximately 48, 26 and 26%, respectively, occurred in the lower (< 1500 m), middle (1500–3000 m) and upper layers (3000–5500 m) of the atmosphere. On a smaller scale, 70–80% of sulphur emissions from the UK were transported eastwards over the North Sea, in the absence of precipitation (Fisher and Callander, 1984).

On a global scale Granat *et al.* (1976) attributed about 60% of the total sulphur emissions into the atmosphere to man-made sources. However in north-western Europe, covering about 1% of the Earth's surface, they found that natural emissions constituted < 10% of the total sulphur emissions, and so were of little importance for the regional sulphur budget.

Primary sulphate emissions

The percentage of total SO*x* released as directly emitted primary sulphates is approximately 1–3% for coal-fired boilers as opposed to 5–9% for oil-fired boilers burning fuels of a similar sulphur content (Homolya *et al.*, 1976). Bituminous coal-fired utility boilers show an average conversion of fuel sulphur to primary sulphate of 1.5% (Surprenant *et al.*, 1981). Non-utility oil-fired boilers, however, have been found to emit significantly more primary sulphate, ranging from 11.7–16% of the total SO*x* emitted, with 95% of the sulphate emitted in the form of H_2SO_4 (Homolya and Lambert, 1981).

Electrostatic precipitators (ESP) reduce sulphate emissions from coal-fired boilers by more than 50%, reducing primarily the metal sulphates component (Nader, 1980). However in some units after solid particle emission rates are reduced to about 5 g/GJ a persistent blue-white plume is seen, characteristic of a stable mist of H_2SO_4. In the past much of this primary sulphate may have been neutralised by basic particles which are now being trapped by the ESP (Anson, 1981).

A side effect of the use of flue gas desulphurisation units (FGD) for SO_2 removal, with consequent reduction in secondary sulphate formation during atmospheric transport, could be a change in primary sulphate emissions. The change to combustion of a high sulphur coal instead of low sulphur, based on SO_2 reductions gained by use of FGD, could produce primary sulphate emissions at the same level as the combustion of low sulphur oil (Nader, 1980). It would also increase the load of aerosol sulphate close to the plant which would otherwise be emitted as gas and only converted to aerosol after atmospheric cooling and dilution during transport.

Homolya and Cheney (1979) investigated the effect of a wet limestone slurry FGD unit dating from 1973 on primary sulphate emissions from a 5% sulphur coal-fired boiler to determine the extent to which emissions are affected by scrubber liquor re-entrainment. The FGD unit reduced SO_2 levels by an average of 76% but total sulphate removal averaged only 29%, with 85% of the flue gas sulphate as H_2SO_4. The acid was in the vapour phase prior to scrubbing but the FGD slurry and demisters, maintained at temperatures below the acid dewpoint, converted it to aerosol. Although sulphate emissions were only 1% of total SO*x* emissions the use of 5% sulphur fuels would produce significant calculated mass emissions of acid and particle sulphates of 785 kg/h and 138 kg/h respectively. Surprenant *et al.* (1981) found that only 28.5% of SO_3 was removed from emissions from an industrial oil-fired boiler using a double-alkali FGD unit although SO_2 removal efficiency was 97.5%. Downs (1981) found a similar result at a unit burning western low-rank coal where a wet scrubber FGD system was used, with average SO_2 removal efficiency of 85%. In all cases the potential amount of secondary sulphate aerosol generated by oxidation of emitted SO_2 was much greater than the primary sulphate aerosol present.

If these results are typical of wet FGD processes then the use of high sulphur coal on the basis of these current FGD processes as controls would reduce SO_2 levels and thus the associated secondary sulphate formation but would also permit a significant local increase in primary sulphate emissions, proportional to the increase in the sulphur levels of the fuel. In a comparison between primary sulphate emissions and secondary sulphate production an ASTRAP model simulation estimated that secondary sulphate accounts for about 85% of surface sulphate concentrations at present (Shannon, 1981a).

Recent oxygen isotope ratio measurements (Holt *et al.*, 1982) have demonstrated that primary sulphates constitute 10–14% of the total ambient sulphates in both aerosol and precipitation at Argonne, Illinois, values much higher than would have been expected from previous estimates based on the percentage of primary sulphates in emissions from large point sources. An interesting seasonal variation was found in aerosol sulphates where the primary fraction varied from about 10% in summer to 30–40% in winter, while the primary fraction in precipitation sulphates ranged from 20–30% during most of the observation period. The results of sampling of rainwater collected beneath a coal-fired power plant by this method implied that, on average, 48% of the scavenged sulphates were emitted as primary sulphates while the remaining 52% originated as SO_2, although typically only 3% of sulphur was emitted as primary sulphates at this plant. This rapid scavenging by rain gives primary sulphates a much more local effect than secondary sulphates (Holt *et al.*, 1983).

Submicron fly ash particles showing a preferential surface concentration of sulphur in comparison with their matrix elements have been found at the Four Corners power plant, New Mexico with physical properties suggesting that sulphur is adsorbed and oxidised on fly ash after volatilisation in the high-temperature zone of the boiler (Pueschel, 1976). Thus cloud condensation nuclei are available before gas-to-particle conversion occurs. Fly ash particles may also play a part in SO_2 oxidation. Mamane and Pueschel (1979), however, found that only up to 6% of fly ash particles were coated with sulphate, equivalent to a conversion rate of 0.03%/h which is negligible in comparison with other mechanisms.

Dimethyl sulphate and its hydrolysis product, monomethyl sulphate, were found in primary airborne particulate matter from several sources of coal and oil combustion including power plants (Lee *et al.*, 1980; Eatough *et al.*, 1981). These compounds constitute about 0.5% of the total sulphur measured but this is the lower limit of the actual amounts present. The results suggest that the concentration of methylated sulphates may be up to 1000 ppm in primary respirable particulate matter and can have a residence time of hours to days in the atmosphere. These species are expected to be produced in combustion of other sulphur-bearing fuels and are of interest because of the proved mutagenic and carcinogenic properties of dimethyl sulphate.

Summary and comments

Sulphur is an abundant element which is essential to life. It is transported through the environment while undergoing transformation from one oxidation state to another. Sulphates, in the fully oxidised state, are the most stable form. On a global scale, the main natural emission sources of sulphur to the troposphere are biogenic emissions (50–100 Tg/y) and sea spray (175 Tg/y ± 50%), although only about 10% of marine emissions affect the land. Volcanic sources contribute 2 to 28 Tg/y of sulphur. Human activities, mainly the combustion of fossil fuels, produce 90–113 Tg/y. Because of the relatively short residence time of SO_2 and sulphate aerosols in the atmosphere, of the order of a few days, they are deposited on a regional scale. Since the bulk of the man-made sulphur emissions are produced over a small area of the planet's surface, human influence on the sulphur cycle varies widely from one region to another. On a global scale human activities produce about 60% of total sulphur emissions but over north-western Europe and eastern North America they contribute about 90 and 99% of total sulphur emissions respectively.

Gaseous SO_2 emissions have increased by almost a factor of thirty since 1860. On a global scale, the present level of combustion of fossil fuels accounts for 82% of SO_2 emissions, with 56% coming from coal. In industrialised areas, about 60%

of SO_2 is emitted from coal combustion. In the developed countries the majority of coal is burned for electricity generation. In eastern USA for example 57% of the total SO_2 emissions are released by electricity generation, with 60–70% of that injected in the 100–700 m height range. About 1–3% of total sulphur is released as directly emitted primary sulphates by coal-fired boilers as opposed to 5–9% from oil-fired boilers burning fuels of a similar sulphur content. The use of particle and SO_2 control systems often do not remove more than a small fraction of the primary sulphate in the flue gas. This is not of great importance since model estimates suggest that about 85% of surface sulphate concentrations originate from atmospheric SO_2 oxidation.

Chapter 20

Atmospheric chemistry of sulphur

Those reactions considered most significant for sulphate production under atmospheric conditions are outlined. Reactions which contribute to SO_2 oxidation under more specialised conditions, such as in stack gases, are also discussed. The formation of particles from gaseous SO_2 is considered together with the characteristics of particle sulphates.

In the atmosphere, sulphur compounds undergo reactions in the direction of increasing oxidation. Reduced sulphur compounds can be oxidised to SO_2, which in turn can be oxidised to sulphates. Sulphur is returned to the earth and oceans mainly in the form of sulphates and dry deposited SO_2. The mechanism of SO_2 oxidation determines the reaction rate and, together with the atmospheric residence time of SO_2, enables predictions to be made of the amount of sulphate likely to be formed from a given amount of SO_2 emitted, the influence of the concentration of SO_2 and other species, and partially even the form of the sulphate product. The relative rates of reaction of the various atmospheric oxidation pathways indicate which mechanisms are likely to be the most important in practice (Kellogg *et al.*, 1972; Urone and Schroeder, 1978).

For control purposes it is necessary to know the order of reaction of the dominant conversion mechanisms. This indicates which factors limit the progress of the reaction. First order reaction the reaction rate is proportional to the concentration of one of the reactants; in a second order reaction the rate is determined by the concentrations of two chemical species; in a zero order reaction the rate is independent of the concentration of the reactants. If the dominant reaction is first order with regard to SO_2 then a reduction in SO_2 emissions will lead to a proportional reduction in sulphate concentration. However, if the conversion rate is dominated by external factors, such as temperature, humidity, insolation or presence of oxidising agents and is only weakly dependent on SO_2, then a large reduction in SO_2 would produce only a marginal improvement in sulphate concentrations. In the latter case it is important to determine the dependence of SO_2 conversion on other pollutants, such as oxidising agents, trace metals or hydrocarbons, which could be more effectively controlled (Altshuller *et al.*, 1980).

In the atmosphere external conditions could also influence the conversion mechanism so that a homogeneous gas phase mechanism depending on the production of the hydroxyl (OH) radical by photochemical activity could be dominant on a warm sunny day while an aqueous phase reaction could be more important in conditions of high humidity such as fogs or when a power plant plume enters cloud or intersects a cooling tower plume. Such factors as plume age and travel time, the chemical characteristics of the air mass in which a plume is embedded and meteorological conditions all affect the rate of sulphate formation (Wilson, 1981).

Reduced sulphur compounds

Reduced sulphur species found in the atmosphere are hydrogen sulphide (H_2S), dimethyl sulphide ($(CH_3)_2S$), carbonyl sulphide (COS), carbon disulphide (CS_2), methyl mercaptan (CH_3SH) and dimethyl disulphide ($(CH_3)_2S_2$). COS, H_2S and CS_2 undergo oxidation reactions in the troposphere, initiated by the OH radical, with SO_2 as the primary product. Reaction between $(CH_3)_2S$ and the OH radical is assumed to take place but the oxidation products have not yet been determined (Bandy and Maroulis, 1980; Graedel, 1980; Sze and Ko, 1980b).

COS may make a considerable contribution to the stratosphere, by its oxidation to sulphate aerosol (Turco *et al.*, 1980). Sze and Ko (1980a), consider CS_2 an important precursor for COS in both stratosphere and troposphere.

Sze and Ko (1980b) calculated the residence times of H_2S as 1.5 days, $(CH_3)_2S$ as 0.75 days, COS as 160 days and CS_2 as 45 days. Bandy and Maroulis (1980) suggested a residence time of the order of one year for COS, extrapolated from the global atmospheric measurements on tropospheric aerosols and gases programme (GAMETAG) measurements, which was supported by Turco *et al.* (1980). Davis *et al.* (1982) gave approximate photochemical lifetimes of one day for $(CH_3)_2S$ and H_2S based on GAMETAG programme data. Khalil and Rasmussen (1984) estimated that the lifetime of COS was about 2 years and of CS_2 about 12 days.

Gas phase oxidation of SO_2

Figure 20.1, after Cox and Penkett (1983), summarises schematically the current thought on homogeneous oxidation mechanisms of sulphur compounds. The predominant oxidising agent involved is the OH radical. Sulphuric acid is the main sulphur species initially produced, which can be wholly or partially neutralised by atmospheric ammonia or other bases present on particles.

Under normal atmospheric conditions direct photo-oxidation of SO_2 is so slow as to be relatively unimportant. The uncatalysed reactions of SO_2 with O_2 and O_3 are also so slow as to be insignificant. However, for the special case of stack gases, at relatively high concentrations of impurities, a burst of SO_2 oxidation from reaction with excited oxygen transient species is anticipated during the early stages of dilution.

Indirect photo-oxidation is the major route for tropospheric oxidation. A variety of strongly oxidising radicals formed in the atmosphere from the photo-oxidation of ozone, nitrogen oxides and hydrocarbons were considered important in oxidation of SO_2. Recent research shows that the hydroperoxy radical (HO_2) reacts only very slowly with SO_2 under atmospheric conditions and that organic peroxy radicals (RO_2) could only become significant in highly polluted atmospheres. It is now generally accepted that the reaction of SO_2 with the hydroxyl radical (OH) is

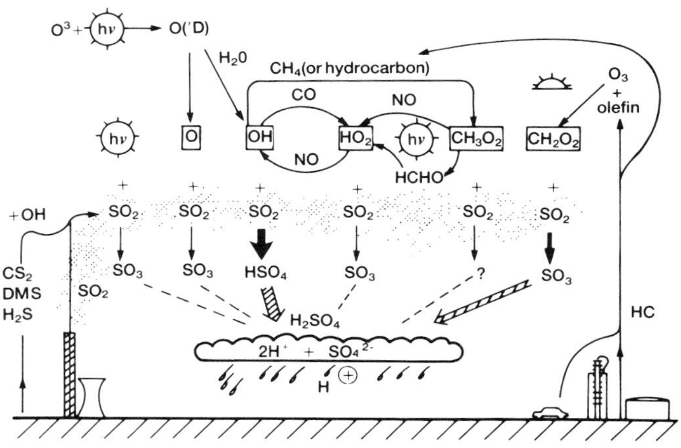

Figure 20.1 Gaseous phase oxidation mechanisms of sulphur compounds in the atmosphere (after Cox and Penkett, 1983)

the major rate-controlling step in the gas phase oxidation of SO_2 in the troposphere (Calvert and Stockwell, 1984; Cox and Penkett, 1983). The dominant pathway is:

$$OH + SO_2(+M) \rightarrow HSO_3^-(+M) \qquad (20.1)$$

This is a pressure dependent third body reaction where M is a molecule of oxygen, nitrogen or another neutral atmospheric gas which carries off the excess energy of the reaction to prevent immediate reversal of the reaction. At tropospheric pressures the reaction is in a transition region between third and second order kinetics, with the rate becoming less sensitive to the concentration of the third body M as the pressure increases (Calvert and Stockwell, 1984; Cox and Penkett, 1983).

The subsequent reactions of the transient HSO_3 radical are still uncertain although, under simulated atmospheric conditions, very efficient conversion of this radical to sulphuric acid aerosol has been observed. Calvert and Stockwell (1984) summarise the evidence for the participation of the HSO_3 radical in a variety of radical–radical and radical–molecule reactions but find an unambiguous choice between mechanisms impossible at present. It is important to know whether the overall reaction consumes OH radicals and is thus terminated when available OH radicals are used up. This could be important in plume chemistry where oxidation of the relatively high SO_2 levels found could be limited by a low level of OH radicals. The assumption of a reaction terminating mechanism in SO_2 transformation and transport models accentuates the non-linearity of response in acid formation to cutbacks in SO_2 emissions. Kinetic studies of the SO_2–OH reaction showed that the OH steady state concentration was insensitive to the concentration of SO_2 in the mixture. This was so even though up to one half of the reacting OH radicals were removed by reaction with SO_2 to form sulphuric acid aerosols. The termination of OH–HO_2 chain reactions by SO_2 was relatively unimportant under the experimental conditions, and would also be relatively unimportant for the usual atmospheric conditions. The most likely mechanism is:

$$HSO_3^- + O_2 \rightarrow SO_3^{2-} + HO_2 \qquad (20.2)$$

$$SO_3^{2-} + H_2O \rightarrow H_2SO_4 \qquad (20.3)$$

$$HO_2 + NO \rightarrow NO_2 + OH \qquad (20.4)$$

More than 80% of HSO_3 radicals regenerated HO_2 radicals (Stockwell and Calvert, 1983; Calvert and Stockwell, 1983).

The theoretical rates of SO_2 oxidation at 50% relative humidity are shown in Figure 20.2 as a function of time of day for a number of polluted air scenarios (Calvert and Stockwell, 1983). The conditions were simulated to match the diurnal variation of sunlight over two days near sea level for 40° N latitude in midsummer. In every case the maximum rates occur near the noon solar maximum. The rate for highly polluted urban atmospheres rich in NO*x* and hydrocarbons, shown as cases A and B, is a maximum of 1.5%/h. This rises to 3.4%/h as the air mass is diluted by a factor of ten in case F, representing the moderately polluted, non-urban troposphere, and to 5.4%/h maximum with dilution by a factor of 100 in

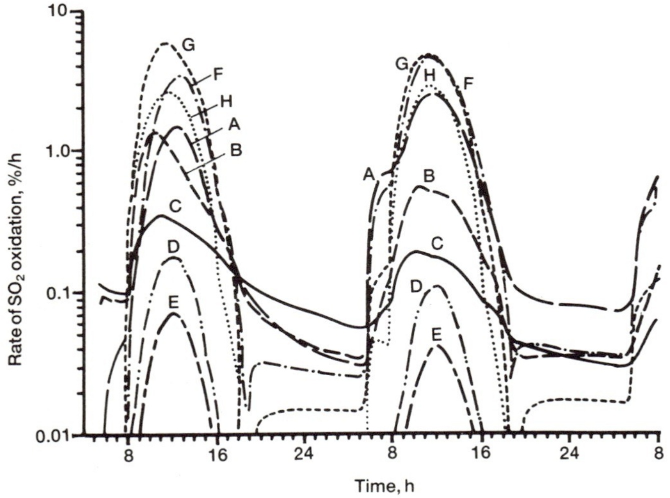

Figure 20.2 Theoretical rates of sulphur dioxide oxidation to sulphuric acid with time for the various two-day polluted air scenarios A–H at 50% relative humidity (after Calvert and Stockwell, 1983)

case G, representing the mildly polluted non-urban case. Further dilution by another factor of ten in case H, representative of the lightly polluted non-urban troposphere, results in a decreased rate of SO_2 oxidation of 2.6%/h maximum. Cases D and E, representing stack gas plumes in the early stages of dilution, show a maximum rate of 0.2%/h. This is similar to the maximum rate of 0.3%/h seen in case C, representing a hydrocarbon rich but low NO*x*, polluted air mass. In the less polluted air masses shown by cases F, G and H, which are the type commonly encountered during long range transport of pollutants, increasing relative humidity increased the maximum rates of SO_2 oxidation from 3.4, 5.4 and 2.6%/h for 50%

relative humidity to 5.4, 6.1 and 3.6%/h respectively for 100% relative humidity. The theoretical average rates of SO_2 conversion to sulphuric acid for a clear 24 hour period in the summer at 50% relative humidity are 16, 24 and 13%/24 h period respectively for cases F, G and H. At 100% relative humidity these rates increase to 23, 30 and 17%/24 h period respectively. These model results are in good agreement with field measurements of SO_2 oxidation presented in Chapter 21.

Since changes in available UV solar intensity with latitude and season result in large variations in OH radical concentrations, the gas phase oxidation of SO_2 will also vary widely. Gas phase SO_2 oxidation producing sulphate is important all year below latitude 35° N but at higher latitudes (above 45° N) it appears significant only in summer. At other times of the year heterogeneous reactions must be responsible for the observed rates of oxidation (Altshuller, 1979).

Other oxidants which undergo reaction with SO_2 are the oxides of nitrogen NO_2, NO_3 and N_2O_5. NO_3 and N_2O_5 are important products in photochemical smog. In a smog chamber study the rate of new particle formation from photo-oxidation in the SO_2/NO*x*/propylene system (compounds of the type found in photochemical smog) was observed to increase with the rate of SO_2 consumption and to decrease as the amount of pre-existing aerosol increased (McMurry and Friedlander, 1979). From current data these reactions are too slow to make a significant contribution to SO_2 oxidation in the lower troposphere (Calvert and Stockwell, 1984; Eggleton and Cox, 1978).

The oxidation reactions of SO_2 with the methoxy radical, the tert-butylperoxy radical and the acetylperoxy radical (including peroxyacetyl nitrate (PAN)) are considered unimportant. The ozone–alkene reaction has been studied in detail and it is thought that oxidation involves formation of a reactive intermediate which rapidly oxidises SO_2, even in the dark. This reaction is not thought to be of great importance in the general atmosphere but may be significant in areas where a localised concentration of reactive hydrocarbons is found, for example near oil refineries (Calvert *et al.*, 1978; Penkett, 1979; Calvert and Stockwell, 1984). Modelling studies of the UK by Derwent and Hov (1980) show that natural hydrocarbons can account for the formation of only 10% of sulphate aerosol.

Miller (1978) concluded, as a result of smog chamber studies which attempted to simulate conditions in polluted atmospheres containing NO*x*, NMHC (non-methane hydrocarbons) and SO_2, that the rate of SO_2 oxidation to sulphate aerosol was directly proportional to the SO_2 concentration. Since NO*x* suppressed or delayed SO_2 oxidation and NMHC promoted the reaction, the ratio of NO*x*/NMHC, rather than their absolute concentrations, was the significant factor. A model produced by Atkinson *et al.* (1982) for polluted atmospheres, with the OH radical the dominant oxidant in SO_2 conversion, fitted this data quite well. These results indicate that daytime conversions of SO_2 in urban areas may be limited to 10–20%. Diluted and filtered coal-fired power plant flue gas and air mixtures of similar composition (containing NO*x*) were used by Luria *et al.* (1982) in a smog chamber study to show that the rate of SO_2 conversion in clean air is slower than the rate measured in the atmosphere, in filtered stack gas or in synthetic mixtures simulating stack gas.

Theoretical and laboratory studies reviewed by Freiberg (1983) suggest that gas phase oxidation of SO_2 is positively correlated with temperature but the correlation is weaker in clean air than in polluted air. Oxidation of SO_2 is sensitive to changes in relative humidity, especially at low relative humidities, but the dependence on humidity is weaker in polluted air containing NMHC than in clean air.

Aqueous phase oxidation of SO_2

The atmosphere is a heterogeneous system with the gas phase loaded with wet and dry particles. Oxidation occurs via a sequence of physical and chemical stages. These are gas phase diffusion, mass transfer at the gas/water interface, hydrolysis and ionisation of dissolved S(IV) species, liquid phase diffusion and oxidation. Any of these steps may be the limiting factor which controls the overall rate of reaction. The rate limiting step may also vary with external conditions such as pH and temperature (Schwartz and Freiberg, 1981).

In an atmosphere containing gaseous SO_2, SO_2 interacts with water to produce three species: the sulphite ion (SO_3^{2-}), bisulphite ion (HSO_3^-) and physically dissolved, or hydrated, SO_2($SO_2 \cdot H_2O$) in equilibrium with SO_2 in the gas phase. The fraction in each of the three species is known and the equilibrium is shifted from sulphite to bisulphite and dissolved SO_2 as pH decreases. In the pH region most relevant for atmospheric chemistry, pH 3–6, the predominant S(IV) species is the bisulphite ion. All these species can be oxidised to sulphate (Gravenhorst *et al.*, 1980).

The importance of uncatalysed oxidation of SO_2 by O_2 is debatable. Beilke and Gravenhorst (1978) reviewed this reaction, based on the assumption that the sulphite ion (SO_3^{2-}) alone was acting as the oxygen carrier. This is believed to be a zero order reaction, although pseudo first-order rate constants are available in the literature to provide an indication of rate of formation of sulphate ions. They believe this mechanism to be unimportant unless droplet pH is higher than about six. Observed cloud water pH values range from 3.7 to 6.8 with a mean of 5.1 ± 0.9. Cloud water pH levels of 5 to 6 are associated with the non-urban environment whereas pH values below 5 tend to be associated with polluted air (Hegg and Hobbs, 1981, 1979).

Hegg and Hobbs (1978) however believed that oxidation by molecular oxygen, via a free radical chain involving sulphite and bisulphite radicals was of importance in cloud droplets. Hegg and Hobbs (1979) used the Easter–Hobbs interactive cloud chemistry model to predict cloud water pH and the amount of sulphate formed at these levels using rate constants from the literature. The model predicted significant amounts of sulphate formed in the pH range 5–6. However, this could reflect a low level catalysis of the reaction rather than a strictly uncatalysed reaction.

Oxidation in droplets by the strong oxidising agents O_3 and hydrogen peroxide (H_2O_2) are possible mechanisms. Both O_3 and H_2O_2 are products of photochemical reactions, often of pollutants. In a laboratory study, Maahs (1983a,b) found that SO_2 oxidation rates by O_3 were competitive with those by H_2O_2 at pH levels above about 4.5 and that this mechanism dominated at pH levels above about 5.5. Since this pH range is typical for non-urban tropospheric cloud water, O_3 is a potentially important contributor to sulphate production in the non-urban troposphere. The reaction does not proceed by a free radical mechanism but by direct ozone attack catalysed by the hydroxyl ion. The measurements of Hegg and Hobbs (1982) in natural clouds provide support for the occurrence of this reaction in the atmosphere at rates similar to those derived in the laboratory (Martin, 1983). H_2O_2 reacts with the HSO_3^- ion, produced by dissolved SO_2, in the following way:

$$HSO_3^- + H_2O_2 \rightleftharpoons \begin{matrix} {}^{-}O \diagdown \\ \quad S\text{–}OOH + H_2O \\ O \diagup \end{matrix} \qquad (20.5)$$

$$\begin{matrix} {}^{-}O \\ \quad\diagdown \\ \quad\quad S\text{–}OOH + H^+ \rightarrow H_2SO_4 \\ \quad\diagup \\ O \end{matrix} \qquad (20.6)$$

With its extremely high solubility in water, the concentration of H_2O_2 in cloud droplets could be as high as the dissolved oxygen concentration. An H_2O_2 concentration of 1 ppb leads to a calculated oxidation rate for SO_2 of 700%/h by H_2O_2 (Penkett, 1985, 1979; Altshuller *et al.*, 1983; Penkett *et al.*, 1979a). Measurements of gaseous H_2O_2 in the non-urban troposphere of eastern USA during the autumn showed that sufficient quantities were available for H_2O_2 to be major oxidant for S(IV) in cloud water. Gaseous H_2O_2 should be more abundant in summer (Calvert *et al.*, 1985). However, Daum *et al.* (1984) found that concentrations of H_2O_2 in cloud water exhibited a strong inverse correlation with SO_2 in cloud interstitial air. Appreciable concentrations of SO_2 and H_2O_2 only rarely coexisted, for the most part only one or the other being present above the limit of detection. This suggests that either of these species can be a limiting reagent for in-cloud SO_2 oxidation.

Penkett *et al.* (1979a) investigated oxidation reactions with O_2, O_3 and H_2O_2 in the laboratory, and used the results to extrapolate to atmospheric conditions. The reaction rate with H_2O_2 was almost independent of pH, due to its positive catalysis by hydrogen ions, unlike all other aqueous phase SO_2 oxidation reactions. The H_2O_2 and O_3 reactions were first order. The reactions of these three oxidising agents with SO_2 proceed much faster in the aqueous phase than in the gaseous phase, so the liquid phase processes may be thought more important than the gas phase processes even though the extent of the liquid phase in the cloud is only one millionth that of the gas phase.

Chameides and Davis (1982) demonstrated theoretically that gaseous HO_2 and OH radicals, generated within cloud by gas phase photochemical reactions, may be dissolved in droplets within the cloud. The dissolved HO_2 radicals are transformed to H_2O_2, forming a major daytime source of H_2O_2 in cloud water. Hydroxyl radicals, dissolved by cloud droplets, also react with HSO_3^- and SO_3^{2-} ions forming sulphates. This shows that H_2O_2 can be rapidly regenerated in both the gas and aqueous phases.

Martin and Damschen (1981) agreed that SO_2 oxidation with H_2O_2 was independent of pH between pH 1 to 5. They calculated that for 1 ppb H_2O_2 in a typical cloud containing 5 ppb SO_2 the conversion rate of SO_2 to sulphate would be 3% per minute but cautioned that the rate may be unrealistically high for high H_2O_2 concentrations. Their laboratory studies of this reaction showed that the reaction rate did not respond to catalysis or inhibition by iron (Fe) or manganese (Mn) ions, hydroquinone, toluene, pinene or hexene.

Hegg and Hobbs (1981) compared measurements of pH and ionic content of water collected in clouds over rural and urban areas in the USA with results from the model presented by Hegg and Hobbs (1979). Reasonable agreement was found which led the authors to conclude that measured sulphate production could be explained by a combination of O_2, O_3 and H_2O_2 oxidation in cloud droplets, based on an estimated 0.3 ppb concentration for H_2O_2 which was not directly measured. Their investigations indicated that rates measured over short time periods could not be extrapolated to longer time scales since in some cases the sulphate production rate fell sharply with increasing time, possibly due to pH dependence of some part of the reaction mechanism.

Shaw and Rodhe (1982) compared observations of SO_2 oxidation during long range transport with predictions from a model using gas phase oxidation by OH radicals and liquid phase oxidation by H_2O_2, and concluded that even in winter time an additional rate of SO_2 oxidation of about ten times that of OH and H_2O_2 combined was needed to explain the observations. Liquid phase oxidation by O_3 was examined but considered an unlikely mechanism unless cloud water pH stayed above a lower limit ranging from 4.6 to 5.5.

The reactions with strong oxidants are the most important pathways in remote areas and during long range transport but, in urban fog conditions where the concentration of catalysts and metal ions is likely to be high, oxidation by O_2 in the presence of catalysts may play an important role (Beilke and Gravenhorst, 1978). The most widely studied promoters have been NH_3 and transition metal ions of the fourth period, especially Mn and Fe.

Ammonia is not itself a catalyst for SO_2 oxidation. It enhances the reaction rate by acting as a buffer to keep solution pH high, preventing termination of the reaction by high hydrogen ion concentrations. There is the possibility that the presence of ammonia may retard the reaction when the ammonium ion forms complexes with a heavy metal catalyst, as is thought to happen with cobalt (Hegg and Hobbs, 1978; Overton *et al.*, 1979).

Beilke and Gravenhorst (1978) and Hegg and Hobbs (1978) reviewed extensive earlier work on the effect of Mn and Fe catalysts on oxidation reactions by oxygen, including evidence of a synergistic effect when both metal ions were present. Most of the mechanisms suggested involve free radical chains similar to those found for the uncatalysed reactions, but the chain initiating mechanism is uncertain, possibly involving the formation of a metal sulphite or bisulphite complex. The mechanism is thought to be very complex and to depend on the concentrations and chemical forms of both reactants and catalysts.

Penkett *et al.* (1979b), using stored rainwater samples, found that the oxidation rate was directly proportional to the Mn content (with Fe having no effect) and depended upon the total dissolved SO_2 concentration. Lindberg (1981) also found a significant correlation between the concentrations of sulphate and Mn ions in rain samples, but concluded that this was at least partially due to a co-dependence on rainfall amount.

Kaplan *et al.* (1981) found only a weak dependence of sulphate production rate on Mn concentration, but consider that this may be an initial artifact of their novel experimental method. In a fog chamber experiment Anderson *et al.* (1979) demonstrated a dependence of oxidation rate on the concentration of trace amounts of heavy metals (catalysts used included hydrated compounds of vanadium (V), Mn and Fe) but did not attempt to ascertain which metals were active. They also postulated a dependence on SO_2 concentration, but found no synergistic effect between irradiation and trace concentrations of heavy metals. A cloud chamber study by Steele *et al.* (1981) showed an uncatalysed conversion rate range of 0.09–4.4%/h in cloud drops comparable in pH with atmospheric clouds. The presence of Mn in the cloud water produced oxidation rates over two orders of magnitude higher than that of the uncatalysed reaction at pH values between 4 and 5. A substantial dependence of the conversion rate on the pH of the cloud water was noted. Holt *et al.* (1981) used an oxygen isotopic labelling method to determine the mechanism of sulphate formation from samples of sulphate in precipitation. The oxygen isotope relationship indicated the origin of oxygen atoms in the compound. The Fe catalysed reaction showed the closest relationship to the values found in field tests, although none of the catalyst mechanisms studied (Fe, H_2O_2 and charcoal) either singly or in combination, could account for the observed isotopy of precipitation sulphates. Martin *et al.* (1979) also found evidence of Fe catalysis and, under the experimental conditions used, found a synergistic effect with Mn.

Harrison *et al.* (1982) pointed out the lack of reproducibility of aqueous measurements between different studies and between different tests by the same investigators. They reported no effect of Fe ions on the oxidation rate of sulphites by ozone, except in the region of pH 4.5. Mn ions did exhibit a catalytic effect on this system with rate maxima near pH 4.5. No effect greater than additive was found when both Mn and Fe ions were present, in contrast to findings reported for oxidation by oxygen. The authors consider that the reaction proceeded by the formation of ligand complexes between metal and bisulphite ions, with hydroxyl and possibly even chloride ions taking part.

Catalysis by the trace metals Mn and Fe has been shown to be important in the fogwater of polluted atmospheres. Under these conditions Mn and Fe catalysis is the main contributor to SO_2 oxidation at night. This mechanism may be more important than oxidation by H_2O_2 in polluted atmospheres. This fog droplet chemistry is similar to that responsible for producing the great urban smoke fogs previously prevalent in Europe. The effectiveness of Mn and Fe catalysts is greatly reduced as their concentration in the aqueous phase decreases so their importance in cloud water remains debatable (Martin, 1984; Hoffmann and Jacob, 1984; Cox and Penkett, 1983).

Fassina and Lazzarini (1981), studying the Venetian environment, consider that the main oxidation mechanism in this area is due to the catalytic action of trace metals dissolved in water droplets but ascribe a primary importance to the presence of NH_3. The percentages of SO_2 oxidised were higher in summer when NH_3 concentrations were high and SO_2 low and lower in winter with low NH_3 and high SO_2 concentrations.

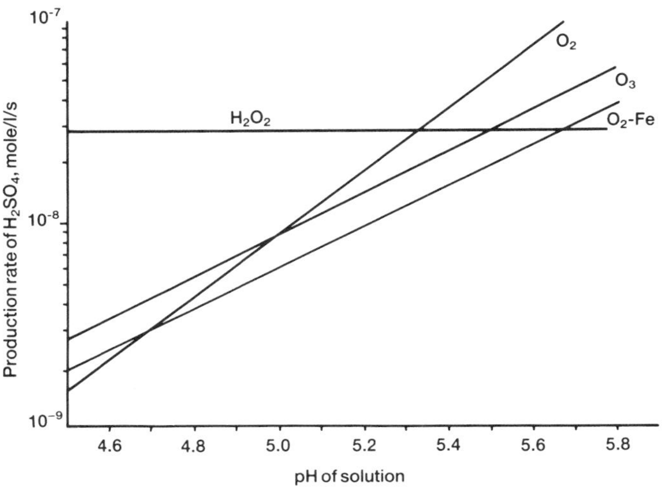

Figure 20.3 Sulphuric acid production rates for various oxidants over the pH range of mid-level clouds (after Altshuller *et al.*, 1983a)

Nitrogen oxides (NO*x*) have been considered as oxidants for SO_2. Martin *et al.* (1981) studied the three reactive species likely to be present in aqueous atmospheric aerosols, which are typically acidic. They found that only the nitrite ion (NO_2^-) or nitrous acid in solution (HNO_2) were important as they reacted rapidly with S(IV) species, with the reaction uncatalysed by Fe^{3+}, Mn^{2+} or VO^{2+} ions. However this reaction is not a major pathway in the general atmosphere but could be important when the lifetime of fog or clouds is long and the droplet pH low. Observations from Florida, USA (Leslie, 1980; Winchester, 1980b), suggest that liquid phase SO_2 oxidation in polluted air occurs at a rate controlled by the supply of available water, which is proportional to the ambient vapour pressure of water. Oxidation is expected to begin on the surfaces of pre-existing particles, producing a coating of H_2SO_4 by surface catalysed reactions. Further oxidation could take place in the coating solution leading to solution droplets containing the nucleating particles.

Figure 20.3, after Altshuller *et al.* (1983a), shows a comparison of the theoretical production rates for the various oxidants over the pH range of typical mid-level clouds (~800 mb pressure level, temperature ~5°C, liquid water content ~1 g/m^3) situated in a moderately industrialised region. The H_2O_2 reaction is the single most important reaction over much of the pH range. However the figure shows that all of the oxidants can contribute significantly to oxidation of SO_2 above a pH of about 5.2. Rate studies and field data also indicated that no one oxidant dominated sulphuric acid production in all atmospheric situations.

Theoretical calculations and laboratory observation show that the effects of temperature and humidity on SO_2 oxidation rate vary considerably from one mechanism to another, so a conversion path can become more or less significant relative to other paths as temperature and relative humidity vary. This suggests that temperature and relative humidity can be major factors in the temporal (seasonal, diurnal) and spatial (coastal, desert) variation of the rate and yield of SO_2 oxidation. In general, the rate of SO_2 conversion in aqueous droplets is negatively correlated with temperature and positively correlated with relative humidity, especially at high relative humidity. Aqueous phase reactions are more sensitive to temperature than gas phase reactions (Freiburg, 1983).

Hegg (1985) used the SO_2 oxidation rate expression of Maahs (1983) and a global cloud model (to calculate the tropospheric liquid water burden) to estimate the extent of aqueous phase SO_2 oxidation. For the troposphere as a whole the SO_2 flux through the aqueous phase was estimated to be about 10–15 times the flux through gas phase mechanisms. Even with the most conservative assumptions, aqueous oxidation of SO_2 was at least as important a sink for SO_2 as was gas phase oxidation. However the major source areas of man-made emissions are highly atypical of the troposphere as a whole.

Gas/particle interactions

Carbon soot and other types of graphitic particles can catalyse SO_2 oxidation in the presence of liquid water but the sulphate production rate has a complex dependence on the concentration of S(IV) species and is not very sensitive to change in SO_2 concentration. The reaction order appears to change with environmental conditions. Siliceous coal fly ash samples containing a low percentage of carbon have also been found to exhibit catalytic behaviour (Benner *et al.*, 1982; Brodzinski *et al.*, 1980; Benner *et al.*, 1980; Novakov, 1979). Incomplete combustion is a source of primary gas phase oxidants, primarily H_2O_2 and organic peroxides, as well as of soot particles. Sulphates are formed from SO_2 oxidation by H_2O_2, following 1:1 stoichiometry, in the conditions following combustion. Catalysis of SO_2 oxidation by soot could take place by a combination of surface catalysis and solution oxidation by H_2O_2 co-emitted with the soot (Benner *et al.*, 1985). Tartarelli *et al.* (1978) characterised the sorption of SO_2 on carbonaceous particles and Liberti *et al.* (1978) examined SO_2 sorption on various dusts including urban particles and stack emissions from industrial plants. Both groups found that relative humidity influenced sorption behaviour. Haury *et al.* (1978) found that oxidation on siliceous fly ash particles was more dependent on SO_2 concentration than on relative humidity, with the reaction being first order with respect to SO_2.

Schryer *et al.* (1980) observed that the presence of NO_2 greatly enhanced the chemisorption of SO_2 on carbon particles. NO_2 also accelerated SO_3 formation (Britton and Clarke, 1980).

At high NO_2 concentrations the final amount of SO_3 was depressed, due to poisoning of the active sites of the carbon by NO_3 formation. This reaction may be important in low temperature flue gases or plumes. Both NO_2 and O_3 enhanced sorption and desorption and increased oxidation on carbon black at concentrations greater than about 0.07 ppm up to 20–40 ppm of SO_2 compared with air. NO_2 alone was a much less efficient oxidant. At low concentrations little, if any, enhancement was found with any oxidant when compared with equivalent SO_2 in air. This reaction, while probably not significant through most of the atmosphere, could be highly significant in power plant flue gases and urban air where SO_2 is found in close association with NO*x* and particulate matter. An activated surface-catalysed reaction could be important in the plumes of coal fired plants, even in the absence of carbonaceous soot (Cofer *et al.*, 1981; Brodzinsky *et al.*, 1980).

Nucleation

Whatever the chemical processes involved, gas–aerosol interactions are subject to certain physical constraints. These restrictions are related to the thermodynamic stability of condensed particles of small diameter, particle nucleation and condensation processes and gas phase diffusion limited rates of absorption or adsorption on particles. Nucleation is a phase transformation process where water molecules in the vapour phase condense onto aerosol particles forming droplets. These particles are known as cloud condensation nuclei (CCN). Aerosol particles may also nucleate ice crystals and these are known as ice nuclei (IN). Condensed particles must be close to or at equilibrium with the surrounding medium to exist in air for any length of time so the partial pressure of condensed species on particles must be less than or equal to the saturation vapour pressure at atmospheric temperature for stability. This condition can normally be met by sulphate salts and H_2SO_4 but severely constrains the ability of nitric acid (HNO_3) to exist either as a pure compound or as an acid diluted in water. Aerosols may be formed by condensation of supersaturated vapour or chemical reaction leading to the spontaneous formation of new particles, or by condensation, absorption or reaction on or in existing particles. H_2SO_4 can undergo nucleation at atmospheric concentrations to produce new particles in the absence of nuclei (Hidy and Burton, 1980; Hales, 1982).

Middleton and Kiang (1978) review a number of studies which consider, both theoretically and experimentally, the nucleation of H_2SO_4 particles. New particle formation in the smog chamber depends on H_2SO_4 vapour concentration, relative humidity and temperature and is very sensitive to supersaturation (the ratio of H_2SO_4 vapour pressure to the solution vapour pressure) and surface tension. Uncertainties in product measurement techniques in many of the parameters required by the available models can affect calculated nucleation rates by several orders of magnitude. The more recent H_2SO_4 vapour pressure data of Roedel used by Middleton (1980) (higher than previously assumed but in good agreement with the results of Chu and Morrison (1980)) give an increase in both the calculated nucleation rates and growth on pre-existing particles. This enhances the perceived importance of sulphur gas to particle conversion as a source of submicron particles.

Since a large proportion of the atmospheric aerosol is in the form of sulphate salts, H_2SO_4 particles must react with other impurities in the air, chiefly NH_3. The presence of NH_3 can also give rise to more complex molecules such as nitrosyl sulphate (SO_4HNO), which has been found in atmospheric particles (Boulaud *et al.*, 1978).

Scott and Cattell (1979) measured the vapour pressures of ammonium sulphates and found that under atmospheric conditions in the troposphere binary nucleation of $(NH_4)_2SO_4$ should be preferred to nucleation of H_2SO_4 solution drops in all instances of water under saturation, provided there is any NH_3 present since $(NH_4)_2SO_4$ exhibits the lower vapour pressure. At stratospheric temperatures ($-50°C$) there is a greater tendency for nucleation of acid aerosols, chiefly NH_4HSO_4.

Depending on the extent of neutralisation and relative humidity, sulphate aerosols may exist in the atmosphere in a completely dissolved, partially dissolved or crystalline state as either pure $(NH_4)_2SO_4$ or mixed salts such as NH_4HSO_4 and $(NH_4)_3H(SO_4)_2$. Tang (1980) found that the effect of HNO_3 on the formation of mixed ammonium nitrate and sulphate aerosols is dependent on the atmospheric HNO_3 concentration and the relative humidity rather than the H_2SO_4 concentration.

Atmospheric sulphates are hygroscopic substances and therefore have the potential to act as CCN and so influence cloud and precipitation processes. Ono and Ohtani (1980) demonstrated that atmospheric sulphates are active as CCN at supersaturations of less than 1%, which is representative of clouds in the troposphere.

Particle size

Whitby (1978) presents a wide ranging review of the physical characteristics and size distributions of aerosols, based on the trimodel model shown in Figure 20.4, adapted from Whitby (1978). The two size ranges usually observed are fine particles, with a particle diameter $<2\,\mu m$, and coarse particles, with a particle diameter $>2\,\mu m$. Coarse particles are readily removed by rainfall and sedimentation and so tend to remain in the atmosphere or for only a few hours. The fine particle mode is formed by chemical processes and is usually chemically different from the mechanically generated coarse particle mode. Formation, transformation,

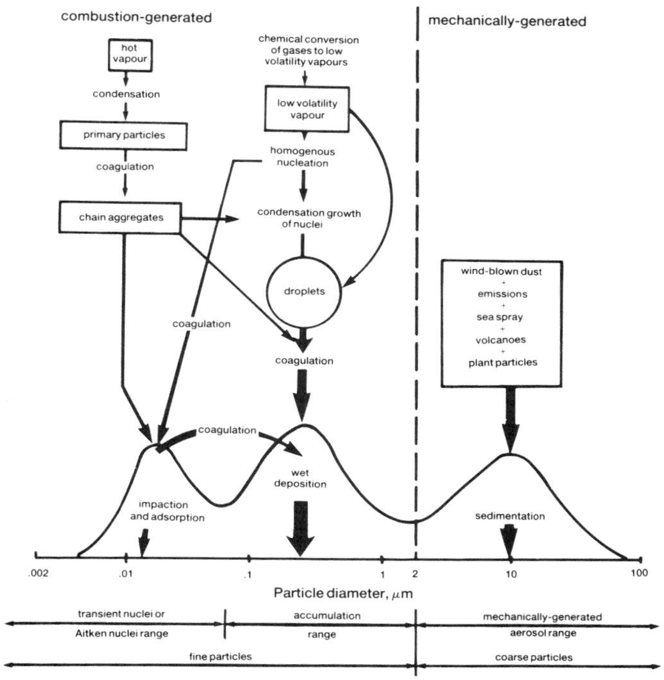

Figure 20.4 An atmospheric surface area distribution showing the three modes, main source of mass for each mode, the principal processes involved in inserting mass into each mode and the principal removal mechanisms (after Whitby, 1978)

transport and removal of the two modes are normally quite independent. Almost all the sulphur generated by fossil fuel combustion and found in atmospheric aerosols is in the fine particle fraction. The fine particle range can be divided further by the size and behaviour of the particles. Transient nuclei mode particles (also known as Aitken nuclei) have a particle diameter $<0.08\,\mu m$ and are usually the recent products of combustion processes formed by condensation from the gaseous phase. They disappear within minutes by coagulation with each other and with slightly larger particles.

Accumulation mode particles, of particle diameter $0.08\,\mu m$ to $2\,\mu m$, grow by condensation and coagulation onto existing nuclei and have an airborne lifetime of some days. In the atmosphere the accumulation mode is usually present in background air and is very persistent since the particles are poorly removed by both settling and meteorological processes and coagulate only slowly (Lodge *et al.*, 1981). Growth by coagulation and condensation of the transient nuclei mode always occurs in the presence of the accumulation mode. Aged particles are those that have had enough time to coagulate and settle for at least a few hours, removing most of the transient nuclei mode. Coagulation of the transient nuclei mode with the accumulation mode exceeds that of the transient nuclei mode with itself. Nucleation is general throughout the atmosphere with gas to particle conversion of sulphur providing most of the nuclei and with much of the sulphate formed ending up in the accumulation mode. This mode is removed only slowly from the atmosphere and so may be transported for long distances (Whitby, 1978).

Knowledge of the relationship between the growth rates of individual particles and aerosol size distributions has been used to draw conclusions on the chemical conversion mechanism for aerosol formation since theory predicts that secondary aerosol mass formed by homogeneous chemical reactions accumulates in smaller particles than that formed heterogeneously (McMurry, 1980). Analysis of field study and smog chamber data showed that sulphate aerosol formation during the daytime in the Great Smoky Mountains, Tennessee, at relative humidities of 65–95%, was consistent with liquid phase reactions while growth of smog chamber aerosols at relative humidities around 30% was consistent with the condensation of low vapour pressure species formed by gas phase reactions (McMurry and Wilson, 1982). In the St Louis urban plume (relative humidity 48–83%), 75% of the secondary aerosol volume formation was due to gas phase reactions and 25% to liquid phase reactions (Whitby, 1980).

Aerosol properties vary with particle size and composition. In the atmosphere the size range 0.1 to 30 μm is predominant for the volume and mass of aerosol present. Although the coarse particles are not very numerous individually they have a much greater mass than fine particles and so can make up as much as 90% of the total suspended particulate matter. Of this total aerosol mass, originating from both natural and anthropogenic sources, 30% is inorganic and water soluble, consisting of the following ions: SO_4^{2-}, NO_3^-, Cl^-, NH_4^+, Ca^{2+}, Mg^{2+}, Na^+ and K^+ (Jaenicke, 1980a). The majority of studies published measure single ions while many properties of sulphates, such as water solubility, particle size distribution, acidity and toxicity depend on the molecular composition. However, it is known that the most common compound in the fine particle range of the continental aerosol is $(NH_4)_2SO_4$ with an estimated residence time in the atmosphere of five days (Jaenicke, 1978; Blokker, 1978).

Summary and comments

Gaseous reduced sulphur compounds are either oxidised to SO_2 in the troposphere, by a mechanism involving the OH radical, or passed into the stratosphere where they undergo photolytic reactions to form H_2SO_4 particles. The probable average residence time of SO_2 in the lower troposphere is about 1.5 days. Removal takes place by dry deposition or by a variety of chemical reactions all of which involve the oxidation of SO_2 to H_2SO_4 and the incorporation of the acid into cloud droplets and aerosols. Although there are many uncertainties in the transformation process, the dominant mechanism in gas phase oxidation is undoubtedly reaction with OH radicals which are generated in the atmosphere by photochemical reactions. In the aqueous phase, the reaction with H_2O_2 is the single most important reaction over much of the pH range. However O_3 oxidation rates may be competitive with H_2O_2 rates above about pH 4.5 and dominate at pH levels above 5.5. Other mechanisms also contribute above a pH of about 5.2 and, in the atmosphere, if one path is not available the reaction will go by another route. Gas phase HO_2 and OH radicals and O_3 are formed by photochemical reactions and may then dissolve in cloudwater, regenerating the aqueous phase reactants. Aqueous phase reactions are not entirely independent of gas phase photochemistry. Dependence on solar radiation to produce oxidant radicals leads to variations in sulphate formation rate with season and latitude.

Although these are the dominant reactions in the background atmosphere other mechanisms can contribute in more polluted areas. Catalytic oxidation may occur in urban fog conditions where the concentration of metal ions is likely to be high. The heavy metals Mn and Fe are thought to promote SO_2 oxidation. The presence of NH_3 can affect aqueous phase reactions by increasing the pH of the solution. NO*x* could be a significant oxidant in power plant flue gases and plumes, where the concentrations reach high levels. Reactions with carbonaceous and siliceous particles may make significant contributions to sulphate production but only under special circumstances which have not been well defined. It has not yet been conclusively determined whether the level of oxidants in the atmosphere is a limiting factor in the oxidation of SO_2.

Sulphates are active as cloud condensation nuclei in the atmosphere and so can influence cloud formation and precipitation processes. After gas phase oxidation combustion generated sulphates form transient nuclei which quickly grow into accumulation mode particles (diameter $<2\,\mu m$). These are removed only slowly from the atmosphere and so may be transported long distances. Sulphates dominate the accumulation mode and the most common compound found is $(NH_4)_2SO_4$.

Chapter 21

Sulphate formation in plumes

The evidence is presented for sulphate formation from SO_2 in plumes, to determine the conditions under which it takes place, the rate and extent of SO_2 transformation and how far conversion can be traced in the life of a plume. Sulphate formation in coal-fired power plant plumes is compared with that seen in other point source and urban plumes to determine whether the circumstances of combustion affect the conversion rate. The effect of atmospheric conditions is considered to determine how environmental influences affect sulphate formation.

Plume dispersion

Sulphur from coal, mainly in the form of SO_2, is emitted from large stationary sources, through only a small number of stacks for any source, in the form of a plume. Tall stacks utilise the dispersion capability of the atmosphere to meet local SO_2 emission standards, leading to an effective dilution, but not removal, of SO_x. The behaviour of SO_2 in a plume is dependent on both the meteorological processes affecting the parcel of air into which it is injected and the chemical reactions occurring within it. During the day the plume is normally emitted into the mixing layer, the layer nearest the ground in which the air is uniformly mixed, where turbulent eddies transport pollutants through the depth of the unstable layer. The mixing layer, or atmospheric boundary layer, is the region of the atmosphere where the direct effect of the ground is detectable. When the boundary layer is stable, or not subject to turbulence, it ceases to be a mixing layer (Smith and Hunt, 1978).

The depth of the boundary layer varies from a few metres during windless nights to several thousand metres in the middle of the afternoon, due to various interactions between the atmosphere and the surface which are the sources and sinks of turbulent energy.

This diurnal cycle results in the observed diurnal changes which occur in the wind and temperature profiles of the boundary layer. At night the boundary layer is very shallow, often associated with a low level inversion, but will begin to deepen and start to become mixed a couple of hours after dawn as the sun heats the ground. It reaches maximum in the early afternoon and then remains constant or decreases slightly until just before sunset when the stability of the layer may change quickly and a stable layer form close to the ground. These nocturnal inversions are common throughout the world, occurring, for example, on about three quarters of

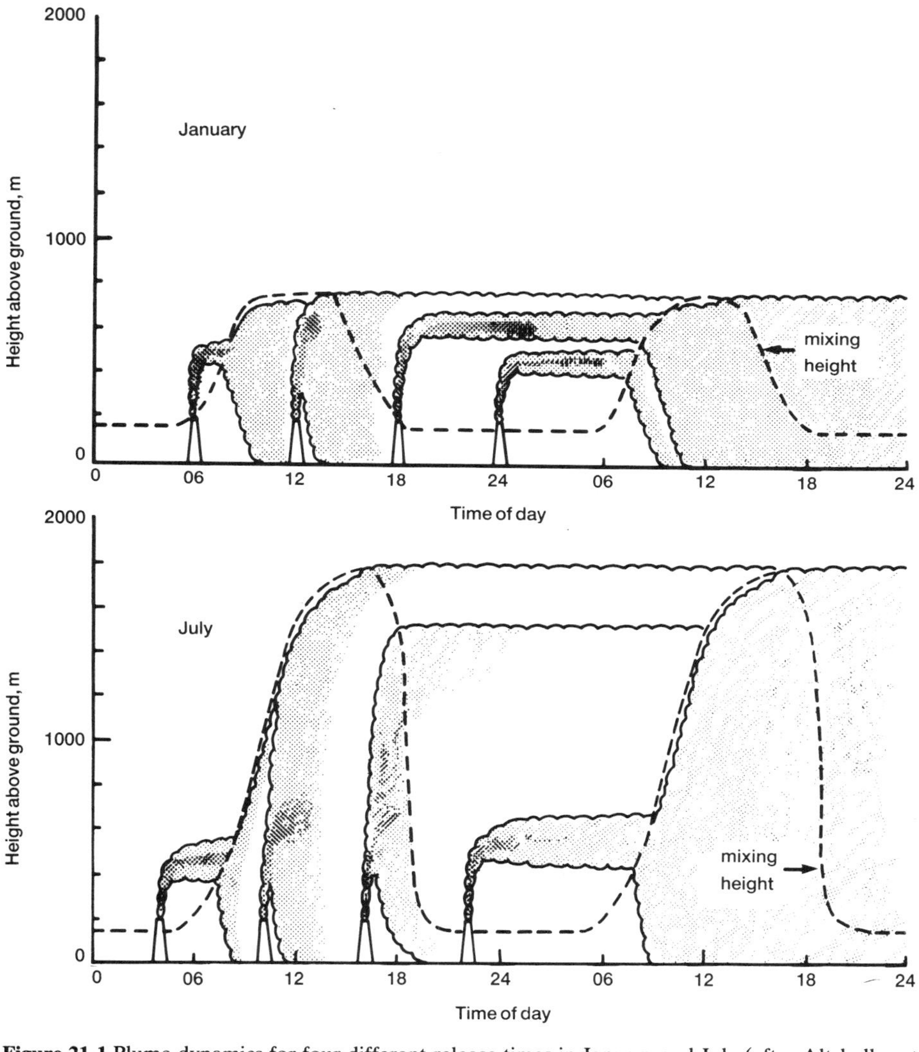

Figure 21.1 Plume dynamics for four different release times in January and July (after Altshuller *et al.*, 1983a)

the nights in western North America and on about half the nights in eastern North America. The height of the mixing layer also varies with season, as it is partially driven by temperature related processes. Peak daytime mixing heights average about 1800 m in July and only about 700 m in January, in the mid-west of the USA. Figure 21.1, after Altshuller *et al.* (1983a), illustrates the plume dynamics for different plume release times during 48 h of transport in January and July, based on data from the plume of the Labadie power plant, Missouri, USA (Altshuller *et al.*, 1983a; Smith and Hunt, 1978; Smith, 1979; Pueschel and Mamane, 1979).

The heat of the plume (which can be up to 120°C) causes it to rise from the stack with an appreciable exit velocity, effectively increasing the stack height from between two to ten times the actual release height for typical elevated, buoyant sources. Turbulence causes the plume to expand both vertically until it fills the mixing layer, when dry deposition of pollutants can start to take place, and laterally as it is diluted. Wind dilutes the pollutant at source and pollutant concentration varies as the inverse of wind speed. For given loss rates the stronger the wind the greater the downwind area affected by surface deposition and the smaller the average deposition per unit area (Briggs, 1984; Smith, 1979). However, an increase of pollutant concentration with wind speed can sometimes be seen associated with emissions which have been transported some distance from their origins in medium or high chimneys (Bower and Sullivan, 1981). Injection of a pollutant into the upper part of the boundary layer leads to longer pollutant residence times and longer transport distances for tall stack effluents than for those of comparable near-surface sources (Husar and Patterson, 1980). Model calculations showed that a low stack height (50 m) increased the local dry deposited mass by a factor of 1.6 and thus decreased the total deposition at 1800 km from the source to about 75% of total deposition from an elevated source (300 m) for both summer and winter months (Gislason and Prahm, 1983).

Much long distance plume travel occurs above the boundary layer, particularly at night. Tall stacks can inject a pollutant plume above the boundary layer, into a layer essentially decoupled from the surface both during daytime hours and at night. Here it experiences much less dilution and vertical diffusion of pollutants effectively does not occur. Plumes can also become isolated from the surface when warm air flows over a surface layer of relatively cold air. When the mixing layer contracts closer to the surface late in the day 90% of the pollutant remains isolated above a nocturnal surface based inversion. Pollution local to a source is then greatly reduced since dry deposition cannot begin until the plume re-enters the lower layer the next morning. If surface heating is weaker than the day before, the maximum boundary layer height may be below some of the polluted layers from the previous day and these layers can remain essentially intact. Because of this decoupling, wind speed increases above the inversion, producing nocturnal jet streams which are a common occurrence over much of the world. Pollutant plumes can be carried unchanged for long distances in these jet streams. Well defined plumes have been tracked by lidar during the night time stable period. With 8–12 h of transport at 5–15 m/s, SO_2 can travel 150–650 km before dry deposition can occur, although bursts of turbulence of several minutes duration, caused by internal wind gradients, can bring pollutants close to the ground (Slinn, 1982; Uthe and Wilson, 1979; Sisterson *et al.*, 1979; Smith and Hunt, 1978; Smith *et al.*, 1978; Smith, 1981).

Because of this common diurnal variation in atmospheric processes it is not possible to treat a plume as a flow reactor with time as the only variable. The chemical reactions in a plume will depend on both the time since the plume left the stack and the time of day at which it was emitted (Wilson, 1981).

Coal-fired power plant plumes

The chemical processes occurring in plumes have been extensively studied. Published values for SO_2 oxidation rate vary widely, as do the calculation methods and model assumptions used to derive these rates, as is shown in Table 21.1. The SO_2 oxidation rates seen in point source plumes vary from nearly 0 to more than 16%/h while rates of more than 30%/h are seen in urban plumes. Some of this variation may be due to uncertainties in data collection and interpretation but the wide variation in plume composition and concentration and the ambient conditions, both chemical and meteorological, encountered by the plume are also important. It is generally conceded that no single factor is dominant under all atmospheric conditions, with reaction mechanism as well as reaction rate varying with prevailing conditions. The use of aircraft to obtain samples from plumes as they are progressively diluted by air has been proved effective, and plumes have been sampled up to 1000 km and over 12 h travel time from the stack. By accounting for plume dispersion and loss to the ground by dry deposition it is possible to estimate the SO_2 conversion rate. Plume dispersion is well characterised and parameterised but loss to the ground is not very well characterised and a large uncertainty remains in the dry deposition velocity for both SO_2 and sulphate which creates a similar range of uncertainty in estimated transformation rates when the plume touches the ground (Sievering *et al.*, 1981).

Mathematical models applied to the binary system of H_2SO_4–H_2O gases by Yue and Hamill (1980) to evaluate theoretically the formation of sulphate aerosols by various nucleation processes showed that the concentration of water vapour was the dominant factor in gas phase nucleation processes. In general, the condensation of aqueous sulphates on pre-existing solid particles was dominant over gas phase nucleation. However the majority of aircraft studies were performed in conditions where aqueous phase reactions were not favoured.

Based on Project MISTT (Midwest Interstate Sulfur Transformation and Transport) aircraft measurements of the plume of the Labadie coal-fired power plant, situated near St Louis in the Midwest of the USA, Gillani *et al.* (1978) concluded that sulphate formation occurred at rates of less than 3%/h with the highest rates corresponding to peak sunlight hours. The total amount of sulphate formed was seen to be linearly related to the total dose of sunlight received by the sampled plume. This study tracked and sampled the plume during day and night for over 300 km with plume transport extending for 10–12 h. Very little oxidation of SO_2 was observed during the night, presenting a strongly diurnal pattern.

Husar *et al.* (1978) analysed the Project MISTT data, using a simple two-box model, and found an SO_2 conversion rate of 1 to 4%/h around noon with night rates below 0.5%/h. However plume transport during the night is faster than at midday by a factor of two, so sulphate formation takes place during the day while long range transport occurs mainly at night, with the plume travelling 300–400 km from the stack. About 30–45% of SO_2 is converted to sulphate aerosol, about half within the first day, with the rest removed as SO_2. In a re-examination of the Project MISTT data, Gillani and Wilson (1980) concluded that the condition of the background air mass which receives the plume, the extent of plume/background interactions and photochemical processes are the most important factors in SO_2 oxidation on warm sunny days because a close link exists between the formation of ozone and sulphate aerosols. The entrainment of non-methane hydrocarbons (NMHC) from the polluted background air mass can provide organic peroxide (RO_2) and hydroperoxyl (HO_2) radicals which can promote oxidation. If the atmosphere is 'clean', containing relatively few hydrocarbons, hydroxyl radicals (OH) are the predominant oxidants. Ozone is produced as a by-product of the photolytic degradation of NO_2 to nitric oxide (NO). NO reacts rapidly with ozone in a plume leading to an ozone deficit, which gradually disappears as NO_2 is produced photolytically until a photostationary state is reached. An excess of ozone is generated if reactions with RO_2 and HO_2 radicals consume NO to produce RO

Table 21.1 Oxidation rates in plumes

Location of source sampled	*Plume age* (h)	*Stack distance* (km)	*Travel time* (h)	*Month and year of sampling*	*Oxidation rate range* (%/h)	*No. of data points*	*Average rate* (%/h)	*Rate derivation method, comments*	*Reference*
Coal-fired power plants									
Cumberland TN, USA	1.5–10	8–160	1.67–12	Aug 1978	<0.1–7.5	19	3	Particulate S/total S ratio	Gillani *et al.* (1981)
Johnsonville TN, USA	6–8	56–160	6–8	Aug 1978	0.8–8.5	6	3	Dry conditions only	
Cumberland TN, USA	1.3–10.7		2.15–4.77	Jul/Aug 1976	1.1–8.5	5	5.5	Average daytime rate	Zak (1981)
	0–7		1.35–7.03	Aug 1978	0.4–16.7	5	6.1	Stack to first measurement	
								Particulate S/gaseous S ratio	
Keystone PA, USA			0.5–2.67	Apr/May/Sep/Oct 1978	0.01–5.92	13	0.05	At r.h 42–64%	Dittenhoeffer and De Pena (1980)
							0.78	At r.h 65–90%	
							3.31	At r.h 91–100%	
								Particulate S/total S ratio	
Paradise KY, USA			0.2–3.8	Jun	0–1.3	5		Particulate S/total S ratio	Meagher *et al.* (1982)
Colbert AL, USA		7–50	0.28–5.20	May/Jun	0.78–2.79	12	1.3	Morning rate	Meagher *et al.* (1981)
Widows Creek AL, USA		2–49.3	0.43–6.01	Aug	0.34–3.41	17	1	Morning rate	
							2.4	Afternoon rate	
								Particulate S/total S ratio	
Cumberland TN, USA	0.76–9	11–200		Aug 1978	0.1–7	21	3	Late morning & afternoon	Forrest *et al.* (1981)
							0.5	Night & early morning	
							2	Average diurnal rate	
								Particulate S/total S ratio	
Bowen GA, USA			0.12–3.33	Dec 1979	0–2.3	7	<0.2	Particulate S/total S ratio	Liebsch and De Pena (1982)
Breed IN, USA	1.3–6.8			Jun/Nov 1977	0–3.7	7	1	Particulate S/total S ratio	Easter *et al.* (1980)

Cobb MI, USA	1.1–3			May/Nov 1977	0.2–8.4	6	2.6		
Labadie, MO, USA	0.8–12	12–320	0.83–12	Jul 1976	0.3–3.2	17	1–3	Particulate S/ total S ratio	Gillani and Wilson (1980)
Labadie MO, USA	0.7–12.5	14–360		Aug 1974/ Jul 1976	0.1–4.8	50	1.6	Particulate S/ total S ratio rate 1–4%/h day, <0.5 night	Husar *et al.* (1978)
Leland-Olds, ND, USA			0.38–1.08	Jun 1978	0–0.06	4		Change in total particle volume & particulate S/ total S ratio	Hegg and Hobbs (1980)
Big Brown, TX, USA			0.63–1.32	Jun 1978	0.15–5.7	4			
Sherburne Co. MN, USA			0.17–2.7	Jun 1978	0–2.2	4			
Centralia, WA, USA			0.03–1.42	Mar/Oct 1976 Sep/Oct 1977	0.03–0.56	5		Change in total particle volume	Hobbs *et al.* (1979)
Four Corners, NM, USA			0.78–0.87	Jun 1977	0.34–6.6	3			
Navajo, AZ, USA	3–6.1			Jun/Jul 1979	0.7–13	22	1.9 0.9	Rate at noon Diurnal average Particle volume/ total S ratio	Wilson and McMurry (1981)
Great Basin NV, USA	0.36–10.5			Jul/Aug 1979	1–7	16		Particulate S/ total S ratio	Eatough *et al.* (1981b)
Navajo AZ, USA	2.5–11	25–115	2.67–10.92	Jul/Dec 1979	0–0.8	13		Particulate S/ total S ratio	Richards *et al.* (1981)
Four Corners, NM, USA		2–90	0.3–12.5	Jun 1978	0.15–0.5	3		Varies with stack distance CN production/ SO_2 ratios	Mamane and Pueschel (1980)
Nanticoke Ontario, Canada	0.15–1.93	3–43		Jun 1978	0–8.7	7	4	Plume age <2 h, with fumigation Particulate S/ total S ratio	Anlauf *et al.* (1982)
Nanticoke Ontario, Canada	0.07–2.9	2–93		Nov 1975	0.32–12.6	19		Varies with stack distance Particulate S/ total S ratio	Melo (1977)

Table 21.1 continued

Location of source sampled	*Plume age* (h)	*Stack distance* (km)	*Travel time* (h)	*Month and year of sampling*	*Oxidation rate range* (%/h)	*No. of data points*	*Average rate* (%/h)	*Rate derivation method, comments*	*Reference*
Oil-fired power plants									
Anclote FL, USA		0.50	0–1.67	Aug 1976/ Feb 1977			⩽0.25	Steady state rate, Particulate S/ total S ratio	Forrest *et al.* (1979b)
Andrus MS, USA	1.2–6.6			May/Oct 1977	0–5.1	9	2.2	Particulate S/ total S ratio	Easter *et al.* (1980)
Northport NY, USA	0–3.3		0–2	various, over 3 years		60	<1	Rate essentially incalculable. Particulate S/ total S ratio	Garber *et al.* (1981)
Metal smelters									
Mt Isa Queensland Australia	0.08–14.83	2–256		Jun 1977	0.06–0.45	65	0.25	Diurnally averaged rate Gaseous S/ total S ratio	Roberts and Williams (1979)
Mt Isa, Queensland Australia	2.2–42.5	60–1001		Jul 1979			0.15	Diurnally averaged rate	Williams *et al.* (1981)
Urban plumes									
St Louis MO, USA				Aug 1975	10–14	18		Gaseous S/ total S ratio	Alkezweeny and Powell (1977)
Milwaukee WI, USA			0–3	Aug 1976/ Jul 1977	1–9		4	SO_4^{2-} & light scattering measurements	Miller and Alkezweeny (1980)
St Louis MO, USA				Jun 1976	0–4			Particulate S/ total S ratio	Forrest *et al.* (1979a)
Budapest Hungary		50	~3h	Jul/Aug 1978	3–31	8	10		Horvath and Bonis (1980)
Long-range transport trajectories									
Sweden		900–1900	23–61	Summer (Apr–Sep)	0.3–5.2	12	1.4	Particulate S/ gaseous S ratio	Traegaardh (1980)
		1300–2500	27–55	Winter (Oct–Mar)	0.4–1.3	4	0.8		

and OH radicals. These peroxy radicals participate in SO_2 oxidation as well as in the generation of excess ozone, and are formed by the reaction of NMHC from the background air with OH radicals. The SO_2/NOx ratio is an important parameter because SO_2 and NO*x* compete for the same radicals and oxidation of NO_2 by OH is more than an order of magnitude faster than SO_2 oxidation by OH (reaction 20.1) so the presence of NO*x* inhibits sulphate formation close to the

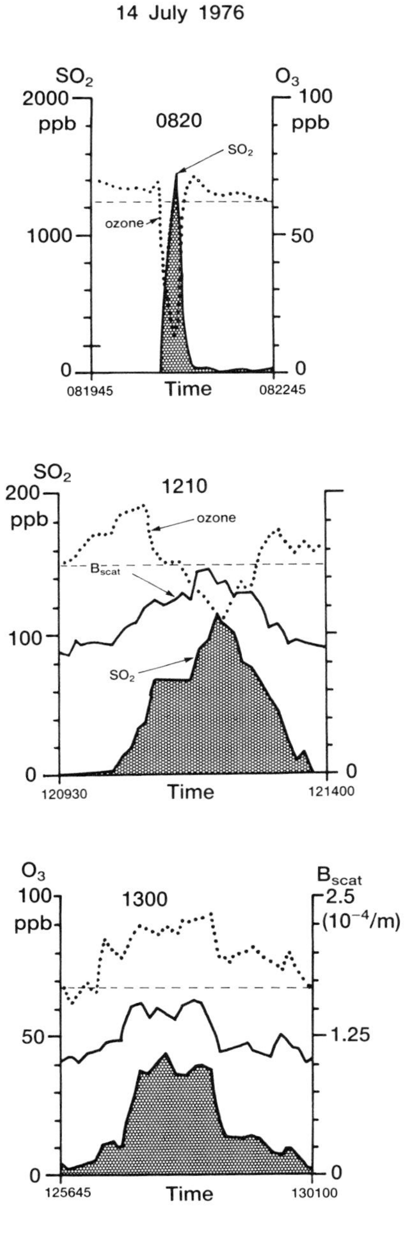

Figure 21.2 The early, intermediate and fully-developed stages of secondary pollution formation (after Gallani and Wilson, 1980)

stack. If the plume is in clean air the OH radical is generated from photolytic decomposition of ozone. A moderate amount of sunlight, ground level temperatures above 28°C and relative humidity above 50% are considered sufficient conditions for rapid formation of sulphate aerosols.

These plume data show three chronological stages of daytime plume development. Figure 21.2, after Gillani and Wilson (1980), illustrates crosswind plume profiles of SO_2, ozone and aerosol light scattering coefficient (B_{scat} which is highly correlated with sulphate) obtained 50 km from the stack of the Labadie power plant at different times of the day. The early stage is dominated by primary emissions and an ozone deficit with the SO_2 transformation rate about 1%/h. The intermediate stage exhibits enhanced formation of sulphate aerosols (conversion rate about 1.3%/h) and ozone in plume fringes as background species begin to be entrained into the plume. The fully developed stage is found when background species have diffused through to the plume core and an excess of ozone has formed over the whole of the plume profile. The average rate of SO_2 transformation is now ~0.8%/h. The time the plume takes to pass through these stages is dependent on the availability of reactive hydrocarbons in the background and intensity of mixing, a matter of a few hours during daytime in summer. Once the plume is mature, secondary pollution (including sulphates) dominates until UV insolation is insufficient to drive the photochemistry. During the night the plume is carried onward, with no new formation of sulphate, but carrying the concentration of pollutants formed the day before unabated, unless there is strong wind shear.

Gillani *et al.* (1981) used this data to construct a model of sulphate production by gas phase reactions, under relatively dry (relative humidity $<75\%$) sunny summer conditions. Plume dilution, background reactivity (measured in terms of ozone formation) and UV insolation were the principal reaction determinants. Model results provide a good match with observations. Plume dilution is important only in the early stages of long range transport. Under these dry conditions the gas phase mechanism is probably dominant, providing 90% or more of SO_2 conversion. Where the plume top was in contact with clouds conversion was much enhanced, presumably due to liquid phase reactions. Instantaneous conversion rates in clouds could exceed 10%/h. Variability over the plume cross section was observed in the Cumberland, Tennessee plume where a conversion rate of 7.5%/h was seen in the side plume while the rate was 3%/h in the corresponding section of the main plume.

This enhancement in the plume edges was also observed by Zak (1981), from balloon studies of the Cumberland plume, who found an average daytime conversion rate of 5.5%/h compared to the average rate of 1.3%/h found by Gillani and Wilson (1980). The balloon data is derived mainly from plume edge measurements with high background air to plume effluent mixing ratios while the aircraft data is normally averaged over the whole cross section. Where the balloon measured a lower background air to plume effluent mixing ratio the conversion rate was only 0.4%/h. The rate of plume dilution is thought to vary over the plume cross-section which would also affect conversion rates. During these measurements the power plant plume was embedded in an urban plume on some occasions which may also have contributed to an enhanced conversion rate. Hobbs *et al.* (1979) observed a decrease in concentration of particles in the centre of a plume with travel time, but an increase with travel time at the edges of a plume in both the Centralia, Washington, plume and the Four Corners, New Mexico, plume.

Figure 21.3 shows the calculated monthly average sulphur budget of the Labadie

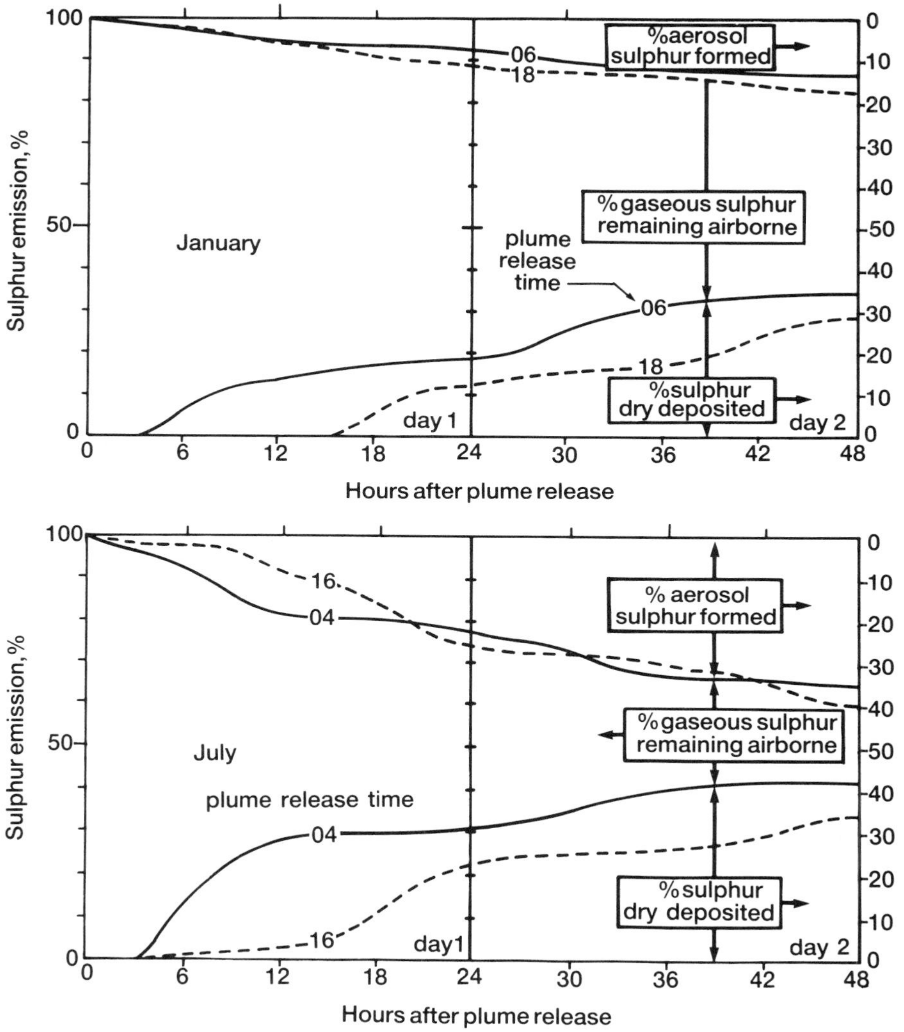

Figure 21.3 Plume sulphur budgets for different plume release times during 48 hours of transport in January and July (after Altshuller *et al.*, 1983a)

plume (stack height 214 m) in January and July, during two days of transport, in the absence of wet deposition. This is based on the results of a time varying plume transport and diffusion model. Production of sulphate is about 25% of total sulphur after 24 h and 40% after 48 h in the July simulation. In the January simulation, sulphate production is about 10% after 24 h and 15% after 48 h. Ground removal of SO_2 is about 18% on each day in January and 30% on the first day in July with an additional 10–12% on the second day. In July, about 40% of the total sulphur emitted is dry deposited in 48 h. On average another 20–40% may be wet deposited in the same period. The model results predict that about two-thirds of the sulphur emitted from a tall stack in the Midwest may be deposited within two days during summer. During this time the plume is likely to have been transported about 1000 km. In January the atmospheric residence time of sulphur is expected to be significantly longer and the potential for long range transport significantly greater. The effect of increased transport distance with greater stack height is also likely to be enhanced more in winter than in summer (Altshuller *et al.*, 1983a).

From the concentration of particles in the plume of the Four Corners power plant, the calculated particle production rate found by Mamane and Pueschel (1980) indicated an SO_2 oxidation rate of about 0.15%/h in the first 0.3 h, increasing to 0.5%/h for the 0.3 to 2.5 h of plume transport then decreasing to 0.3%/h at 12.5 h travel time. Up to 6% of fly ash particles were coated with sulphate (Mamane and Pueschel, 1979) and so may contribute to SO_2 oxidation up to a calculated conversion rate of 0.03%/h, negligible in comparison with other mechanisms.

Direct measurements of the OH radical have been made in the vicinity, although not in the plume, of the Four Corners power plant, by Davis *et al.* (1979) who calculated the SO_2 conversion rate under summer atmospheric conditions via reaction (20.1) as the initiating step. The rate of reaction (20.1) is dependent on the OH concentration, which has been shown to be a function of the solar zenith angle, the intensity of UV light and irradiation of O_3 (Davis, 1977). Assuming an average insolation the calculated conversion rate ranges from about 0.2%/h in the early morning and evening to 2.3%/h around noon. The calculated conversion time for SO_2, taking into account the OH diurnal cycle, gave a lifetime of about four days for SO_2, corresponding to an average conversion rate of 0.7%/h. An induction period of about 2 h would be required, but this period could vary significantly with plume characteristics. Since the hourly conversion rate is so low the majority of SO_2 is converted to sulphate under very dilute plume conditions, nearly equal to that found in ambient air.

Dittenhoeffer and De Pena (1978) in a study of particle formation at the Keystone, Pennsylvania, power plant, in spring and autumn, also found that under conditions of low relative humidity and intense insolation, gas phase photochemical reactions forming small new particles were the dominant sulphate formation mechanism. However, where the plume merged with a cooling tower plume (relative humidity near 100%) the formation of sulphate on pre-existing particles, by SO_2 absorption and liquid phase oxidation, predominated. The highest rate of particle growth occurred when average plume temperature was lowest and the lowest rate when temperature was highest, possibly indicating that a metal catalysed reaction was taking place. In a later study of the Keystone power plant under a wide range of meteorological and plume conditions Dittenhoeffer and De Pena (1980) found an average conversion rate of about 1%/h during the first 1–2 hours of plume transport, with a distinct increase in rate at about 65% relative humidity possibly due to deliquescence of sulphate particles in the plume encouraging aqueous phase reactions. The highest conversion rate (5.9%/h) and the largest sulphate particles ($\sim 2\ \mu m$) diameter) were produced when the plume merged with clouds.

In a winter study of the Bowen, Georgia power plant plume in conditions of relative humidity $<50\%$, Liebsch and De Pena (1982) found that the highest conversion rate was 0.2%/h. However on the single occasion when relative humidity rose to 88% the conversion rate rose to 2.3%/h. On this occasion the results correlated better with plume water vapour content than with relative humidity. Transient nuclei concentrations were strongly correlated with solar flux and with plume dilution which, with evidence of enhanced nuclei production at plume borders, implied that a gas phase reaction with photochemically produced species entrained from the background air was taking place.

The merging of cooling tower and power plant plumes was studied by Meagher *et al.* (1982) at Keystone, who observed a high activity near the plant under these conditions, with 3% of plume sulphur as sulphate before sampling started and with no oxidation occurring later, possibly due to plume dilution or evaporation.

Where the plumes did not merge, 1% of plume sulphur was found to be sulphate prior to the onset of measurable SO_2 conversion, with little oxidation near the plant, increasing to about 2%/h after the plume had aged 1.5 to 2 h. Sulphate formation in the merged plumes may take place by a catalysed aqueous phase mechanism, possibly with aqueous solutions of nitrogen compounds acting as catalyst since the cyclone furnaces at this plant emit 3–4 times as much NOx as conventional units for the same power output.

Meagher *et al.* (1981) compared conversion rates at the conventional Colbert, Alabama plant with the Widows Creek, Alabama plant which uses a wet limestone SO_2 scrubber, to evaluate the effect of an FGD system on the chemistry of production of secondary pollutants. Although the SO_2 concentrations in the scrubbed plume were much reduced, sulphate formation rates at both plants were almost identical, the average morning rate being 1.2%/h with an average rate of 2.4%/h for afternoon measurements. During an episode of high humidity the morning rate reached 2.8%/h, otherwise no relation was found between meteorological conditions and oxidation rates. Net ozone production was observed.

A summer plume study in a dry region (relative humidity ~5–20%) was conducted by Wilson and McMurry (1981) at the Navajo power plant, Arizona. A diurnal conversion pattern was observed with conversion beginning only after emissions had been exposed to sunlight for about an hour. A noon oxidation rate of 1.9%/h with a diurnal average of 0.9%/h together with the formation of new particles and condensation on to small particles (0.01–0.32 μm diameter) depending on SO_2 concentrations and time of day implied that the mechanism was a gas phase reaction involving the OH radical. Since the SO_2 plume concentrations were low the calculated conversion rates would also apply in dilute plumes, on plume edges and in background air in regions with similar climatic conditions. Richards *et al.* (1981) also sampled the Navajo plume, in both winter and summer, finding measurable sulphate formation only in the dilute plume during the day. The highest rates found were 0.8%/h in summer and 0.2%/h in winter with formation rates increasing with increasing distance from the stack, suggesting a gas phase reaction involving the OH radical.

Hegg and Hobbs' (1980) study on five coal-fired power plants in the west and midwest of the USA was conducted under conditions favouring gas phase photochemical reactions. The dominant mechanism found was oxidation by OH radicals with conversion rate varying with travel time from the stack and UV light intensity. Differences in sulphate concentrations observed by the three different measuring methods used gave an uncertainty in the calculated oxidation rates of a factor of two. A later study (Hegg and Hobbs, 1983) in Arizona on the plume of a modern plant fitted with wet scrubbers found a measurable conversion rate in only one out of five flights. Other results, such as the low nitrate production rates in the plume, suggest that relatively low concentrations of the OH radical were present. Mean rates of sulphur gas to particle conversion in the plume of the coal-slurry-fired Mohave power plant were about 0.6%/h. The rates showed a tendency to increase with travel time (Hegg *et al.*, 1985).

Forrest *et al* (1981) found a summer conversion rate varying with time of day in a study of the Cumberland plume. Insolation and other related parameters such as turbulence may account for the variations. Late morning and afternoon rates averaged 3%/h; night and early morning rates averaged 0.5%/h. Excess ozone was observed in the plume suggesting an OH induced mechanism of SO_2 oxidation. Acidic bisulphate particles were more prevalent than ammonium sulphate.

During fumigation conditions in the mixing layer, involving entrainment of ambient air into the plume, an average rate of 4%/h was seen by Anlauf *et al.* (1982) at relative humidities of 30–50%. Calculation of back trajectories showed that the level of pollutant concentration was related to the passage of high pressure weather systems.

Melo (1977) found a generally low level of oxidation, normally below 3% with the highest measurement of 8.5%, at plume ages up to 2.9 h and distances up to 90 km from the stack. Oxidation rate depended on distance from the stack with 3–10%/h found at plume ages up to 10 min reducing to less than 1%/h at plume ages >50 min. In a later study, Melo (1979) found a level of oxidation of only 1–3% at stack distances up to 17 km with plume age varying up to 1.12 h.

Oxidation rates of 0–3%/h were found by Easter *et al.* (1980) from data collected as part of the SURE study. Reduced oxidation during the first hour of plume aging was sometimes found. Rates appeared sensitive to hydrocarbon levels with indications of greater oxidation in the plume fringe than in the core. Losses of SO_2 by dry deposition were on occasion found to be significantly larger (about 10%/h) than losses due to oxidation.

The chemical species formed by atmospheric oxidation in plumes in the arid west of the USA were studied by Eatough *et al.* (1981b) with ground-based sampling procedures. They found evidence of conversion of SO_2 to particulate sulphate, organic and inorganic S(IV) compounds. Neutralisation of H_2SO_4 was limited only by the rate of introduction of basic material (NH_3, metal oxides or carbonates depending on the location) from the ambient air. NH_3 was responsible for neutralising about half the H_2SO_4 produced in one of the coal-fired power plant plumes studied. The primary sulphate emitted accounted for 66% of the total ambient concentration close to one of the plants, since the air was clean. Secondary sulphate increased with plume age to become comparable with background levels. At a plume age of 5 h, 40% of sulphate in the atmosphere was due to power plant emissions, present mainly in a particle size $<0.5\,\mu m$. The relative amounts of sulphate, organic and inorganic S(IV) species formed were dependent on ambient temperature and humidity. In the dry, hot daytime conditions of south-western USA the sulphate conversion rate was up to 6%/h, and the conversion rate to organic S(IV) species up to 1.7%/h. The concentration of the latter ranged from 25 to 75% of the secondary sulphate concentration in aged plumes. During cold, dry periods the conversion rate to organic S(IV) was still about 1.7%/h but the sulphate conversion rate was down to below 1%/h and organic S(IV) was the main secondary sulphur species present. The chemical composition of secondary sulphur particles may vary with region since secondary sulphur in eastern USA is predominantly found as sulphuric acid and ammonium sulphate with only about 10% present as organic S(IV) compounds (Altshuller *et al.*, 1983a).

Two sets of measurements were made in January over the North Sea in the plume from the Eggborough, UK, power station. The same parcel of air was sampled off the coast at about 105 km from the station and again the next day at about 650 km, when the boundary layer was shallow and filled with moist cloudy air. Clark *et al.* (1985, 1982) estimated a conversion rate of about 1%/h at 5 h travel time, with a calculated 15% of sulphur in the form of sulphate. This is comparable with a calculated 4.3%/h SO_2 oxidation rate in rural air over the same period. Oxidation in the plume may have been suppressed both by low cloud water pH and the unavailability of oxidants in the plume (due to low lateral dispersion). These occurrences may be common in winter over the North Sea, although the measurements were taken in unusual meteorological conditions (Cocks *et al.*, 1983). Similar sets of measurements taken over England and Wales in July gave calculated oxidation rates of 2%/h in a plume embedded in clear dry air and rates in the range 5.3 ± 1.5%/h to 21 ± 8%/h in a plume which entered cloud. The much higher in-cloud oxidation rate seen in the summer study may be due to the higher temperatures or greater insolation in July than in January or to higher oxidant availability due to greater mixing or to higher ammonia levels over land (Bamber *et al.*, 1984).

Models of SO_2 oxidation in plumes generally propose a conversion rate of between 1 and 5%/h for conditions equivalent to eastern USA. An approximate 25 ± 10% conversion of SO_2 to sulphate during the day in summer is suggested for power plant plumes. The extent of SO_2 conversion during the night is 5% or less (Gillani, 1978; Husar *et al.*, 1978; Isaksen *et al.*, 1978; Hov and Isaksen, 1981; Gillani *et al.*, 1981; Meagher and Luria, 1982; Bottenheim and Strausz, 1982). A similar range is found in the simple dispersion model of Cocks and Fletcher (1982), illustrating the interdependence of reacting species in the plume on expansion into ambient air of various compositions. A model for the arid west of the USA (Seigneur, 1982) gives maximum conversion rates of 0.5 and 0.6%/h for July and December respectively, in fair agreement with observations.

Oil-fired power plant plumes

The chemical composition of emissions from oil-fired power plants is different from those from coal-fired power plants, especially as regards possible trace metal catalysts, so the atmospheric behaviour of SO_x species could also differ because of the diversity of possible SO_2 oxidation mechanisms (Garber *et al.*, 1981). Earlier investigations found that the extent of oxidation could be as much as five times greater than in a coal-fired power plant plume but this has not been borne out in more recent studies (Newman, 1981).

A study on the Anclote power plant, Florida (which operates without particle controls), by Forrest *et al.* (1979b) showed the extent of SO_2 oxidation was generally between 1 to 3% up to 50 km from the plant in a travel time of 1.67 h, with conversion rates <0.25%/h. About half of this sulphate may be due to primary emissions.

The Northport power station on Long Island, NY, USA, was extensively sampled by Garber *et al.* (1981) over a wide range of meteorological conditions to give a similar picture with the extent of oxidation generally ranging from 0 to 5% for plume ages up to 3.3 h. The oxidation rate was essentially unmeasurable but certainly <1%/h. Primary emissions accounted for a major portion of plume sulphate observed during the first few hours after emission. The plume aerosol was generally acidic.

Eatough *et al.* (1981b) found no significant difference in SO_2 oxidation between the coal- and oil-fired power plants studied. Sulphate, organic and inorganic S(IV) compounds were found in all these plumes, with the relative amounts of each species dependent on ambient temperature and humidity rather than fuel burned. Insolation, plume dispersion and ambient hydrocarbon concentration were the main factors influencing oxidation rate in the Easter *et al.* (1980) study of two coal-fired and one oil-fired power plants in eastern USA The level of oxidation was normally $< 10\%$ and conversion rate ranged between 1 to 3%/h for all plants.

Direct measurements of the rate of SO_2 oxidation in the plume of a Pacific coast power plant as it passed through a fog bank were reported by Eatough *et al.* (1984). The average conversion rate in dry air of $3.1 \pm 0.8\%$/h agrees well with other studies on midday oxidation in power plant plumes. However, the average rate of oxidation in the fog bank was $30 \pm 4\%$/h, a factor of ten higher. This may be due to aqueous phase reactions controlled by H_2O_2 or organic peroxides present in the primary emissions from the oil-fired plant.

In Sweden, Enger and Hoegstroem(1979) found that 70% of the sulphur from the coastal Karlshamn power plant plume was in the form of sulphate at 30 km from the stack, on a day with relative humidity close to 100%. However on a much drier day only about 10% was transformed to sulphate at 30 km. A rapid conversion rate was seen with high relative humidity but a low rate in drier weather.

Smelter plumes

Smelters emit SO_2 together with a variety of particles different from those emitted by power plants. Plumes can be followed for long distances more easily since smelters are often located in remote areas which lack other SO_2 emission sources. Newman (1981) reviewed a variety of smelter plume studies finding oxidation rates at a large nickel smelter to be generally below 1%/h with less than 10% of the sulphur transformed to sulphate. In a study of copper and lead smelters at Mt Isa, Australia, during winter Roberts and Williams (1979) found a strong diurnal effect, related to insolation, on SO_2 conversion up to 256 km from the stack with plume age up to 13 h. A later study (Williams *et al.*, 1981) followed the plume 1000 km from the source at plume age up to 42.5 h. The extent of oxidation of plume sulphur was up to 4%, giving a diurnally averaged oxidation rate of 0.15%/h. At longer distances condensation nuclei 'wings' were seen in the plume. The calculated lifetime of SO_2 over the dry northern part of Australia was 14 days, with atmospheric oxidation accounting for half the loss. The conversion mechanisms were gas-phase photo-oxidation reactions involving the OH and HO_2 radicals.

Eatough *et al.* (1981) sampled a copper smelter plume among mountains, from the ground, and found that not all of the H_2SO_4 formed was being neutralised, because of a lack of basic material to be incorporated in the plume. H_2SO_4 was neutralised only by metal carbonates or oxides since no NH_3 was available. At short travel times the plume was dominated by primary sulphate emissions but, due to the high SO_2 emissions, after 5 h 80% of the sulphate was secondary and only 6% primary. Sulphate formation in the plume during periods of high insolation was temperature dependent and first order in SO_2. The percentage of sulphate in the $< 0.5\,\mu m$ particle size range increased regularly with plume age until 75% was in this range at 60 km from the source. No correlation was seen between particle formation and plume expansion, particle acidity, metal content or S(IV) species (Eatough *et al.*, 1982).

Urban plumes

All the plumes so far considered have been produced from point sources but urban areas can also form plumes in which the combined air pollutants from the major emission sources are carried, with air masses moving across the city, out over non-urban land or water. Urban plumes differ from point source plumes in the much lower effective plume release height and in the greater relative load of NO*x* and reactive hydrocarbons, largely from automobiles. The plume can be distinguished by its chemical composition and high concentrations of certain pollutants and can affect ground sites at distances on the order of 100 km downwind of the city. There is evidence that the plume is detectable up to 350 km from an urban area. In areas such as the eastern USA, where urban source areas are closely spaced, the intervening non-urban areas are often under the combined influence of several urban plumes, as well as other man-made SO*x* sources such as power station plumes. In general, higher and more variable oxidation rates have been reported for urban plumes than for point source plumes, possibly due to the more highly polluted conditions that occur in urban areas (Winchester, 1980a; Ellestad, 1980).

Alkezweeny and Powell (1977) found conversion rates of 14%/h and 10%/h in the St Louis plume on two successive summer afternoons, although the uncertainty was such that the rate could have been the same on both days. A number of groups took part in the 1976 Da Vinci measurement programme which used a manned balloon tracking an urban air mass to measure chemical species evolution in a parcel of air as it travelled. From the Da Vinci II data on a stagnant urban atmosphere Forrest *et al.* (1979a) concluded that the oxidation rate could be as low as zero and no greater than 4%/h. Chang (1979) derived an upper limit of 3.6%/h during the daytime from the same data. From the Da Vinci III study which tracked the St Louis plume, although influenced by the plume from the Labadie power plant embedded in it, a daytime conversion rate of 4.2%/h was calculated. However Zak *et al.* (1981) concluded that the daytime conversion rate from Da Vinci II data was 8.5 ± 4%/h, with a nocturnal rate of 1.1 ± 0.5%/h, and from Da Vinci III 4.4 ± 2%/h. The Da Vinci II aerosol was highly acidic, indicating that the sulphur present was mainly H_2SO_4 although partial neutralisation to NH_4HSO_4 took place over and downwind of the city.

In a study of the Milwaukee urban plume over Lake Michigan, Miller and Alkezweeny (1980) found a rate in the range of 6 to 8%/h on one afternoon while for the same period on the following day the average rate was $<$1%/h although the meteorological conditions and precursor levels were similar on both days.

Horvath and Bonis (1980) in dry summer weather found an average rate of 10%/h in the Budapest urban plume, from ground-based measurements. They concluded that the reaction rate was dependent on distance, implying that different reactions are controlling the reaction rate at different distances from the source.

Particle formation

The SO_2 gas to sulphate particle conversion process produces hygroscopic sulphate aerosols which can act as cloud condensation nuclei (CCN). As the SO_2 conversion time is of the order of hours to days and the residence times of fine particles in the atmosphere are of the order of days to weeks production of CCN in plumes could have an effect on cloud structure and precipitation downwind from plants in the regional scale and possibly the mesoscale. Mamane and Pueschel (1980) observed a formation rate of the order of 10^{16} to $10^{17}\,s^{-1}$ CCN, corresponding to an oxidation rate ranging from 0.15 to 0.5%/h in the first 12.5 h of travel time in the plume of the Four Corners power plant. Dry deposition of plume particles reached a maximum of 36% in the first 2.5 h of travel time. Comparison with natural sources suggested that the power plant caused an estimated doubling of the CCN production rate in the area. This could have major effects on regional cloud structure and precipitation (Pueschel and Mamane, 1979; Pueschel and Van Valin, 1978).

A sulphate particle generation rate of 10^{15}–$10^{16}\,s^{-1}$ was found in the plume of the coal-fired Colstrip, Montana, power plant between 0.5 and 45 km downwind. The plume was deflected as much as 20° away from the prevailing wind direction by the terrain, a factor which could not be taken into account using current models (Van Valin *et al.*, 1980; Van Valin and Pueschel, 1981). Lukow and Cooper (1980) report a CCN flux of $10^{15}\,s^{-1}$ at the Colstrip plant and the Bridger coal-fired plant in Wyoming with the highest concentrations of CCN found furthest downwind from the plants (up to 25 km) as total particle concentration decreased. Hobbs *et al.* (1980) found a similar rate of CCN production from a coal-fired power plant plume with most CCN produced by gas-to-particle conversion rather than directly emitted, after a travel time of 1 h. The concentrations of ice nuclei did not differ significantly from those in ambient air.

Summary and comments

In the atmosphere the fate of a plume is dependent on meteorological and chemical conditions. The depth of the mixing layer is governed by input from solar radiation causing atmospheric turbulence and creating a diurnal cycle varying from stable to unstable, well-mixed conditions. Plumes rapidly expand to fill the vertical dimension of the mixing layer as they experience dilution and dry deposition of SO_2 starts as soon as they intersect the ground. Conversion to sulphate takes place during plume transport and all species are subject to removal during transport either directly or by precipitation. Tall stacks can inject a plume above the mixing layer, during both night and daytime hours, where it experiences more rapid transport, less dilution and very much less deposition of sulphur species than is the case in the mixing layer.

The average sulphate formation rate seen in point source plumes is usually less than 4%/h with the extent of SO_2 oxidation typically about 25 ± 10% in conditions equivalent to those in the eastern USA. Variations are due to external conditions rather than type of fuel burned. No single factor is dominant under all atmospheric conditions but some of the most important influences are insolation, relative humidity, oxidant and NO*x* levels. Little photochemical oxidation occurs from late evening to early morning but the rate rises considerably around mid-day. There is a large seasonal variation, with the average conversion rate higher in summer than in winter. Both gas and liquid phase mechanisms probably contribute to observed rates. Most aircraft studies take place in the dry, stable weather conditions which are most favourable to a gas phase mechanism, but possible liquid phase reactions are seen where a plume intersects clouds, fog or a cooling tower plume and here rates are up to an order of magnitude higher. The theoretical basis of predicting oxidation rates consistent with measurements for the more complex liquid phase mechanisms is not as far advanced as for gas phase mechanisms, which can also lead modellers to pay more attention to the latter. The generally higher rates seen in urban plumes are probably due to the higher levels of pollutants present to provide the necessary free radicals or catalysts.

Chapter 22

Atmospheric aerosol concentration patterns

In this chapter the background concentrations of sulphates in the atmosphere are compared with the levels seen in areas affected by industrialisation to indicate the likely levels of combustion generated sulphates. The variation in ambient sulphate concentrations over short periods of time and with season are considered. The relative importance of meteorological and pollution factors in determining ambient sulphate concentrations is discussed.

This discussion will be confined to fine particles ($<2\,\mu m$) in the transient nuclei and accumulation modes since secondary sulphates are mainly found in this range. The majority of the mass of sulphate aerosol is concentrated between 0.1 and 1 μm in diameter. The natural background concentration of sulphate is still only partially resolved. Sulphate concentration as a function of latitude is estimated at a maximum of $2\,\mu g/m^3$ around 40° N where SO_2 also has a maximum, but decreases below $1\,\mu g/m^3$ above 50° N. An average sulphate concentration of $1.3\,\mu g/m^3$ over the Atlantic Ocean, possibly affected by sulphur emissions from the land, and about $0.3\,\mu g/m^3$ over clean areas in the northern hemisphere was estimated by Meszaros (1978). To put the later sulphate data into perspective aerosol characteristics in remote and regional background locations will be summarised.

The air above the sea up to the top of the mixing layer is dominated by coarse sea salt aerosols mechanically produced by the sea surface. Above the mixing layer fine particles can be found which, in the main, have been transported from land. Over land, variations in ground cover (which affects surface reflectivity), topography and precipitation cause great variation in particle generation and removal rates so the natural background concentration varies with surface characteristics. Where the usual ground cover is snow or ice the fine aerosol concentration can be as low or lower than over the oceans. Studies over Antarctica and the Arctic have shown that transient nuclei counts of $100/cm^3$ are common, and counts as high as $10\,000/cm^3$ have been observed at altitudes of 1 km, with accompanying accumulation mode mass concentrations of $10\,\mu g/m^3$. In remote dry and desert areas mainly coarse particles are produced and the resulting dust can be advected to high altitudes. A submicron volume of $1.03\,\mu m^3/cm^3$ with a transient nuclei count less than $1\,000/cm^3$, observed in the western USA is the lowest value likely to be commonly found in most dry areas. The amount of natural accumulation mode aerosol that exists in areas of average precipitation and vegetative cover, such as is found in western Europe and eastern USA, is not clear since these areas all seem to be affected at some time by man-made aerosol sources. A volume concentration of 1 to $3\,\mu g/m^3$ was measured in the USA when the wind was blowing over uninhabited regions of Canada and the US. There is some evidence that the background sulphur concentration has increased in recent years (National Research Council, 1979).

The fine particle size fraction is dominated by sulphate aerosol particles but the percentage of sulphate present varies from location to location. For example Weiss *et al.* (1977) found that sulphate formed more than 90% of the particles seen in the 0.1–1.0 μm size range at three sites in Missouri, Michigan and Arkansas, USA. Richter and Granat (1978), in a rural area near Stockholm, Sweden, found that sulphate comprised 30% of the total bulk aerosol weight below 5 μm. Sulphate made up 39% of the fine particle fraction in studies of the south-western USA by Macias *et al.* (1980).

Measurements of sulphate levels often do not show a close correspondence to SO_2 levels when annual means from the same site are compared. This may be because SO_2 has a short residence time in the atmosphere (about 1.5 days). Regional sulphate concentrations show the combined effects of local emissions, transformation from SO_2 in the atmosphere by a variety of pathways and transport from distant SO_2 sources, giving an average residence time for sulphate of 5–7 days with a possible transport distance 1000–2000 km from the source. Hence the relationship of sulphate to SO_2 and to meteorological, climatological and pollution variables is a function of the particular locale (Henry and Hidy, 1979, 1982; Husar and Patterson, 1980; Rodhe, 1978). Table 22.1 shows recent measured sulphate concentration levels mainly in Europe and North America, since little measurement is carried out elsewhere except to provide data on clean background air. Sample collection and analysis, and the time basis used for calculation may differ therefore values may not be directly comparable.

The average levels in Table 22.1 do not show the range of variability observed in sulphate levels. Measurements made at the summit of Whiteface Mountain, New York (a remote site), in July 1982 gave monthly mean sulphate concentrations of 5.3 μg/m^3, with values in clean air of about 1.5 μg/m^3 and in polluted air up to 80 μg/m^3. The SURE program, monitoring a variety of air pollutants at 54 stations in north-eastern USA, produced data displaying the highly episodic nature of widespread high concentrations of sulphate, lasting for 2 to 6 days, mingled with periods of low sulphate concentration. These regional pollution episodes take place over an area of a million km^2 (Hidy *et al.*, 1979; Mueller *et al.*, 1980; Lavery *et al.*, 1981; Kelly *et al.*, 1984).

Altshuller (1980) analysed US air quality data for sulphates and SO_2 from 1963 to 1978. The monitoring sites were grouped into five sets (north-east urban, south-east urban, east non-urban, mid-west urban and mid-west non-urban) showing different characteristic seasonal trends. Sulphate concentration levels in the winter quarter decreased at all urban sites but increased slightly at non-urban sites, indicating a decoupling of processes between urban and rural areas. This decoupling was also seen between the north-east urban and the east non-urban sites in the spring quarter. The urban excess of sulphate over regional sulphate levels, especially in the north-east, can be accounted for by local primary sulphate emissions. In all groups of sites regional scale processes were most significant in the summer quarter. Increases seen in summer sulphate concentration levels were related to increases in regional scale emissions of SO_x and photochemically induced reaction processes, and were often associated with regional scale meteorological processes. Summer sulphate levels were strongly influenced by episodes of elevated sulphate concentrations. The oxidation rate of SO_2 varies seasonally, with faster

Table 22.1 Atmospheric sulphate aerosol concentrations

Location	*Type of location*	*Date*	*Fine* SO_4^{2-} (μg/m³)	*Comments*	*Reference*
North America					
Ohio Valley	non-urban	1977	<1–30	24 h averages	Mueller *et al.* (1980)
West Massachusetts	non-urban	1977	<1–30	24 h averages	
Allegheny Mt., PA	non-urban	1977	1–32	12 h averages	Pierson *et al.* 1980)
Tennessee	non-urban	1978	6.2–17.3	12 h averages	Stevens *et al.* (1980)
Nashville, TN	urban	3 years data	5–11		Reisinger and Crawford (1980)
Lake Michigan		1976/1978	0.17–0.2	clean air mass	Alkezweeny and Laulainen (1981)
			14–28	polluted air mass	
Chicago, IL	urban	Apr–Jun	1.7–30	29 days tested	Cooke and Wadden (1981)
Lexington, KY	urban	spring–summer 1979	10–40	24 h averages	Hazrati and Peters (1981
Washington DC	urban	summer 1976	7.6–9.8	averages of 37 samples	Kowalczyk *et al.* (1982)
Buffalo, NY	urban	Sep 1978–	1–15.8		Rao and Sistla (1982)
	non-urban	Mar 1979	1–10.3		
Virginia	non-urban	Jul–Aug 1980	11.1–14.2	12 h means	Weiss *et al.* (1982)
Michigan	non-urban	Jun 1979	0.6–2	averages, clean air mass	Alkezweeny *et al.* (1982)
Tidewater region VA		1979–1981	17.92	mean value	Brooks and Salop (1983)
New York City, NY	urban	Jul–Aug 1977	22.6	highest observed	Lioy *et al.* (1980)
High Point, NJ	non-urban	”	34.6	daily average	”
Brookhaven, NY	non-urban	”	20	concentrations	”
New Haven, CT	urban	”	25.7	”	”
New York City, NY	urban	Jul–Sep 1977	11.58	6 h average	Leaderer *et al.* (1982)
High Point, NJ	non-urban	”	16.67	concentrations	”
Brookhaven, NY	non-urban	”	9.47	”	”
New Haven, CT	urban	”	10.62	”	”
New York City, NY	urban	Aug 1976	7.3–42.2	average for highest and lowest days	Tanner *et al.* (1979)
Whiteface Mt, NY	non-urban	Jul 1975	1.1–23.4	daily samples, 6 days	Husain and Samson (1979)
St Louis, MO	urban	Jul 1977–Jun 1978	5.1–16.5	monthly averages	Cobourn and Husar (1982)
St Louis, MO	mostly urban	1975–1977	3.12–21.72	monthly averages	Altshuller (1982)
St Louis, MO	urban	Jul–Aug 1976 Sep 1978	15–20	averages, 37-7d	Dzubay (1980)
Steubenville, OH	urban	fall 1978–	11–23	24 h means	Colome and Spengler (1982)
Portage, WI	urban	fall 1979	4.5–5.5		

S Dakota	non-urban	1978	1.9	background air	Kelly *et al.* (1982)
Denver, CO	urban	Nov–Dec 1978	0.7–7.7		Countess *et al.* (1980)
South Coast Air Basin, CA	mostly urban	Jul–Oct 1972 & 1973	5.7–21.8	24 h averages	Appel *et al.* (1980)
Colorado R area, NV, AZ	rural, desert	winter 1976–1977	0.09–18	24 h averages	Hoffer *et al.* (1979)
Salt Lake City, UT	urban	1971–1973	1–81	measurement range	Murray and Farber (1982)
Los Angeles air basin, CA	urban	1976–1977	3–52		Hering and Friedlander (1982)
North America					
S Ontario, Canada	non-urban	Nov 1978–Jun 1979	<1–18		Barrie *et al.* (1981)
Long Point, Canada	non-urban	1979	4.5–15.5	monthly averages	Whelpdale and Barrie (1982)
Chalk River, Canada	"	"	1.5–7	"	
Kejimkujik	"	"	1–4	"	
Ela-Kenora	"	"	0.5–3.5	"	
Europe					
Austria	non-urban	1973–1974	8.8–13.3	2 stations	OECD (1979)
Switzerland	"	"	1.6–9.1	3 stations	annual means
FR Germany	"	1974	2.1–4.3	5 stations	"
Denmark	"	1973–1974	0.4–7	6 stations	"
France	"	"	2.6–12.7	6 stations	"
Iceland	"	"	1–1.1	1 station	"
Norway	"	"	1.1–4.8	7 stations	"
Netherlands	"	"	6.7–11	4 stations	"
Sweden	"	"	2.1–5.3	10 stations	"
Finland	"	"	1.1–3.2	5 stations	"
United Kingdom	"	"	1.6–7.6	7 stations	"
Arnham, Netherlands	urban	Jun–Aug	1–16	weekly means	Elshout *et al.* (1978)
Rotterdam	urban	1976	3–28	"	"
Massvlakte	urban	"	2–19	"	"
Venice, Italy	urban	1972–1973	8–55	monthly averages	Fassina and Lazzarini (1981)
London, UK	urban	Apr–Sep 1976	3–38	24 h means	Barnes and Lee (1978)
Czechoslovakia	urban and non-urban	Apr–Jun 1979	0–28.9	averages of flight data	Prokop (1979)
Czechoslovakia	background	1978	0.2–5.8	monthly means	Jaroslav and Dusan (1979)
Irafoss, Iceland	non-urban	Aug 1979	0.04–1.34	measured values range	Borys and Rahn (1981)
Chilton, UK	non-urban	1957–1974	2.5–9.7	daily samples, quarterly means	Salmon *et al.* (1978)

Table 22.1 continued

Location	*Type of location*	*Date*	*Fine* SO_4^{2-} (μg/m^3)	*Comments*	*Reference*
Other areas					
S America	background	1976–1977	0.06–0.27	means	Lawson and Winchester (1979)
Sao Paulo, Brazil	urban	”	6.75–30.6		
Perth, Australia	urban		0–5.5	measurement range	O'Connor *et al*. (1981)
Remote regions					
Antarctic	background	summer only 1974–1979	0.15–0.25	measured values	Shaw (1980)
Canadian Arctic	background	1979–1980	0.05–3.5	weekly averages, 2 stations	Barrie *et al*. (1981)
Spitsbergen, Arctic	background	winter 1979	1.85	average value	Heintzenberg *et al*. (1981)
Barrow, Alaska, Arctic	background	Sep 1976–Sep 1977	0.08–2	monthly means	Rahn and McCaffrey (1980)
N Atlantic	background		0.4–1.4	average excess sulphate	Gravenhorst (1978)

oxidation during warmer months. The proportion of total atmospheric sulphur present as sulphate particles was about 30–40% during summer but only 10–24% during winter in the SURE study area. However, this proportion varied with region, with the maximum occurring within 80–200 km of the major SO_2 source areas (Mueller and Hidy, 1983). Altshuller (1984) found that, in a number of studies in the USA and Canada, the percentage of total sulphur particle sulphate present as particulate sulphur was highest in the third quarter, followed by the second quarter and then the first or fourth quarters. Sulphur budget calculations indicated that the rate of SO_2 oxidation was also about a factor of two larger in summer than in winter in Europe. As a result the residence time of sulphur is about 60% longer in winter than in summer (Rodhe and Granat, 1983; Saltzman *et al.*, 1983). Sulphate episodes are considered further in Chapter 23.

Data taken from 1957 to 1974 in Chilton, UK, show that over this period the sulphate concentration was rising steadily but the rise was almost entirely confined to the summer months; 4.4% per annum compared with $<0.1\%$ per annum in winter. In 1982–1983 in the UK, the average proportion of the total sulphur burden present as sulphate was 10–15% (Martin and Barber, 1985; Clarke *et al.*, 1984; Salmon *et al.*, 1978).

Since the appearance of elevated sulphate concentrations depends on a complex sum of chemical and meteorological factors, statistical methods have been used on air quality data to attempt to find the variables which best predict sulphate levels. Normal regression methods require independent variables but air quality and meteorological data are highly correlated because the data are measurements of different aspects of the same underlying physical and/or chemical processes. For example temperature, ozone and sulphate levels tend to rise and fall together in part because of their common relationship to photochemical activity.

Henry and Hidy (1979, 1982) used the method of principal component analysis (PCA), a type of factor analysis, to replace the set of intercorrelated variables on air quality, not including sulphate, with a set of uncorrelated variables which are linear combinations of the original variables. The sulphate values are then regressed on these independent parameters to determine which of the derived components were most important in describing sulphate variability. The first principal component is that linear combination of variables which explains a maximum of the variability of the original variables. The second principal component is the combination, uncorrelated with the first principal component, which explains a maximum of the total variability not already accounted for by the first component. The third, fourth, etc, principal components are defined similarly. These principal components are statistically independent and normally the first few components explain most of the variability of the whole data set.

Sufficient aerometric data were available to apply the method to Los Angeles, New York, Salt Lake City, and St Louis. The results demonstrated that the relationship of sulphates to SO_2 and other meteorological and pollution variables was a function of the particular locale. In New York, Los Angeles and St Louis the principal factor was associated with photochemical activity; explaining up to 25% of the sulphate variability in St Louis. The second component in New York and St Louis was identified with atmospheric dispersion, relating local combustion sources to regional transport of polluted air. In Los Angeles the second component was related to atmospheric moisture and dominated by noontime relative humidity with the atmospheric dispersion component, although similar to that of New York, having only a small effect. St Louis results resembled those of Los Angeles in that the air chemistry factor appeared to outweigh dispersion in determining the sulphate levels. Salt Lake City is situated in a semi-arid region subject to greater normal temperature extremes than the other cities studied and isolated from most transported pollution by mountains. A copper smelter, and its associated coal-fired power station, dominated SO_2 air quality close to the city. The first principal component here represented seasonal changes in, for example, temperature, absolute humidity and mixing layer height. Air stagnation meteorological conditions were the second component, with primary emissions from the smelter a possible fifth component. Photochemistry in moist air was not important, unlike at the other cities, possibly indicating that the predominant SO_2 oxidation process was heterogeneous. The results indicate that high sulphate levels were mainly associated with air stagnation conditions which trapped local SO_2 emissions in the area. These conditions were most common in winter and possibly autumn.

Lioy *et al.* (1982) applied PCA to measurements from Lewisburg, West Virginia, using 36 h upwind air parcel trajectories and estimated SO_2 emissions and climatological data from Cincinnati, Ohio, and Salem, Virginia, to identify upwind relationships. They found that variation in summer sulphate was associated with upwind and local photochemical smog processes but was not simply related to either local SO_2 concentrations or to total upwind SO_2 emissions.

The elemental characteristics of five source classes of fine particles were employed together with PCA techniques to identify and quantify the major sources of fine particles affecting a site in Boston, Massachusetts. Coal combustion related aerosols, transported from the midwest, accounted for about 45% of fine particle sulphur with only 20% attributed to local emission sources (Thurston and Spengler, 1985).

PCA techniques were also applied to non-urban data by Mueller and Hidy (1983) using SURE data for greater north-eastern USA From PCA analysis of daily average air quality and meteorological data over a year, sulphates were most strongly related to SO_2 concentration. The association with SO_2, which was also seen in the 3 h data, was stronger than that found in urban data. Seasonality and diurnal climatological factors were weakly associated with sulphate as were surrogates for both photochemical and heterogeneous oxidation processes. At some sites 25% of annual sulphate variability was associated with high pressure stagnation conditions. PCA analyses were performed on 3 h averaged aerometric data for January 1978, July 1978 and August 1977. The most important component at most stations for the winter month was associated with high values of primary pollutants, low wind speeds and low mixing heights; interpreted as high pressure stagnation. Between 62% and 20% of sulphate variability was accounted for by this type of component. Three components were found to explain the majority of sulphate variability in the summer months and these were identified with the diurnal temperature cycle, air circulation around high pressure cells, and warm, moist air mass conditions. These principal components were thought to be representative of stagnation and transport of warm moist air, similar to the conditions dominant in the winter data. Concentration of SO_2 was an important variable in the winter stagnation components but it was missing from the warm summer stagnation components, possibly due to an increase in SO_2 dry deposition rate in summer.

PCA was used to examine the relationship between many different variables at one site. To seek relationships between observations of one variable at many different sites Mueller and Hidy (1983) applied empirical orthogonal function analysis (EOFA) to the SURE data base. This factor analysis technique is mathematically similar to PCA, deriving EOFs (or eigenvectors) from the estimated covariances of sulphate concentrations between pairs of stations. Spatial patterns are derived for each EOF by plotting the value of the eigenvector for each station, interpolating the values onto a grid and then drawing contours from the grid. Two spatial EOFs explained a large fraction of sulphate variability in summer (74%) and winter (68%), although the regional patterns differed with season. In summer the largest amount of sulphate variability was related to long-range transport while the next most important was related to air mass stagnation episodes. In winter an EOF identified with air mass stagnation in the mid-west over the zones of high emission density, combined with a sub-regional EOF concentrated further to the north west, accounted for the largest amount of sulphate variability. The geographic agreement of calculated distribution of sulphate production corresponded well with the distribution of high SO_2 emission density. In all cases the patterns were consistent with the dominant weather experienced during the periods examined. Therefore in any given month or year the zones of maximum source influence should be predictable and should be related to synoptic scale meteorological behaviour (Mueller and Hidy, 1983; National Research Council, 1983).

Ashbaugh *et al.* (1984) used this method to determine spatial patterns of fine particle sulphur concentrations in the western USA. Two large regions were identified which accounted for 33% of the variance in the data. The first included all sites in the southern part of the network and was attributed to copper smelter emissions in Arizona and New Mexico. The second included sites in the northern Great Plains and was attributed to episodic incursions of sulphur from the east.

Summary and comments

Secondary sulphate particles occur predominantly in the fine particle range ($<2\mu m$). They are the dominant species in this range although the percentage present varies with location. The background concentration is still uncertain but there is evidence that it has increased in recent years. In the areas most affected by man-made sulphur pollutants there is a wide variability in atmospheric sulphate concentration of a highly episodic nature, involving sequential periods of 2 to 6 days of high concentration amidst periods of lower concentration. The highest sulphate concentrations are seen generally in the warmer months of the year. Statistical studies have shown that the relative importance of meteorological and pollutant loading factors in determining sulphate concentration varies with the locality.

Chapter 23

Transport

The scale of sulphate transport is discussed in this chapter, together with evidence for long distance transport of sulphates in the atmosphere. The role of meteorological processes in the transport of sulphates and in the build up of episodes of high aerosol sulphate concentration are considered.

The scale of sulphate transport is defined by the distance over which the pollutants travel in concentrations that are identifiable above background concentrations. The emission of primary sulphate aerosol is negligible and little secondary sulphate aerosol is formed up to about 10 km from an SO_2 source so local sulphur pollution is largely caused by wet or dry deposition of SO_2. Between ten and a few hundred kilometres, sulphate transport is associated with mesoscale weather systems and urban or large point source plumes. Large scale, (or synoptic scale) transport occurs over several hundred to several thousand kilometres, associated with weather systems such as migrating high and low pressure areas. In mid-latitudes, sulphur pollution sources would have a range of influence of about 500 to 200 km (Rahn and Lowenthal, 1985; Wolff, 1980). AIRSOX modelling simulations gave an average distance from source to receptor of 585 km for ambient air concentrations of sulphate and 357 km for SO_2 in the SURE programme area of eastern USA. In the Adirondacks, the average distance from source to receptor was 932 km for sulphate and 810 km for SO_2 (Kleinmann, 1983).

Mesoscale transport

Transport of pollutants is seen within mesoscale weather systems such as sea–land breezes, mountain–valley breezes and urban heat island systems (Wolff, 1980). Cass and Shair (1981) observed the sea–land breeze circulation in the Los Angeles area, which transports pollutants seaward at night followed by return of aged material inland the following day. This wind reversal pattern increases the retention time for oxidation of SO_2 to sulphate and causes individual air parcels to make multiple passes over large coastal emissions sources, increasing the sulphate burden. As a result the Los Angeles atmosphere exhibits high peak day and high annual mean sulphate concentrations in spite of low background concentrations. A summer SO_2 oxidation rate of about 6%/h is seen.

In the Tampa Bay area, Florida, Young and Winchester (1980) identified a sea–land breeze circulation pattern which amplifies the effect of local pollution sources for short periods of time under certain meteorological conditions. Although sulphate transported by synoptic-scale air masses is thought to be the dominant pollutant on average, sulphate from local pollution sources may be significant over short time periods.

Plume studies were introduced in Chapter 21 with reference to sulphate formation processes and most of these observations were made during transport in the mesoscale. The results are interpreted by modelling the twin problems of transport and deposition. It is in the mesoscale that the plume becomes dispersed through the mixing layer and first comes into contact with the ground. Source height has its greatest effect, and the amount and distribution of sulphate which become available for long-range transport on the synoptic scale are determined by mesoscale effects on emission (Gillani, 1978).

Long-range transport

Transboundary transport of sulphate has been recognised for many years. Attempts to quantify long-range transport are based partly on experimental data and partly on mathematical modelling, including air trajectories, and budget calculations (Cowling, 1982; OECD, 1979).

North American evidence

The results of principal component analyses of atmospheric aerosol concentration in Chapter 22 serve to underline the importance of long distance transport of aerosols and their precursors which has been observed in some regions of the United States, notably in the collaborative New York Summer Aerosol Study (NYSAS). The results showed that on some summer days most sulphate and a significant fraction of total suspended particles (TSP) observed in New York City, originated from distant upwind regions (Wolff *et al.*, 1979a; Lippmann *et al.*, 1979; Lioy *et al.*, 1979b). Lioy *et al.* (1980) concluded that periods of high sulphate $> 15\ \mu g/m^3$ (6 h average) in New York City were associated with air parcels which had passed over the major SO_2 sources in the mid-west, while periods of low sulphate $< 5\ \mu g/m^3$ were associated with wind trajectories that had not passed over the Ohio Valley. During an episode recorded sulphate concentrations appeared to be related to residence time over the main source regions. Leaderer *et al.* (1982), analysed NYSAS data and concluded that in summer, sulphate aerosol transport was more important than local generation in producing observed sulphate levels. However a major source of sulphate in the New York City subregion in winter was local emissions of SO_2 from combustion of oil for heating (Lioy and Morandi, 1982; Tanner and Leaderer, 1982). In both summer and winter the sulphate ions were closely associated with NH_4^+ ions and strong acid at all sites. Concentrations of strong acid were high at rural and semi-rural sites and low at urban sites suggesting that a large flux of NH_3 was available at urban sites to neutralise acid sulphates (Lioy *et al.*, 1980; Tanner and Leaderer, 1982; Leaderer *et al.*, 1982).

Transport of sulphates into the state was observed in Florida (Annegarn *et al.*, 1980; Leslie *et al.*, 1978; Ahlberg *et al.*, 1978). During the winter regional scale transport occurred with the passage of cold fronts to create a climatic sulphate gradient down the Florida peninsula. During the summer cyclonic circulation carried polluted air from north-eastern USA into Florida and then on over the Atlantic Ocean for several days. The principal compound present was $(NH_4)_2SO_4$.

The impact of man-made emissions in the south-west deserts of the USA was studied by Eldred *et al.* (1983). Sulphate concentrations were recorded at 12 sites between 100 to 650 km from a group of 11 copper smelters near the US–Mexico border (the largest source of atmospheric sulphur in western USA) from August 1979 to September 1981. The majority of smelters were closed by a strike from July to September 1980, providing data from summers of normal emission both before and after the strike summer, when only 20% of the normal sulphur load was emitted. During the strike the mean sulphate concentrations at these remote sites in Arizona, western New Mexico and southern Utah were approximately 1 $\mu g/m^3$, or less than a third of the maximum level of normal summers (4 to 6 $\mu g/m^3$). The smelters increased mean sulphate concentrations by 2 to 3 $\mu g/m^3$ at sites within 100 km and by about 1 $\mu g/m^3$ at sites between 200 and 600 km distant. On average the smelters were responsible for about 70% of the sulphate at near sites and 50% throughout the rest of the region.

The SURE Program documented regional pollution episodes in north-eastern USA which took place over an area of a million km^2 and time periods up to a week, analogous to synoptic scale meteorological processes. Over the north-eastern USA they are related to two meteorological situations. In the first, which can occur in all seasons, regional air mass stagnation takes place under a broad high pressure belt with low mixing height, creating light surface winds. No evidence has been found for the transport of pollution beyond 400 km from the major sources under these conditions. The second occurs when polluted air accumulates on the west of a high pressure area between the Appalachian Mountains (or a slow moving cold front to the south) and a second quasi-stationary front across the Great Lakes north-west of an anticyclone. This situation has only been seen in summer and its behaviour provides circumstantial evidence of sulphate transport more than 1000 km from SO_2 source regions. During these episodes moist warm air is channelled inland from the south-east over the Ohio valley then north-east across New England and New York or New Jersey. In all seasons, relatively high sulphate concentrations consistently occurred on the back side of both continental and maritime high pressure cells temporarily stationary over the mid-eastern US, especially those cells which had drawn warm, moist maritime air over regions of high emissions. Regional-scale events took place 30% of the time in 1977 and 1978 (Mueller and Hidy, 1983; Lavery *et al.*, 1981; Mueller *et al.*, 1980; Hidy *et al.*, 1979).

Vukovich (1979) also reported high concentrations of sulphate, TSP and ozone in the western half, or back side relative to the direction of motion, of high pressure systems in eastern US, with low concentrations in the eastern half. Reisinger and Crawford (1982), in a comparison of model predictions with measurements, showed that sulphate pollution episodes in lower mid-west USA were caused by the same high pressure system that ultimately led to high sulphate levels in the north-east. Results from the 1977 NYSAS, reported by Lioy *et al.* (1980), indicated that high sulphate values occurred simultaneously with high ozone concentrations on the back side of high pressure systems in summertime episodes. Morandi *et al.* (1983), in a two month study of sulphur species in a sparsely populated area of New York state, confirmed this finding. Canadian studies (Whelpdale, 1978; Whelpdale and Barrie, 1982) have shown that the occurrence of high sulphate episodes is usually associated with southerly air flows behind large high pressure areas to the south and east. Sulphate measurements over Lake Erie were compared to four day back trajectories by Anlauf *et al.* (1982) and these also showed the same pattern.

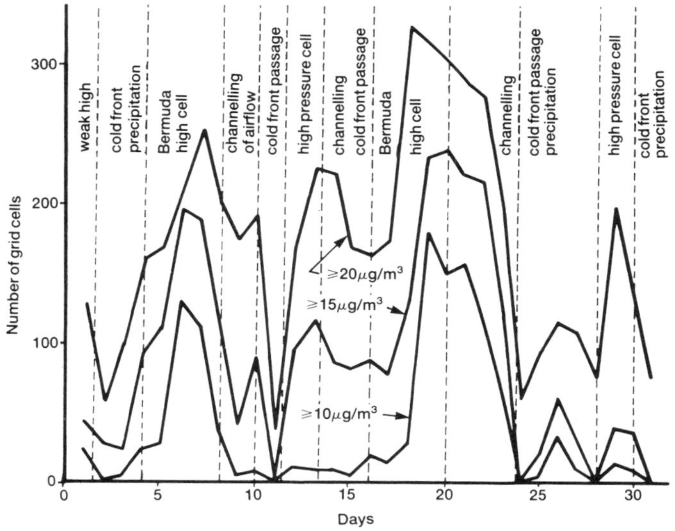

Figure 23.1 Variability of regional sulphate levels with weather conditions over the north-eastern USA for July 1978 illustration the number of 80 × 80 km grid cells experiencing 24-hour average sulphate concentrations of more than or equal to 10, 15 and 20 μg/m³ (after Hidy *et al.*, 1979)

Figure 23.1 (after Hidy *et al.*, 1979) illustrates the variability of regional sulphate levels with weather conditions over north-eastern USA Each cell represents a ground area of 80 × 80 km, covering north-eastern USA in 690 cells. A regional sulphate event is defined as at least 40–50% coverage of land area with calculated 10 μg/m³ concentrations, together with 25% coverage of > 15 μg/m³, and 10% coverage of > 20 μg/m³. Sulphate concentrations during any one month varied from no stations experiencing levels above 10 μg/m³ to more than half the grid area above 20 μg/m³. Bornstein and Thompson (1981), studying the episode illustrated in Figure 23.1, found evidence that SO_2 to sulphate transformation occurred in conjunction with long-range transport.

Regional episodes of high sulphate concentration are correlated with hazy air masses which can be tracked by meteorological satellite allowing synoptic monitoring of the formation, growth and movement of the pollutant masses (Lyons, 1980; Lyons *et al.*, 1978). Wolff *et al.* (1982a) studied the distribution of hazy air masses with elevated sulphate and ozone levels extending over most of eastern USA and concluded that the haze was primarily due to aged sulphate aerosols formed from SO_2 emissions in north-eastern and mid-western USA. The haze formed over and downwind of these source areas, was transported to the Gulf Coast region and, in two out of three episodes, was subsequently transported back to the midwest source region.

Regional sulphate episodes were linked to the season. Summer maxima were observed extending over 1500 km from Indiana north-eastward to New York State. Significant differences were seen between episode development in summer and winter, from analysis of daily 24-hour averaged data. Regional episodes occurred with appreciable frequency in the summer months, when maximum concentrations in rural areas as high as 80 μg/m³ were seen. A summer increase of about 30–50% over the annual averaged level were commonly observed in eastern USA.

Variations in rural levels corresponded to changes in temperature and dew point (Hidy *et al.*, 1978; Husar and Patterson, 1980). Sulphate episodes were also associated with elevated concentrations of ground-level ozone on a synoptic scale. These co-occurrences occurred primarily in warm, fair weather when photochemical activity was enhanced (Tong *et al.*, 1979).

In a study around St Louis, Altshuller (1985) found that during episodes when sulphate concentrations were particularly high, fine particle sulphur constituted an especially large percentage of the fine particle mass. The sum of the non-sulphur species in the fine particle mass did not show similar episodic patterns to the fine particle sulphur, in particular during the late spring and summer months, but elevated concentrations of O_3 and sulphate did tend to occur together. During high sulphate concentration episodes, transported sulphates constituted over 70% of total sulphates at urban locations.

Macias *et al.* (1981b) demonstrated that episodes of haze with high concentrations of sulphate and ozone in the desert south-west USA were due to the transport of southern Californian smog over at least 750 km. However the bulk of the sulphate aerosol was formed in the moist air of the coastal regions and not during transport across the desert. On a smaller scale Hazrati and Peters (1981), in a study of sulphate levels in an area of Kentucky with relatively low levels of SO_2 emission, concluded that sulphate levels were proportional to SO_2 emissions from 23 coal burning power plants situated at distances from 28 to 400 km from the study area and showed a strong inverse dependence on plume travel distance.

Primary pollution aerosols were used as elemental tracers of secondary aerosols by Rahn and Lowenthal (1985, 1984). The technique relies on variation of the proportions of elements between source areas due to the different mixes of major aerosol sources, fuel usage, etc, found in these areas. Regional pollution aerosols had characteristic signatures which were discernable up to several thousand kilometres downwind. In aerosols of mixed origin, regional contributions were resolved by statistical techniques. The seven elements used were As, Sb, Se, Zn, In, noncrustal Mn and noncrustal V. Ratios of these six elements to Se were used as regional signatures to normalise for variable meteorological effects such as dispersion and removal. Near strong sources of primary aerosol, primary constituents cannot accurately trace secondary aerosol components such as sulphate. The system works better outside such areas, where regional aerosols dominate. In remote areas, where a stable SO_2/SO_4^{2-} proportion has been reached, the aged regional aerosols effectively contain a sulphate component linked to the primary signature elements. This has been shown to be the case in measurements from Underhill, Vermont, where predicted sulphate concentrations from tracer techniques reproduced observed values to within ~25%, including good predictions of each of the peaks and valleys. At a less remote site, Narragansett, Rhode Island, however poorer results were seen. The authors concluded that both primary and secondary aerosols in the coastal north-east were more likely to be local in origin than those in interior New England.

On an annual basis, sulphate at both sites came roughly equally from the midwest and the north-east. The apportionments at Underhill varied with season. In summer 62% came from the midwest and 34% from the north-east, while in winter the figures were 21% and 34% respectively. There was much less seasonal variation at Narrangansett, a factor of 2 as compared to a factor of 6 at Underhill, but in the same direction. Only 3 to 5% of the annual average sulphate came from the Canadian smelters. Analysis of aerosol from three sites showed a persistent northeastern 'foreground' upon which pulses of midwestern aerosol were superimposed every few days, in response to large-scale meteorological features. When these pulses were strong enough they formed the classical episodes of elevated midwestern sulphate aerosol. Distant, midwestern sources affected the quantity of pollution aerosol in the north-east USA, and may have dominated episodically but nearer, northeastern sources were comparably important in the long term.

Husain *et al.* (1984) used the ratio of noncrustal Mn to noncrustal V in a similar way to determine the origins of aerosols collected at several rural sites in New York State. During two elevated sulphate episodes out of three the air masses were transported from the midwest. During the third the air masses at the most westerly site had travelled from the midwest but the other sites were affected by air masses from the east coast. During a later episode, when Mn/V ratios and meteorological conditions indicated a midwestern source for the high sulphate air mass, particles with coal fly ash morphology, chemistry and mineralogy were present in the sample. Samples collected before and after the episode were free of coal fly ash particles. Particles with the characteristics of oil fly ash were only seen before the episode when the Mn : V ratio was close to that expected for eastern oil-burning areas (Webber *et al.*, 1985). Tuncel *et al.* (1985) questioned the effectiveness of trace element ratios as regional signatures since they doubted whether submicron particles retained their composition patterns during transit for many hours or days. They observed fractionation of trace elements on submicron particles during long-range transport, resulting in modification of the elemental composition pattern. This implied that ratios changed with distance from the source and possibly also with season, sampling locality and other variables.

European evidence

In Sweden episodes of high sulphate concentration are associated with wind direction. Episodes of black particles (the simultaneous occurrence of high levels of sulphate and soot) are observed only in air transported from the continent. The particles contain soot, sulphates, nitrates, polyaromatic hydrocarbons and metals, and are slightly acidic. Black episodes are often accompanied by increased concentrations of SO_2 and NO_2. During white episodes (high sulphate but normal soot concentrations) smaller colourless particles are seen, consisting mainly of acidic ammonium sulphates. This situation occurs in dry, sunny weather, with winds from either the west in summer or east in spring. The two types of episode provide a useful distinction to represent particles of essentially different origin, and transition stages are seen (Brosset, 1976, 1978, 1980). Rahn *et al.* (1980a) suggested that black and eastern white episodes represent progressive stages of a unified process of atmospheric aging of similar parent air masses, with ages of roughly one and three days respectively, although originating from different regions within Eurasia.

Data from the European Air Chemistry Network covering about 15 years' observations were first analysed in detail in 1968. The results showed that deposition of sulphur, nitrogen and hydrogen ions had increased and the area of highly acidic precipitation (pH 3 to 4) located around the industrial centres of Europe was slowly expanding. Concern about this finding led to the establishment in 1972 of a five-year cooperative research program by the Organisation for Economic Cooperation and Development (OECD) to measure the long-range transport of air pollutants principally SO_2 and sulphates, and precipitation chemistry. A network of 76 ground stations in 11 countries was chosen to be representative of larger regions in the countries and relatively uninfluenced by local emissions. Airborne data collection was used to obtain more information on the vertical and lateral distribution of pollutants than could be provided by the ground stations and to estimate both the relative importance of sulphur sinks and the sulphate formation rate. Sampling was carried out during 111 flights on 92 days. To help determine the relationship between emissions, meteorological data and pollutant concentrations, an emissions survey was produced based on information from the participating countries and a number of mathematical atmospheric dispersion models were developed to interpret the data collected (OECD, 1979).

Aircraft measurements showed that most SO_2 was contained in the lowest 1000 m of the atmosphere while sulphate aerosols were less dependent on height. Generally however there was a peak concentration of sulphur species up to 500 m above ground with a progressive decline above that height. Under certain conditions, distinct plumes were seen several hundred kilometres downwind of major source areas. The mean mixing height was 1200–1300 m. As in North America sulphate episodes were associated with both slow moving fronts and with movement of stagnant air from a polluted area into the frontal area of a depression along the northern edge of an anticyclone. Sulphate aerosols were usually found as $(NH_4)_2SO_4$ with small amounts of NH_4HSO_4, but periods of highly acidic aerosols were seen in some areas (OECD, 1979).

Polar evidence

The results of recent studies demonstrate that sulphates, together with other aerosols and gases, are regularly transported from Eurasia to the Arctic during the winter half-year, over a few principal pathways, causing increase in sulphate concentration of about an order of magnitude over summer levels. At distances ranging from 5000 to 10000 km it is the longest routine transport of sulphate aerosol known (Koerner and Fisher, 1982; Rahn, 1981; Barrie *et al.*, 1981; Borys and Rahn, 1981; Ottar, 1981; Heintzenberg, 1981; Rahn *et al.*, 1980a, 1980b; Rahn and McCaffrey, 1980).

The Arctic aerosol is of an order of magnitude more concentrated in winter than in summer and has a much greater enrichment of sulphate, carbon and other pollution derived constituents. Episodic high concentrations of SO_2 are also seen and sulphate levels usually peak at the same time, (concentrations about 2–4$\mu g/m^3$) suggesting that the sulphur species come directly from polluted areas. The submicron aerosol has an aged continental character. Its continental origin is shown by the high concentration of ^{210}Pb, a decay product of ^{222}Rn, which diffuses into the atmosphere from the coninental crust. Aging is demonstrated by high ratios of secondary constituents, such as sulphate, to primary pollution constituents such as vanadium and manganese aerosols. The SO_4^{2-}/V index increases by an order of magnitude from mid-latitudes to the Arctic. Noncrustal vanadium is associated with sulphates derived from burning fuel oil. The concentration of noncrustal manganese is high in aerosols originating from Europe whereas noncrustal vanadium dominates the aerosol from eastern North America. Both gaseous and aerosol pollutants are thought to have much longer residence times in the Arctic than in mid-latitudes, roughly 20 to 100 days for the Arctic aerosol compared with 3 to 5 days in mid-latitudes. This is probably due to such factors as lower temperatures, reduced rates of wet and dry deposition due to the inherent stability of Arctic air masses, lack of cloud and direct solar radiation (Rahn and McCaffrey, 1980; Shaw, 1981; Rahn *et al.*, 1980; Dovland and Semb, 1980; Rodhe, 1978). The transport of cloud condensation nuclei associated with sulphate aerosol, and possibly largely identified with it, over 3000 to 4000 km from Eurasia to Iceland during summer was observed by Borys and Rahn (1981).

The mean SO_2 oxidation rate between Eurasian sources and the North American Arctic was 0.1%/h in early December, 0.04%/h in late February and 0.1 to 0.2%/h in early April. The residence time of SO_2 was 14 to 20 days in late autumn, 16 to 32 days at mid-winter and 10 to 19 days in April. The sulphate was present as sulphuric acid rather than neutralised sulphates across the Arctic. Accumulation mode sulphate was the principal aerosol component of the visible pollution haze (Cahill and Eldred, 1984; Radke *et al.*, 1984; Barrie and Hoff, 1984). Rahn and McCaffrey (1980) present a simple transport model and description of the chemical changes thought to be occurring during transport of European air masses over European Russia and then north to the Arctic. The most rapid changes occurred during the first few days with a maximum in secondary sulphate after 2 to 3 days. The results are quantitatively consistent with actual concentrations of sulphate found at Barrow, Alaska. Winter transport to the Arctic is controlled near the surface by the Icelandic low-pressure and Asiatic high-pressure systems and is episodic, as is that from central Europe to Scandinavia. Rahn *et al.* (1980b) suggest that the black and white episodes observed in southern Sweden by Brosset (1980) and other workers represent the first few days of aging of Eurasian air masses which may later be transported to the Arctic. They use these and other data to investigate characteristics of Eurasian air masses along a proposed transport pathway at estimated aging times of 0 days in central Europe; 1 day in southern Sweden, black episodes; 3 days in southern Sweden, white episodes; 5 days in northern Norway; 10 days in Norwegian Arctic; and 20 days in Barrow. The results of this first approximation exercise are shown in Figure 23.2 and are similar to the model of Rahn and McCaffrey (1980). The greatest changes in composition from primary to secondary pollutants are seen within the first 4 to 5 days, with a maximum in the sulphate curve. This agrees with OECD (1979) calculations showing a maximum for sulphate about 800 km downwind of its main sources, roughly the distance of the southern Swedish site from the central European source areas. The rate of aging decreases rapidly during transport, reaching very small values in the cold, dry and dark Arctic night.

Joranger and Ottar (1984) reported persistent annual variations in sulphur pollutants from 1977 to 1983, with maximum values in winter. In this season, the polar front was generally far to the south and allowed episodic transfer of pollutants at ground level from industrial areas in northern USSR. In summer the polar front was situated north of all the major industrial sources and the Arctic air was generally very clean.

Sulphur is the most common element in the Antarctic aerosol but, on average, primary sea salt derived sulphate amounts to 5% of the total in summer and to 10% in winter. The remainder is secondary sulphate aerosol, which has a marine biogenic or volcanic origin. Human activity has not been found to affect the measured values. The particles over the polar ice sheet are dominated by $NH(_4)_2SO_4$ and hydrated droplets of H_2SO_4. Marine biogenic activity in this area is responsible for the production of organic sulphur compounds, probably CS_2 and OCS, which can be transported over the central ice cap and oxidised to sulphates

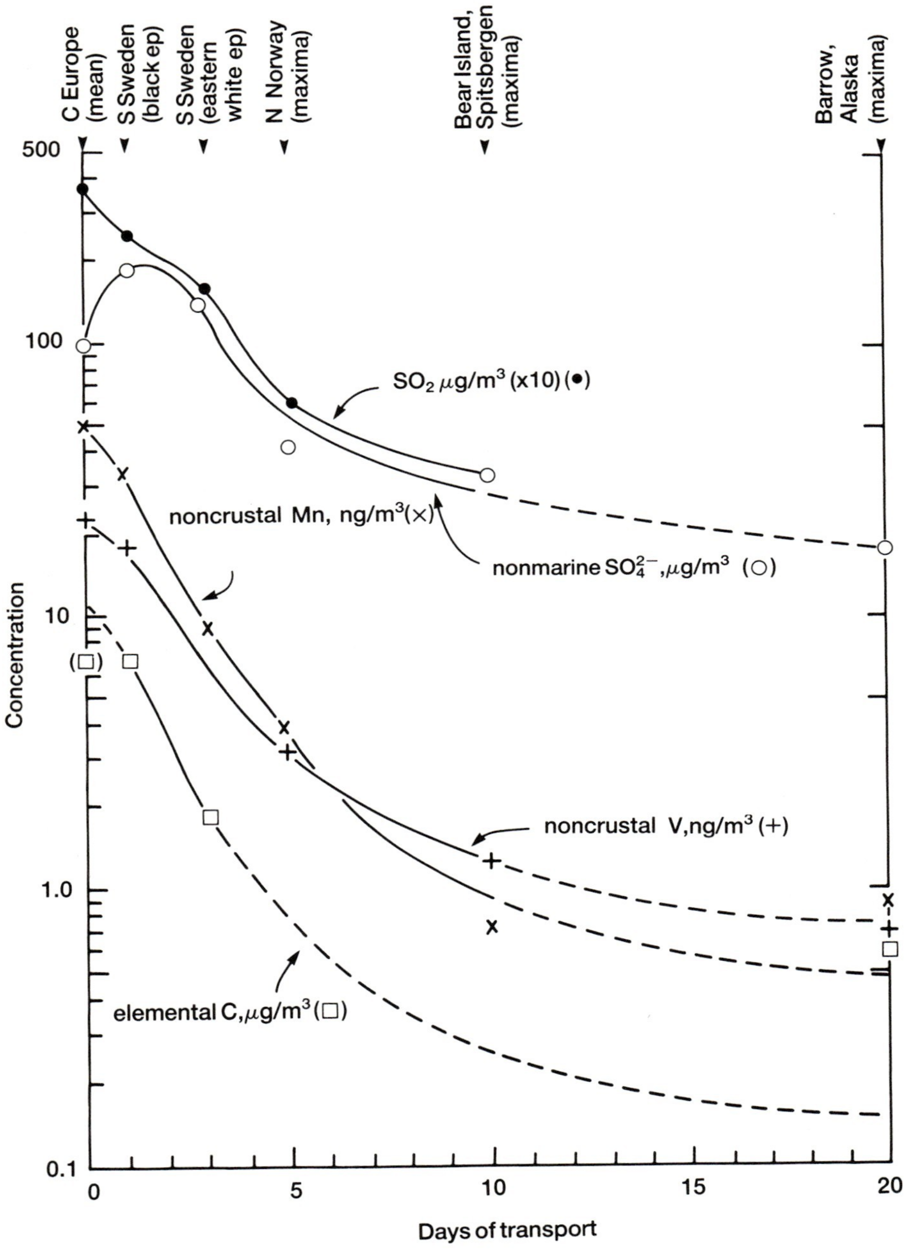

Figure 23.2 Empirical, pseudo-Lagrangian aging diagram for polluted air masses travelling between Eurasia and the Arctic during winter (after Rahn *et al.*, 1980b)

during their long lifetimes. Downwelling air from the stratosphere is not considered to be an important source of sulphate. Sulphate emissions from the active Antarctic volcano Mount Erebus have been estimated to be of the order of 0.42×10^5 t/y. However, this is only a small part of the total sulphur burden over the ice sheet, 0.042×10^9 kg/y as opposed to the total 1.5×10^9 kg/y, and only about 10% of this is deposited in the Antarctic. The rest is transported long distances over sub-Antarctic ocean surfaces where aerosol removal is rapid and efficient (Legrand and Delmas, 1984; Radke, 1982; Delmas, 1982; Shaw, 1980; Delmas and Boutron, 1978).

More recent measurements (Rose *et al.*, 1985) have shown that SO_2 emissions alone from Erebus, excluding the possible contributions of H_2S and particle sulphur, represent about 10% of the total Antarctic sulphur budget.

Summary and comments

Transport of sulphate aerosols over distances between ten and a few hundred kilometres can be seen in plumes and mesoscale weather systems such as sea-land breezes and urban heat island systems. Under certain conditions these weather systems, especially those involving wind reversal patterns, can concentrate sulphates and amplify the effects of local pollution sources. Long-range trans-boundary transport of sulphates and episodes of high sulphate concentration are associated with synoptic scale weather systems. Regional air mass stagnation and migrating high pressure systems are both related to elevated sulphate episodes. The sulphates form hazy air masses which can be tracked by satellite. The formation of sulphates takes place during long-range transport. The majority of regional atmospheric sulphate aerosol accumulation episodes occur in the summer months.

Man-made sulphate aerosols are transported to the Arctic during the winter months where they cause an increase in sulphate concentrations of more than an order of magnitude over summer levels. Aerosols are more persistent in the Arctic than at lower latitudes since the normal deposition processes do not operate as effectively. Sulphur is also the most common element in the Antarctic aerosol, but this is of natural origin.

Chapter 24

Deposition

The next step in the chain from polluted emissions to possible effects is the deposition of aerosol particles from the atmosphere. The mechanisms of sulphate deposition in dry conditions, with precipitation and by interception of water droplets (occult precipitation) is outlined, together with the relative importance of these processes. Geographical patterns of concentration and deposition of sulphates and acidity in precipitation are presented. The ionic composition of precipitation and the relation of precipitation acidity to sulphates and other ionic species is discussed. Episodicity of sulphate deposition is considered in relation to precipitation episodicity. The evidence of historical trends in sulphate deposition is compared with trends in SO_2 emissions to determine if any relationship between the two exists.

The atmospheric lifetime of sulphate aerosols is limited by deposition processes which restrict concentrations and control the transportation distance. Dry deposition is the direct collection of aerosols on land or water surfaces. It is important in the first layer of air close to the surface. Wet deposition refers to pollutant removal in precipitation and comprises the incorporation of compounds in cloud droplets and removal by falling precipitation. Dry deposition is fairly uniform while wet deposition is episodic and highly variable (Garland, 1978).

In the past, wet deposition was considered the main mechanism for removal of sulphur from the atmosphere. Recently more accurate dry deposition observations, mass balance calculations and modelling results have led to the conclusion that dry deposition, largely of SO_2, is probably of equal importance over western Europe and North America (Galloway and Whelpdale, 1980). For example a study of dry deposition in the southern Lake Michigan basin (Sievering *et al.*, 1979) showed that over half the total sulphate delivered to the lake came from dry deposition as opposed to precipitation and surface runoff inputs. ASTRAP model simulations, using meteorology from 1976 to 1981 and 1980 sulphur emissions, estimated the relative contributions of wet and dry deposition processes to sulphur deposition on an annual basis over North America. Dry deposition was more important than wet over the USA (1.34 : 1) while over Canada wet deposition was slightly more important (1.09 : 1) (Shannon, 1985). The trajectory model constructed during the OECD LRTAP study (OECD, 1979) indicated that within Western Europe about 50% of the sulphur emitted was dry deposited, 30% was wet deposited and the remainder left the region. Fowler (1984) showed that over most of England dry deposition of sulphur exceeded wet deposition. For north-west England, Wales and northern Scotland, particularly the high rainfall areas, wet deposition exceeded dry deposition.

The relative importance of wet and dry deposition processes to total sulphate and hydrogen ion concentrations measured in any area is dependent on local conditions. In areas close to sulphur emission sources, where pollutant concentrations are higher, direct deposition of gaseous SO_2 and sulphate aerosol by dry deposition is relatively more important. Wet deposition becomes progressively more important as distance from source regions increases.

Dry deposition

There are considerable experimental difficulties in the measurement of dry deposition so measurements tend to be carried out in conditions most conducive to observation, i.e. in areas of high pollutant flux, and most of the techniques have intrinsic difficulties which means the results have a high level of uncertainty.

The rate of dry deposition for particles is dependent on particle size, the nature of the surface and time of day. Most sulphate aerosol is found in the size range 0.1 to 1 μm where gravitational settling is unimportant. In the free atmosphere turbulent diffusion transports particles close to surfaces. The shallow (~1 mm) layer of air in contact with a surface is known as the laminar boundary layer and here turbulence is suppressed. Sulphate particle transport across this layer is by a number of slow processes; Brownian diffusion, eddy diffusion, interception and impaction. Overall, sulphate aerosol dry deposition is not as efficient a removal process as dry deposition of gaseous SO_2 (Garland, 1978; Sehmel, 1979; Hicks and Wesely, 1980; Fowler, 1980b).

Deposition rate is usually derived by multiplying the air concentration at a specific height by an experimentally derived parameter called the deposition velocity. This is the ratio of the deposition flux divided by the airborne pollutant concentration per unit volume at some height above the surface layer. A deposition velocity of 0.1 cm/s for sulphate aerosol has been commonly accepted but this may be an underestimate. Recent indirect measurements of particle sulphur deposition velocity are: ~1.4 cm/s by Everett *et al.* (1979) on streaker measuring devices; an average 0.7 cm/s by Hicks *et al.* (1982) over a pine plantation; an average 0.7 cm/s on three species of forest tree by Dollard and Vitols (1980); 0.039 and 0.096 cm/s on snow and 0.015 cm/s on a white pine branch by Ibrahim and others (1983); a range of 0.1 to 1.9 cm/s with an average of 0.38 cm/s over agricultural fields by Sievering (1982); a maximum of 0.1 cm/s over grass by Garland and Cox (1982). Hicks *et al.* (1982) and Garland and Cox (1982) also observed an upward flux of particles. During a field intercomparison study of methods of measurement of dry deposition at a common site, sulphate mean deposition velocities on the order of 0.3 cm/s were found by several methods. Night-time values were small, generally near zero, while daytime values of up to 1 cm/s were observed (Dolske and Gatz, 1985). Wesely *et al.* (1985) found a long term mean deposition velocity for sulphate over grass of 0.22 ± 0.06 cm/s with a variation greater than ±50% from day to day, depending on local atmospheric conditions. Peak deposition velocities occurred on windy afternoons.

Models of dry deposition to various surfaces, including vegetative canopies and water surfaces, can be found in Hicks and Wesely (1980); Hicks and Williams (1980); Davidson *et al.* (1982); Slinn (1982); Sehmel (1980) and Sheih *et al.* (1979)

and many of these also indicate variations of over an order of magnitude in sulphate deposition velocity. Long range transport and deposition models are reviewed in Chapter 25.

Occult deposition

This occurs when vegetation intercepts water drops which, because of their form (cloud, fog, frost), are not collected efficiently in standard rain gauges. The water may originate from wind-drive cloud in mountainous areas or from special micrometeorological events. These events generally are the results of a change of state in water vapour in the layer of atmosphere closest to the ground and include fog, dew and frost. They remove sulphates directly from the atmosphere by the processes described for cloud droplet formation and may dissolve previously dry-deposited sulphates on surfaces leading to a potentially severe localised build-up of acidic substances. Light mist may create higher acidities than fog since the smaller amount of moisture present causes less dilution of dry deposited substances. These special events are likely to be significant since they occur more frequently than snow or rain in some areas. Few field measurements have been made. In the worst case, pH values in dew of less than three may result after only half a day of dry deposition. The major influence on potential dew-water acidity is the initial rate of dry deposition. Occult precipitation is known to contribute significant amounts of water and dissolved ions in situations when orographic cloud envelops vegetation covered mountains and when advection fog covers coastal vegetation (Dollard *et al.*, 1983; Wisniewski, 1982).

Measured pH values in cloud and fog vary with location and air trajectory. At Whiteface Mountain, New York, Falconer and Falconer (1980) found the mean pH of non-precipitating cloud was around 3.5 during 1977 and 1979, with ~90% of all observations falling in the pH interval between 2.6 and 4.7. By contrast, rain was seldom observed with a pH <4 at this site. In summer seasons the total number of cloud hours was at least four to five times greater than the corresponding number of rain hours above an altitude of 1100 m. Waldman *et al.* (1982) found fog water from three sites in Los Angeles and Bakersfield, California, ranged in pH from 2.2 to 4.0. This was significantly more acidic and contained higher concentrations of sulphate and other ions than cloud and rain water collected in Southern California. Fog water droplets were thought to coalesce around pre-existing aerosols, when the fog evaporated the fine aerosol content remained in the air. Fog water chemistry in this area was dominated by the composition of the preceding haze-forming aerosol with subsequent condensation and evaporation controlling the observed concentrations. A mathematical model of advection fog formation produced by Hung and Liaw (1980) also supported the contention that fine sulphate aerosols enhance fog formation. Although sulphates provided the majority of condensation nuclei, fog formed more quickly when NO_x were also present.

In upland areas a large proportion of deposition can result from interception of upslope-forced fog (fog drip) and cloud droplets. Dollard *et al.* (1983) measured the rate and chemical content of occult deposition in Cumbria, UK, and concluded that including this pathway could increase wet deposition estimates by up to 20% over values recorded in conventional rainfall chemistry gauges. Rain gauges underestimated total sulphate deposition by about 15% at the site studied. These underestimates would be most significant in forested regions since wind-driven drops were captured more efficiently by trees than by short vegetation. In addition, the very large sulphate concentrations reported in cloud water, of 123 μg/ml, may make the effects of this form of deposition proportionately greater than its contribution to total deposition. The atmospheric transfer processes of turbulent deposition played a significant role in deposition of fog or cloud water to grass, with deposition velocity 2 to 3 times the bulk sedimentation velocity (Dollard and Unsworth, 1983).

Input to subalpine balsam fir forests from occult deposition was modelled by Lovett *et al.* (1982) based on measurements in the northern Appalachian Mountains, USA. In this area of high wet deposition rates, water input from occult deposition made up about 46% of bulk precipitation. Cloud droplet capture contributed 81% of total annual sulphate ion deposition.

Wet deposition processes

An arbitrary distinction is drawn between processes incorporating sulphate into cloud droplets before they begin to fall as raindrops, known as in-cloud precipitation scavenging, and those mechanisms transferring material to falling raindrops, known as below-cloud scavenging. This is for the convenience of investigators since most of the physical processes involved operate throughout the lifetime of the droplet. The whole process of pollutant capture by precipitation elements and deposition to the ground is known as precipitation scavenging or washout. The scavenging ratio, sometimes known as the washout ratio, provides an indication of the collective efficiency of all wet removal processes. The sulphate scavenging ratio is the concentration of sulphate per unit mass of rain divided by the concentration of sulphate per unit mass of surface air (Hales, 1982a; Fowler, 1980b).

Figure 24.1 (modified from Hales, 1982a) illustrates the four stages sulphur pollutants can go through prior to wet deposition. First the pollutant and condensed atmospheric water (cloud, rain, snow) intermix in the same airspace. The pollutant attaches to the condensed-water elements. Within the aqueous phase the pollutant may react physically and/or chemically. The pollutant-laden water elements are delivered to the ground in precipitation. As Figure 24.1 shows reverse processes are also possible; even after deposition resuspension processes can re-emit deposited material to the atmosphere. Although aqueous phase reactions are not essential for the scavenging process this step is influential. Oxidation of SO_2 to sulphate eliminates gaseous desorption of SO_2 from condensed water and since the sulphate reduction reaction is unimportant in the atmosphere the overall scavenging rate is enhanced as a result. Any of these steps may limit the deposition rate if they occur very slowly compared to the others. Precipitation scavenging of pollutants is closely linked to the behaviour of water in the atmosphere, so pollutant wet removal behaviour tends to mimic precipitation patterns (Hales, 1982a, 1978).

Sulphate aerosol may exist in the atmosphere for several hours or days before encountering condensed water and during this time dry deposition can occur. Mixing with condensed water may occur by relative movement or by an *in situ* phase change of water vapour, producing condensed water in the vicinity of the

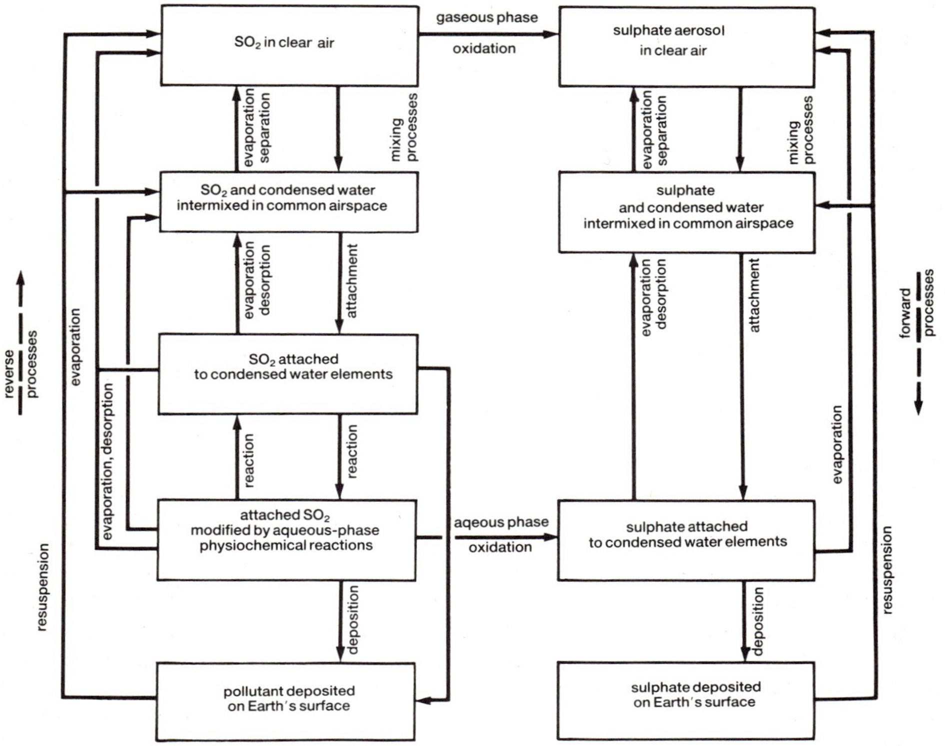

Figure 24.1 Possible interactions of sulphate, sulphur dioxide and water during atmospheric scavenging (adapted from Hales, 1982a)

sulphate particle. The sulphate molecule is then subject to the storm behaviour of the associated water. Reverse processes, such as evaporation, may take place to re-inject sulphate from cloudy to clear air. A single molecule may cycle through these processes in separate cloud systems several times before being deposited with precipitation. This implies that the volumes of air affected by cloud systems are typically much larger than the volumes of the clouds themselves (Hales, 1982a, 1979, 1978).

The attachment of sulphate aerosol to condensed water can occur by a number of mechanisms: nucleation; Brownian diffusion; impaction or interception; diffusiophoresis; thermophoresis; or electrical attachment. If these processes are considered together the overall scavenging coefficient exhibits a strong minimum for aerosol particles between 0.1 and 1.0 μm radius, known as the Greenfield gap. The size gap where particles are inefficiently scavenged corresponds to the size range containing the greatest mass of sulphate particles. However nucleation and the presence of opposite electric charges on drop and particle cause some filling in of the gap. Since H_2SO_4 and its common salts are very hygroscopic, nucleation is the most important mechanism for water attachment to sulphates, and so they tend to be scavenged early in storm systems. Dana (1980) reports that well over 50% of the atmospheric sulphate aerosol exists as condensation nuclei under normal storm conditions. The formation of sulphate aerosol cloud condensation nuclei also occurs in the plumes of coal-fired power plants. These mechanisms occur in a simultaneous and competitive manner (Hales, 1982a, 1978; Fowler, 1980b; Pruppacher and Klett, 1978).

Gaseous SO_2 readily absorbs and desorbs from condensed water and can be scavenged as such. Flack and Matterson (1980) have shown that droplets growing by water vapour condensation are able to concentrate SO_2 at levels far in excess of saturation. However, since most of the sulphur in precipitation occurs as sulphate SO_2 must be oxidised to sulphate either in the gaseous phase or in the aqueous phase after attachment to condensed water. The solubility of SO_2 is such that, at equilibrium, only about 1% of the total SO_2 loading occurs in the aqueous phase. Therefore, oxidation must be rapid for much of the sulphate to be formed in the aqueous phase (Dana, 1980; Hales, 1978).

Atmospheric measurements in clouds are limited. Hegg and Hobbs (1982) measured sulphate production rates ranging between 2 and 1500%/h in wave clouds over western Washington, USA, with the most accurately calculated rate at 50 $\pm$ 20%/h. These rates were measured in the first few minutes of reaction and may not be applicable to longer periods since oxidant concentrations would be depleted. No single oxidation mechanism could account for these findings, measured under comparatively clean conditions, but the observed sulphate production rates were relatively close to those calculated by Penkett *et al.* (1979a) from oxidation by H_2O_2. A comparatively high correlation was found between sulphate production rate and the bisulphite ion suggesting this is an intermediary species. The results suggest that aqueous oxidation mechanisms in natural clouds contribute substantially to sulphate levels in the atmosphere and that the production rate is relatively insensitive to temperature.

The primary means of delivery of sulphate-laden droplets to the ground is sedimentation. Once a sufficient number of precipitation-sized elements have formed the accretion process becomes the dominant producer of precipitation water. Accretion occurs by the sweeping action of large droplets or ice crystals falling through a field of smaller elements and attaching them (Hales, 1982b, 1978).

These scavenging processes are highly interactive and competitive for the available cloud water. Many of the mechanisms are strongly dependent on aerosol particle size. The relative significance of these individual pathways to the whole scavenging process can vary markedly with the meteorological circumstances (Hales, 1982b, 1978; Garland, 1978). Fowler (1980b) estimated the average contribution of these removal pathways to wet deposited sulphur was: diffusiophoresis 2.5%, Brownian diffusion 2.5%, impaction and interception 10%, solution and oxidation of gaseous SO_2 20% and nucleation 65%.

Ion deposition is greater at higher altitudes than at lower and Carruthers and Choularton (1984) suggested a mechanism to account for this. Deposition of ions from orographic rain by the feeder-seeder mechanism, involving collection of small cloud drops formed in low-level polluted air by initially 'clean' raindrops falling from mid-level clouds, could enhance deposition rates at higher altitudes as it is known to enhance rainfall amount. Over hilly areas where this mechanism accounted for over 50% of total rainfall, the washout process accounted for a large fraction of total ion deposition. Where this is a frequent occurrence, total deposition was expected to be approximately proportional to the total rainfall.

Plume scavenging

Scavenging of sulphate by precipitation occurs in plumes in the vicinity of large emitters such as power plants. Chen *et al.* (1982) found a sulphate deposition rate from a coal-fired power plant plume up to 45% above the mean deposition rate in the control area. Calculation showed that 70% of the observed excess sulphate could be attributed to scavenging of aerosol from the plume. Maximum deposition rates were reached after a plume residence time of about 1 h at a site 12 km from the stack. This site did not have the highest rainfall rate.

Millan *et al.* (1982) found that sulphate levels in rainwater collected under the Sudbury smelter plume were 60% higher than background levels at a distance of 12 km from the source, corresponding to a rainwater pH <4.1. However the percentage of total emitted sulphur deposited within 15 km of the stack was estimated to be only about 0.2%. At this distance from the stack primary sulphate was being removed and this formed only 1% of total sulphur emissions so roughly 20% of primary sulphate was removed within 15 km of the source.

Sulphate scavenging could not account for the deposition pattern observed around a coal-fired power plant with short stacks in an area of low background sulphate. Scavenging of dissolved SO_2 was the most probable explanation in these conditions of clean rainwater falling through a concentrated plume at low elevation (Hutcheson and Hall, 1974). Hales and Dana (1979) concluded that dissolved SO_2 made up a minor but significant portion of the total sulphur in rainwater. This could cause errors to arise in sulphate measurement where this factor is not taken into account.

Scavenging of SO_2, with an annual cycle peaking in winter, was seen by Davies (1979a) in the Norwich, UK, urban plume in conditions of both rainfall and snowfall. In a rural area SO_2 scavenging provided about 14% of total precipitation sulphur whereas in the city it was 22%. This percentage was higher still in Sheffield, UK, a heavily industrialised city, at 47%. There was a greater difference between urban and rural SO_2 scavenging levels than between the respective precipitation sulphate levels (Davies, 1979b). Ferguson and Lee (1983) found a similar pattern in a study of rainfall sulphate levels in Manchester and at Holme Moss, a nearby rural site in the southern Pennines, UK. In summer rain pH and sulphate levels were higher in Manchester than at Holme Moss, but the position was reversed in winter.

Pena *et al.* (1982) found that in an area with no local pollution sources the sulphate content in rain was negatively correlated with the SO_2 concentration in air at the surface, suggesting that SO_2 oxidation in rainwater was not an important contributor to rain sulphate concentration. During winter S(IV) species contributed up to 13% of total free acidity while during summer this fell to practically zero.

Precipitation chemistry

The pH of a solution indicates its free hydrogen ion (H^+) concentration, or acidity. The pH scale is a logarithmic scale with each decrease of one pH unit representing a 10-fold increase in acidity. This means that a drop in pH of 0.5 units from pH 6 represents an increase in H^+ ion concentration of about 2 μg/l whereas a 0.5 pH drop from pH 5 represents an increase of around 22 μg/l. A pH of 7 is chemically neutral, for example distilled water. A pH higher than 7 indicates alkalinity, with the scale running to 14. A pH lower than 7 indicates acidity, with the scale running to 0; vinegar has a pH of 3, battery acid has a pH of 1. The pH of unpolluted rainwater is affected by the dissociation of dissolved atmospheric carbon dioxide (CO_2). If the bicarbonate ion (HCO_3^-) were the only species present, rainwater would have a pH of about 5.6. By convention, precipitation with pH <5.6, equivalent to a H^+ ion concentration $>2.5\,\mu g/l$, is referred to as acidic precipitation. One of the most acidic rainstorms on record occurred in August 1977 at the Hubbard Brook experimental forest, New Hampshire, when rain of pH 2.8 fell, equivalent to more than 1400 μg/l of H^+ ions (Likens *et al.*, 1980). An episode of black, acidic snowfall with a pH of 3 and a sulphate concentration of ~20 mg/l occurred in the Cairngorm mountains, UK, an area remote from major pollution sources, in 1984. Coal fly ash particles were identified in the deposit (Davies *et al.*, 1984).

Hydrogen ions in precipitation exist in two states, free and bound, which together give the total acidity. Free H^+ ions constitute the measurable pH as determined with an electrode pH meter. Bound H^+ ions have no influence on this pH reading and can be determined only by titration with a strong base. Strong acids such as H_2SO_4 and HNO_3 dissociate completely leaving no bound H^+, so the total acidity of the solution is equal to the measured free acidity. Weak acids, such as carbonic acid (H_2CO_3) and organic acids, undergo partial and pH-dependent dissociation. The percentage dissociation of most weak acids increases sharply as the concentration decreases. For H_2CO_3 there is a contribution to both the bound and free acidity at pH >5, but at pH <5 the only contribution is to the total acidity with no free acidity exhibited. Brønsted acids (for example dissolved Al and Fe, and NH_4^+) also contribute to total acidity in a titration to pH 9, but they do not contribute to free acidity below pH 5 at the concentrations typical in precipitation (Tyree, 1981; Galloway *et al.*, 1976).

Precipitation contains a complex mixture of ions and the acidity of precipitation is not totally governed by the acidic species present. The major ions normally found are SO_4^{2-}, NO_3^-, Cl^-, NH_4^+, Na^-, K^+, Ca^{2+} and Mg^{2+}. Together with H^+ and HCO_3^- they generally account for $>90\%$ of ions found in precipitation samples. If the pH of a sample is not known it is possible to calculate it reliably from the known concentrations of the major ions, excluding H^+ and HCO_3^-. The free acidity is equal to the excess of anionic charge over cationic charge. The maximum error in pH calculated from the ion balance occurs around pH 5.6 and is about 0.5 pH units. All the major ions should be measured to ensure that the relationships between the species, from both natural and man-made sources, are not obscured. In the past this has not always been done since all the species are present in very low concentrations and there is considerable potential for error (Kallend *et al.*, 1983; Hansen and Hidy, 1982). Because of the low ion concentrations and the extreme care required when collecting and analysing precipitation samples, to avoid sample contamination, the quality of some published data is questionable. However sampling techniques and analytical methodology in general are beyond the scope of this review.

Charlson and Rodhe (1982) concluded that natural variability in the sulphur, nitrogen and water cycles could cause a decrease in rainwater pH, with considerable spatial and temporal variation. Therefore pH alone makes a poor indicator of man-made acidification, because of its large natural variability and the difficulty of measuring it accurately. They calculated that in the absence of basic materials pH values could range from 4.5 to 5.6 in areas unaffected by human activity, due to variability of the sulphur cycle alone. For example rain is likely to be more acidic in marine environments than over the continents. It is therefore not appropriate to use pH 5.6 as a reference value for acidic precipitation. Instead it is necessary to study the natural elemental cycles so that human influences can be identified. However the low pH values sometimes seen in remote regions may not be exclusively due to natural processes since local human influences and long-range transport may also be contributing.

The primary sources of H^+ ions which acidify precipitation in parts of Europe and north-eastern USA are the strong mineral acids H_2SO_4, HNO_3 and hydrochloric acid (HCl). For the most part in these regions, organic acids make a small and irregular contribution; HCl is a relatively minor component and all other components contribute only to the total acidity (OECD, 1979; Marsh, 1978; Galloway *et al.*, 1976). Cogsbill and Likens (1974) found that at several locations in north-eastern USA 65% of the acidity was due to H_2SO_4, 30% to HNO_3 and less than 5% to HCl. The chloride ion, from the combustion of high chlorine coal, can be more significant in some areas. This is the case in eastern England and parts of Wales, UK (Barrett *et al.*, 1983; Martin, 1982). In the Federal Republic of Germany the composition of precipitation, as an annual average, breaks down to sulphate 50–60%, nitrate 25–30% and chloride <15%. However chloride is deposited locally, with higher values found near the sea coast and near polluted industrialised areas (Georgii *et al.*, 1984). Galloway *et al.* (1976) analysed over 1500 precipitation samples from New York and New Hampshire states, USA, and concluded that H_2SO_4 and HNO_3 were the primary sources of H^+ ions. The weak-to-strong acid ratio was higher in southern California, USA, than that reported for eastern USA but the strong acids were still considered to control rainfall pH (Liljestrand and Morgan, 1981). The proportions of 70% of acidity due to H_2SO_4 and 30% due to HNO_3 were roughly true for rural areas, remote from sources, in industrialised countries. However, for rural areas within 50 km of towns or major traffic sources the reverse proportions applied (Martin, 1984; Martin and Barber, 1984; Barrett *et al.*, 1983). Barrie (1981) found significant correlations between observed SO_4^{2-} and H^+ ions in rain at many sites in the Canadian Network for Sampling Precipitation (CANSAP) and the US Multistate Atmospheric Power Production Pollution Study (MAP3S) networks. The molar ratio H^+/SO_4^{2-} in rain was commonly 1.6 to 2 in north-eastern USA and about 1 in acid sensitive areas of eastern Canada. Downwind of major SO_2 source regions H^+ and SO_4^{2-} ions were strongly related at sites where surface geology was non-calcareous.

Studies in the Hubbard Brook experimental forest showed that while these two strong acids were the main contributors to precipitation acidity the relative proportions had changed over time. The proportion of SO_4^{2-} ions dropped from 83% to 66%, whereas HNO_3 increased from 15% to 30% during the years 1964 to 1974 (Likens *et al.*, 1976). A later study (Galloway and Likens, 1981) found that the maximum contributions of H_2SO_4 and HNO_3 to precipitation acidity in this area during summer were 73% and 31% and during winter were 59% and 61% respectively. The data also showed an increase of about 50% in the importance of HNO_3 relative to H_2SO_4 and about a 30% decrease for H_2SO_4 relative to HNO_3 over the 15 year period 1964 to 1979. The ratio of sulphate to nitrate ions varied seasonally, for example Dasch and Cadle (1985) found a ratio of 2.4 in the summer compared to 1.7 in the winter. This was due to variation in SO_4^{2-}, since NO_3^- showed no obvious seasonal trend. The ratio also varied from place to place but over eastern North America as a whole a ratio of 1 : 1 was seen in winter while on an annual basis the ratio was 2. However a SO_4^{2-} to NO_3^- ratio of 5 was seen on a global scale (Whelpdale, 1985).

A decrease of 0.8 pH units, or almost an order of magnitude, over three years was seen near the Continental Divide in Colorado, USA. This was attributed to increasing amounts of HNO_3 in precipitation, even though H_2SO_4 was still the major contributor to total acidity. Significant increases in NO*x* release from multiple, often mobile, sources throughout western USA could be causing widespread changes in precipitation chemistry (Lewis and Grant, 1980).

Brimblecombe and Stedman (1982) studied rainfall records, from non-urban sites in North America and western Europe, over the past century and found that over this period there was a marked increase in the annual deposition of NO_3^- ions. They concluded that the increase paralleled the increases in NO*x* emissions from combustion processes and was large enough for NO_3^- ions to make an important contribution to acidity in rain. For northern Britain approximately 29% of acidity was attributable to HNO_3 and the remaining 71% to H_2SO_4 on an annual basis. However at sites in eastern England over several years the closest correlation of H^+ was with NO_3^- in rainwater. A trend of increasing NO_3^- concentration in rain and air was seen during a period of increasing NO*x* emissions (Martin and Barber, 1984; Barrett *et al.*, 1983; Fowler *et al.*, 1982). The OECD LRTAP study (OECD, 1979) found that in some areas of north-western Europe the contribution of NO*x* to precipitation acidity was as great as that from SO*x*. In Japan a study of cloud and rain water on Mt Tsukuba showed that the NO_3^- concentration, in the majority of cases investigated, was equal to or greater than the SO_4^{2-} concentration (Okita, 1977).

Rain in remote locations shows considerable variability in pH from site to site and from year to year, with a great deal of variation about the mean at each site. The pH value at supposedly background sites commonly falls below pH 5.6, as shown in Table 24.1 from early results of the monitoring programmes at remote sites carried out by the World Meteorological Organization and the US National Oceanic and Atmospheric Administration (reported by the GCA Corporation, 1981). An extensive review of precipitation acidity at background sites, including the Antarctic, is provided by Delmas and Gravenhorst (1983). Conditions at background sampling sites are affected by both the local environment and natural and man-made pollutants which have been transported long distances. Variations

Table 24.1 Average value of pH from remote NOAA/WMO sites (after GCA Corp 1981)

Location	*pH*		*Years*	*Number of values*
	Mean	*Range*		
Mauna Loa, Hawaii	5.30	3.84–6.69	1973–1976	28
Pago Pago, Samoa	5.72	4.74–7.44	1973–1976	30
Prince Christian Sound, Greenland	5.73	4.70–6.81	1974–1976	13
Valentia Island, Ireland	5.43	4.2–6.8	1973–1976	48
Glacier Park, Montana	5.78	2.60–7.10	1972–1976	33
Pendleton, Oregon	5.90	4.67–7.60	1972–1976	35
Ft Simpson, NWT, Canada	6.27	5.22–7.14	1974–1976	14
Mauna Loa, Hawaii	5.84	4.76–6.60	1976–1978	24
Pago Pago, Samoa	6.00	5.55–6.51	1977–1978	12

Table 24.2 Sulphate and hydrogen ion concentrations in precipitation

Location	*Sample type†*		*SO_4* mg/l	*H^-* μg/l	*pH*	*Comments*	*Reference*
North America							
USA, Tallahassee, FL	E–R	southern rains	0.21		5.3	mean values, Nov 1978–Jan 1979	Tanaka *et al.* (1980)
		northern rains	0.56		4.3		
USA, Gainesville, FL	T–R	min	1.18		4.51	annual means, 1977–1980	Hendry *et al.* (1981)
		max	1.36		4.66		
USA, Hubbard Brook, NH	Bulk		2.47	70	4.15	volume weighted annual means 1964–1977	Likens *et al.* (1980)
USA, New York and Pennsylvania States	Bulk	min	3.55	20	4.25	1965–1978, 9 sites,	Peters *et al.* (1982)
		max	6.72	56	4.69	mean values	
USA, Walker Branch watershed, TN	E–M		2.8	65	4.19	volume weighted means and	Lindberg (1982)
		min	0.4	6	3.75	ranges, May 1976–Dec 1977	
		max	16.0	178	5.22		
USA, Brookhaven, NY	E–M	min	2.26	38		annual averages	Raynor and Hayes (1982c)
		max	4.32	66		1976–1980	
USA, MAP3S/RAINE network							
Whiteface, Mt, NY	E–R		2.68	59.6		means of data 1977–1979	MAP3S/RAINE Research Community (1982)
Ithaca, NY			3.85	94.4			
Penn. State, PA			4.23	96.4			
Charlottesville, VA			4.07	92.1			
Urbana, IL			4.14	55.9			
Brookhaven, NY			2.53	56.4			
Lewes, DE			2.66	63.0			
Oxford, OH			3.65	77.0			
USA, Adirondacks, NY	E–M	min	0.67	30		7-ILWAS sites, monthly weighted	Altwicker *et al.* (1981)
		max	6.96	151		means, May 1978–Aug 1980	
USA, Florida	T–R	min	0.96*	6.9	4.6	5 sites, volume weighted	Brezonik *et al.* (1980)
		max	2.08*	24	5.6	means, 1979	
USA, Athens, GA	E–R		2.25		4.19	volume weighted means, Oct 1976–Sep1977	Hanes *et al.* (1980)
USA, N Minnesota	E–R	min	1.47		4.64	volume weighted means,	Thornton and Eisenreich (1982)
		max	2.1		5.32	Apr 1978–May 1979, 3 sites;	
USA, N Minnesota	T–S	min	0.67		4.3	snow cores taken	
		max	1.8		6.0	Mar 1979, 7 sites	
USA, Minnesota	E–R		2.96	12.14	5.35	mean, 1977–1980, summers only, 7 sites	Pratt *et al.* (1983)
USA, Tewaukon, ND	E–R		1.74	5.4	5.27	volume weighted means,	Gorham (1980)
Itasca, MN			1.53	10.0	5.00	3 sites, Apr 1978–Jun 1979	
Hovland, MN			1.89	21.5	4.67		

USA, Austin, TX	E–R		1.81	11.7	4.93	volume weighted means,	Feeley and Liljestrand
Prairie View, TX			2.41	9.33	5.03	1980–1981	(1983)
Highlands, TX			3.66	32.4	4.49		
USA, Davis, CA	E–R	min	0.7	1	4.35	volume weighted means,	Leonard *et al.*
		max	3.7	26	6.58	1972	(1981)
USA, Lake Tahoe, CA	E–R	min	0.3	1	4.58		
		max	2.7	26	5.79		
USA, Southern California	E–R	min	0.31	3.8		9 sites, volume weighted	Liljestrand and Morgan
		max	2.68	39		means, 1978–1979 hydrologic year	(1981)
USA, Wasatch Mts, UT	T–S		1.09		6.17	snow cores, Apr 1982, 10 sites, mean values	Messer (1983)
USA, Cascades, WA	T–M	min	0.71		4.68	5 sites, volume weighted	Logan *et al.*
		max	1.2		4.84	means, Jan–Jul 1981	(1982)
Canada, Lake Ontario basin	T–M	min	5.0		4.5	annual means, 6 sites	Chan and Kuntz (1982)
		max	8.0		7.0	1969–1978	
Canada, St Margaret's Bay, Nova Scotia	E–M	min	3.7		4.0	1979, means according to	Shaw (1982)
		max	7.8		4.3	air mass source region	
Canada, Montreal area	E–S	min	1.3	3	4.0	Jan–Mar 1980, means,	Lewis *et al.* (1983)
		max	6.14	69	7.1	6 events, 10 sites	
Canada, Alberta, urban	E–R		1.34		5.3	median values, summers only	Klemm and Gray (1982)
rural	E–R		1.12		5.4	1977 & 1978	
Canada, Ontario	T–M	min	2.83*	26	4.1	8 sites, volume weighted	Melo (1981)
		max	5.47*	84	4.6	means Jul 1978–Jun 1980	
Canada, S central Ontario	E–M	min	1.0	36		5 sites, Aug 1976–Apr 1979	Kurtz and Scheider
		max	6.6	106		weighted means	(1981)
Europe and other areas							
OECD Network	Bulk-M					Annual means 1973, 1974	OECD (1979)
Austria		min	7.3	1	4.84		
		max	14.8	43	6.09	2 sites	
Switzerland		min	1.5		4.2	6 sites	
		max	4.7		4.94		
Federal Republic of Germany		min	3.4*	26	4.13	5 sites	
		max	5.3*	73	4.37		
Denmark		min	1.1*	3	4.3	6 sites	
		max	6.4*	62	5.19		
France		min	3.4	−35	4.65	6 sites	
		max	5.7	58	6.16		
Iceland		min	1.4*	10	4.65	1 site	
		max	2.1	23	4.99		
Norway		min	0.4*	−11	4.24	21 sites	
		max	4.0*	62	5.11		
Netherlands		min	3.5*	61	4.23	4 sites	
		max	6.1*	81	4.4		

Table 24.2 continued

Location	*Sample type*†		*SO_4* mg/l	*H^-* μg/l	*pH*	*Comments*	*Reference*
Sweden		min	0.4*	15		10 sites	
		max	6.1	58			
Finland		min	1.6*	21	4.48	5 sites	
		max	3.5*	45	4.8		
United Kingdom		min	1.7*	19	4.07	8 sites	
		max	4.8*	85	4.72		
UK, Scotland	R	min	2.6*	18	4.09	means, 1975–1977	Miller and Miller (1979)
		max	4.9*	8.2	4.75		
UK, Wales	T–R	min	2.0*	10		monthly volume weighted means	Martin (1982)
		max	6.3*	43			
Ireland	T–R	min	0.4*		4.8	9 sites, mean 1968–1974	Fisher (1982)
		max	2.0*		6.0		
UK, Holme Moss, rural	T–M		4.77		4.09	volume weighted means, Nov 1979–Nov 1980	Ferguson and Lee (1983)
UK, Norwich	E–M		2.03		4.45	volume weighted means	Davies (1979a)
rural Norfolk			1.53		4.87		
UK, Norfolk, urban	E–M		2.25		4.51	volume weighted means	Davies (1979b)
rural	E–M		2.25		4.81	Oct 1977–Aug 1978	
Netherlands,	T–R	continental	9.13*	88.7	4.06	means of ~20 measurements	Asman *et al.* (1981)
Den Helder		maritime	3.94*	77.6	4.11		
South Sweden	T–M		3.3	52	4.3	mean of monthly values, 1973–1975	Granat (1978)
Poland	T–R	min	0.55		4.3	14 stations, annual means	Kasina (1980)
		max	5.76		5.3		
Poland, Cracow region	T–S		17.25		7.1	Mar 1977, Jan, Feb 1978 27 sites, means	Zajac and Grodzinska (1982)
India, near Bombay	E–R	min	3.3			cloud water	Sadasivan (1980)
		max	11.8				
India	T–R	min	0.71		5.53	9 stations, monsoon season, weighted means	Kelkar and Ashawa (1979)
		max	5.38		6.61		
China, Qinghai Province	E–R	min	1.82	0.0013	2.25	Tibetan plateau, June 1981, Urban & non-urban samples	Harte (1983)
		max	16.6	5600	8.9		
USA, Hawaii	T–R	min	0.3		4.3	9 sites, pH varied with site elevation, volume weighted means	Miller and Yoshinaga (1981)
		max	8.0				
Bermuda	E–R		1.03*	21.15	min 4.3 max 5.45	volume weighted annual means	Jickells *et al.* (1982)

*corrected for sea salt sulphate (only significant for coastal sites)
†T wet precipitation, collected over a time interval; E wet precipitation, storm events only; S snow; R rain; M rain and snow

in these factors cause the observed variation in pH from place to place and time to time. Despite this variability pH levels as high as those seen in regions influenced by industrialisation have not been found in remote areas. From this data the authors conclude that the background pH values of precipitation in a region of continental vegetation-covered soil would be in the range pH 4.5 to 5.5 and in a region of arid soil would be pH 5.8 to 7.4.

A great deal of precipitation chemistry data are available. Many of these data cannot be directly compared since sample collection methods, collection intervals and methods of analysis differ. However Table 24.2 summarises some recent information on the concentration of SO_4^{2-} and H^+ ions in precipitation, largely gained using sampling methods which collect wet deposition only. The precision of analysis methods is typically ±0.1–0.4 mg/l for SO_4^{2-} which is well below the levels usually found in the field (OECD, 1979; Kramer, 1978). As can be seen the concentration ranges are often wide but it is noteworthy that the weighted means of rainfall volume tend to vary between 1 to 4 mg/l of SO_4^{2-} regardless of the area studied. The range in pH values does not appear so wide, the majority falling between pH 4 to 6, but since a log scale is used this represents two orders of magnitude difference.

Work on monitoring of precipitation concentration and deposition was carried out by a work group established under the Memorandum of Intent on Transboundary Air Pollution, agreed by the governments of Canada and the USA in 1980. The final reports of the working groups were approved by both the Canadian and US scientists. They used results from two Canadian networks, CANSAP and APN (Atmospheric Precipitation Network), and two US networks MAP3S and NADP (National Atmospheric Deposition Programme). Data from 90 to 100 sites in 1980 were used to construct maps of precipitation-weighted mean concentrations of SO_4^{2-} and pH, shown as Figures 24.2 and 24.3. In the more remote areas of North America, sulphate concentrations in precipitation were 0.5 to 1 mg/l (Figure 24.2). However, concentrations in eastern North America were greater than 2 mg/l with values reaching 4 mg/l in an area approximately 1000 by 250 km, covering Ohio and southern Ontario. The major man-made sulphur emission source regions lie in this area. The high concentrations seen in western Canada were associated with alkaline windblown dust, while the eastern values corresponded to acidity. A similar pattern is seen in pH measurements (Figure 24.3), with highest concentrations covering an area about 600 km wide running from western Illinois and Kentucky, following the Ohio River valley, northwest-ward through New York and southern Ontario. An area of high pH in central Ontario, close to major emissions from smelting operations, was also found (Memorandum of Intent Work Group 2, 1982a). Galloway *et al.* (1984b) calculated that the 1980 volume-weighted means of sulphate concentration in eastern North America were enriched by up to 16 times relative to those observed in remote areas. The area enclosed by the 3.8 mg/l isopleth in Figure 24.2 was enriched by a factor of 16 over concentrations in remote areas, the 2.8 isopleth was enriched by 12 times, the 1.9 isopleth by 8 times and the 1 mg/l isopleth by a factor of 2.

European data from the cooperative Programme for Monitoring and Evaluation of the Long-range Transmission of Air Pollutants in Europe (EMEP) were used to prepare mean SO_4^{2-} and pH concentration fields in precipitation for Europe based on data from January 1978 to December 1982, as shown in Figures 24.4 and 24.5 after EMEP (1984). All EMEP monitoring stations were located in remote areas, away from local influences as far as possible. Sulphate concentrations of less

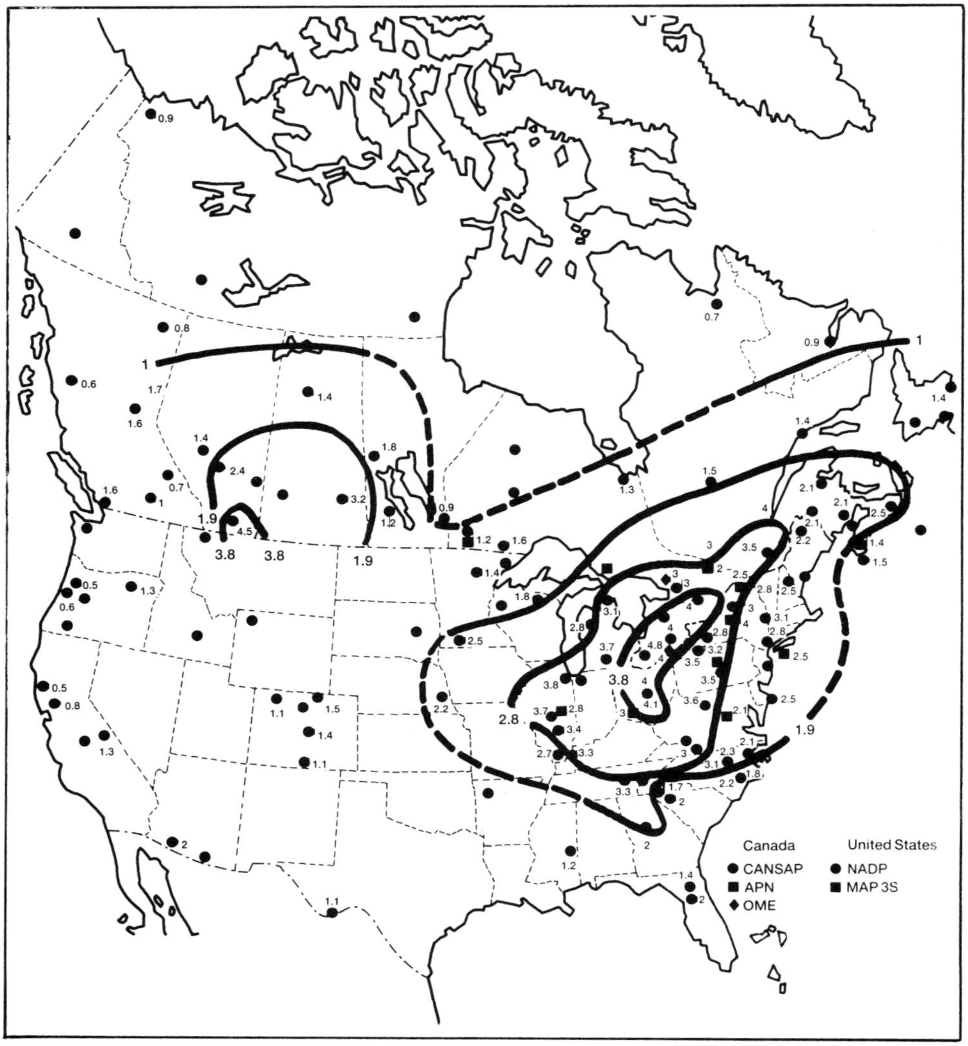

Figure 24.2 Spatial variation of precipitation-amount-weighted mean sulphate ion concentration (mg/l) in North America in 1980 (after Memorandum of Intent Work Group 2, 1982a)

than 1 mg/l occurred in western and northern Europe with the highest concentrations, above 2 mg/l, covering a large part of central Europe. The maximum acidity was also seen in central Europe, but the lowest mean pH was found in the Federal Republic of Germany while the highest mean sulphate concentration was in Yugoslavia. The pH isopleths were a different shape from the sulphate isopleths in the North Sea area and in the southeast, possibly due to the presence of alkaline substances such as Ca^{2+} and NH_4^+. The areas of highest ion concentration were over the main emission source areas but with a slight displacement in the direction of the prevailing wind.

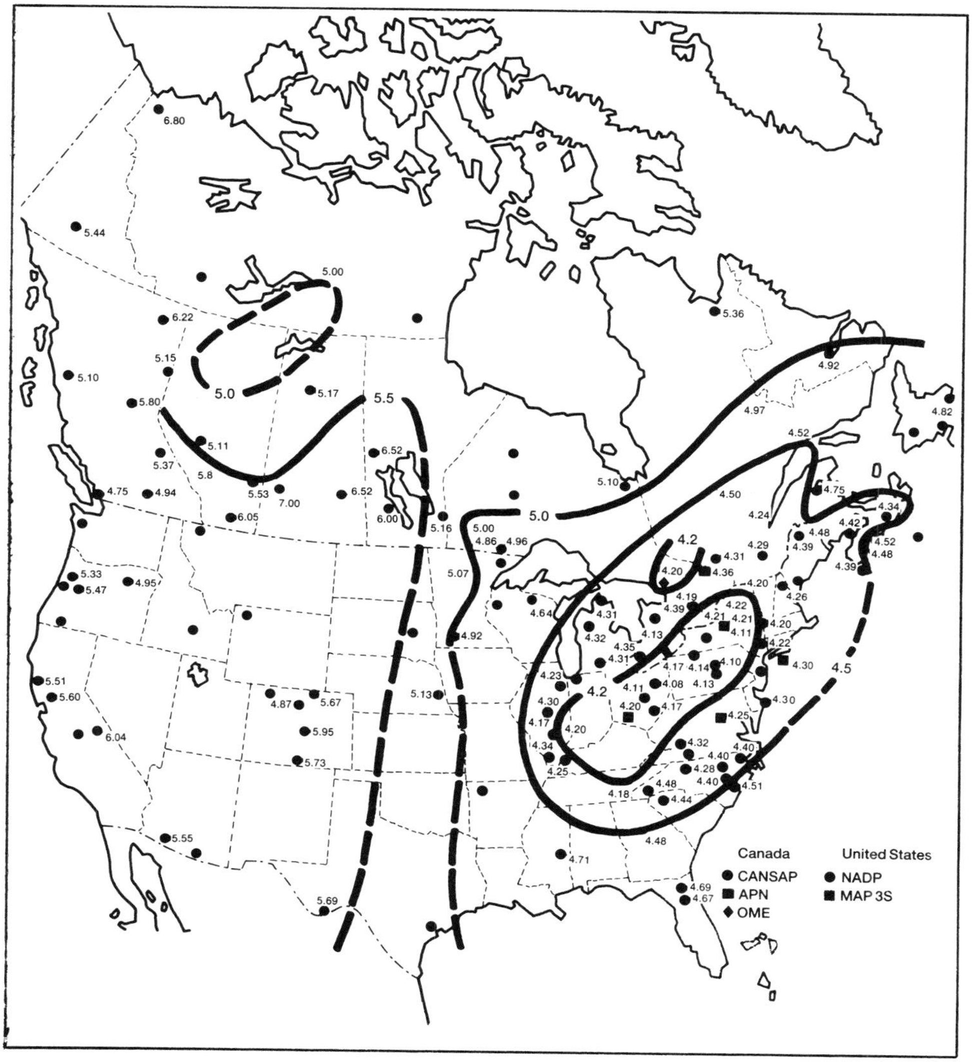

Figure 24.3 Spatial variation of precipitation-amount-weighted mean pH in North America in 1980 (after Memorandum of Intent Work Group 2, 1982a)

The flux of ions can often show a clearer picture than ion concentrations, which are difficult to interpret since the concentration is influenced by meteorological transport and transformation processes and the amount of rainfall. The concentration of SO_4^{2-} and H^+ ions in precipitation is inversely related to total rainfall on an event basis. For example SO_4^{2-} concentrations in rainwater are low in mid-Wales compared with eastern England but the larger rainfall amounts in Wales led to greater deposition of ions there (Kramer, 1978; Semonin *et al.*, 1981; Martin, 1982). Table 24.3 is a compilation of estimates of SO_4^{2-} and H^+ ion deposition rates from the recent literature. Sample collection and analysis methods and the

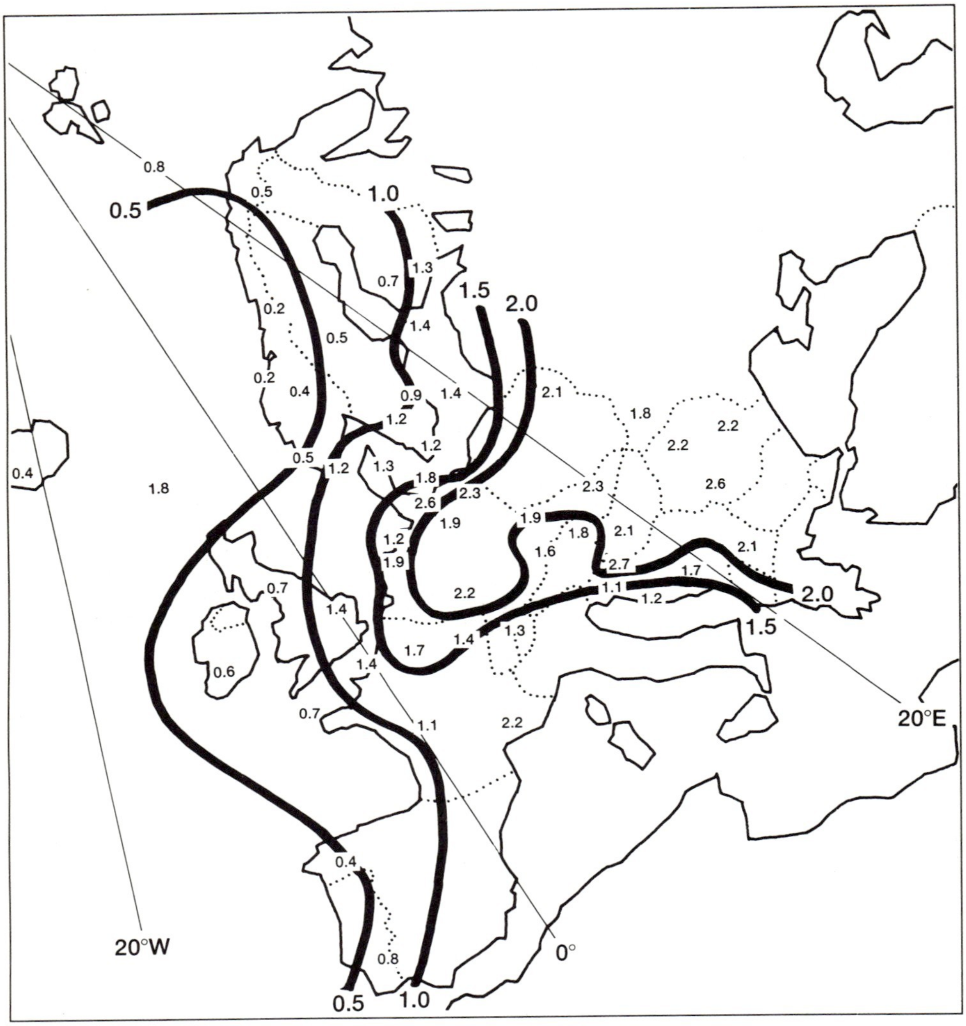

Figure 24.4 Mean concentration of sulphate in precipitation in mg/l January 1978–December 1982 (after EMEP, 1984)

time basis used in calculation may differ therefore values may not be directly comparable. Deposition of H^+ ions near emissions sources can be up to two orders of magnitude higher than that seen in more remote areas, and SO_4^{2-} deposition shows a similar pattern.

Semonin *et al*. (1981) found that although the highest sulphate ion concentrations are seen in Canada, the deposition patterns have their maximum in the USA. This could be due to the higher precipitation levels in the USA. As can be seen in Figure 24.6, the deposition of sulphate ions on a regional scale has a similar pattern to the distribution of sulphate concentration. This is also the case for hydrogen ions.

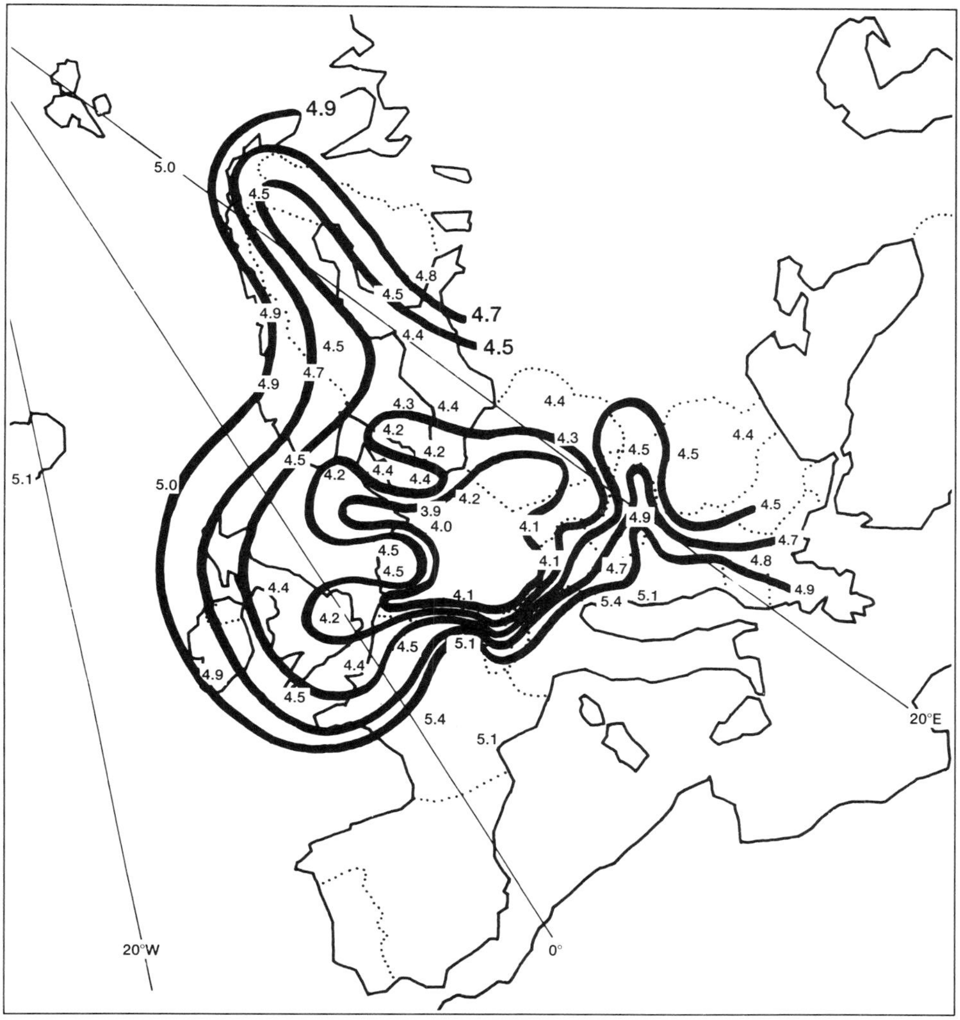

Figure 24.5 Mean pH values January 1978–December 1982 (EMEP, 1984)

The wet deposition patterns may be more variable from year to year than concentration patterns because of the added variability of annual precipitation patterns. The areas of maximum deposition in Figure 24.6 were located over and immediately downwind of the regions of maximum emissions. The zone of elevated deposition extended for a considerable distance in the direction of the prevailing winds, to the east and northeast. The pattern of hydrogen ion deposition corresponded closely with the sulphate distribution. The industrialised areas of the south-east could account for the extension of high deposition rates into the Carolinas. Mixing processes on a variety of scales may act to distribute pollutants

Table 24.3 Sulphate and hydrogen ion deposition in precipitation

Location	*Sample type†*		*S* g/m²/y	*H* mg/m²/y	*pH*	*Comments*	*Reference*
Ireland	T–R	min	0.9*		4.8	8 sites, 1968–1974	Fisher (1982)
		max	2.3*		6.0		
UK, Scotland	R	min	1.39*		4.09	means 1975–1977	Miller and Miller (1979)
		max	1.66*		4.75		
Poland	T–R	min	0.8		4.3	14 sites, annual means	Kasina (1980)
		max	3.55		5.3		
USA, Whiteface Mt, NY	E–R	min	1.91	41.2		1977 to 1979, MAP3S data	Wilson *et al.* (1980)
		max	2.33	47.1			
Ithaca, NY		min	2.23	41.5			
		max	2.43	62.0			
Penn State Univ., PA		min	2.45	53.7			
		max	3.19	75.0			
Charlottesville, VA		min	1.62	31.5			
		max	2.28	55.7			
USA, Adirondacks, NY	E–M		3.7	81.7		7 sites, monthly May 1978–Aug 1979	Johannes *et al.* (1981)
ILWAS Watersheds						Jun 1978–May 1979 (1981)	Johannes *et al.*
Woods Lake			3.88	85.9			
Panther Lake			3.93	85.5			
Sagamore Lake			3.11	69.4			
USA, Davis, CA	E–R		1.41	6.3		1972	Leonard *et al.* (1981)
Lake Tahoe, CA	E–R		1.44	16.97			
USA, Boulder Co, CO	T–R	average	0.94	min1.9	4.63	1 site, 150 weeks of data	Lewis and Grant (1980)
				max9.3	5.43		
Canada, Ontario	Bulk					monthly measurements Jan 1976–Apr 1979	Scheider *et al.* (1981)
Values for smelter working and shutdown Sudbury S (10–12 km S of smelter)		work	2.34	47.83		2 sites	
		shut	2.24	54.17			
Sudbury Centre (4–5 km S of smelter)		work	3.25	58.63		2 sites	
		shut	1.7	53.31			
Sudbury N (30–50 km N of smelter)		work	2.66	47.83		3 sites	
		shut	2.67	59.49			
Muskoka–Haliburton (225 km SE of smelter)		work	2.15	47.14		8 sites	
		shut	3.06	96.51			

*corrected for sea salt sulphate (only significant for coastal sites)
†T wet precipitation, collected over a time interval; E wet precipitation, storm events only; S snow; R rain; M rain and snow

evenly from the major emission regions, when the data are considered over a year or longer (Wilson and Mohnen, 1982).

A similar picture is seen in Figure 24.7, after UNECE (1985), showing the deposition field in western Europe. The highest values were concentrated over emission sources with a fairly uniform gradient in almost every direction. The regions of highest deposition were in Poland, Czechoslovakia and the USSR, with areas of elevated deposition (6 g/m/y of sulphur) extending into East and West Germany. Other zones of elevated deposition were seen in the UK, Northern Italy and Yugoslavia.

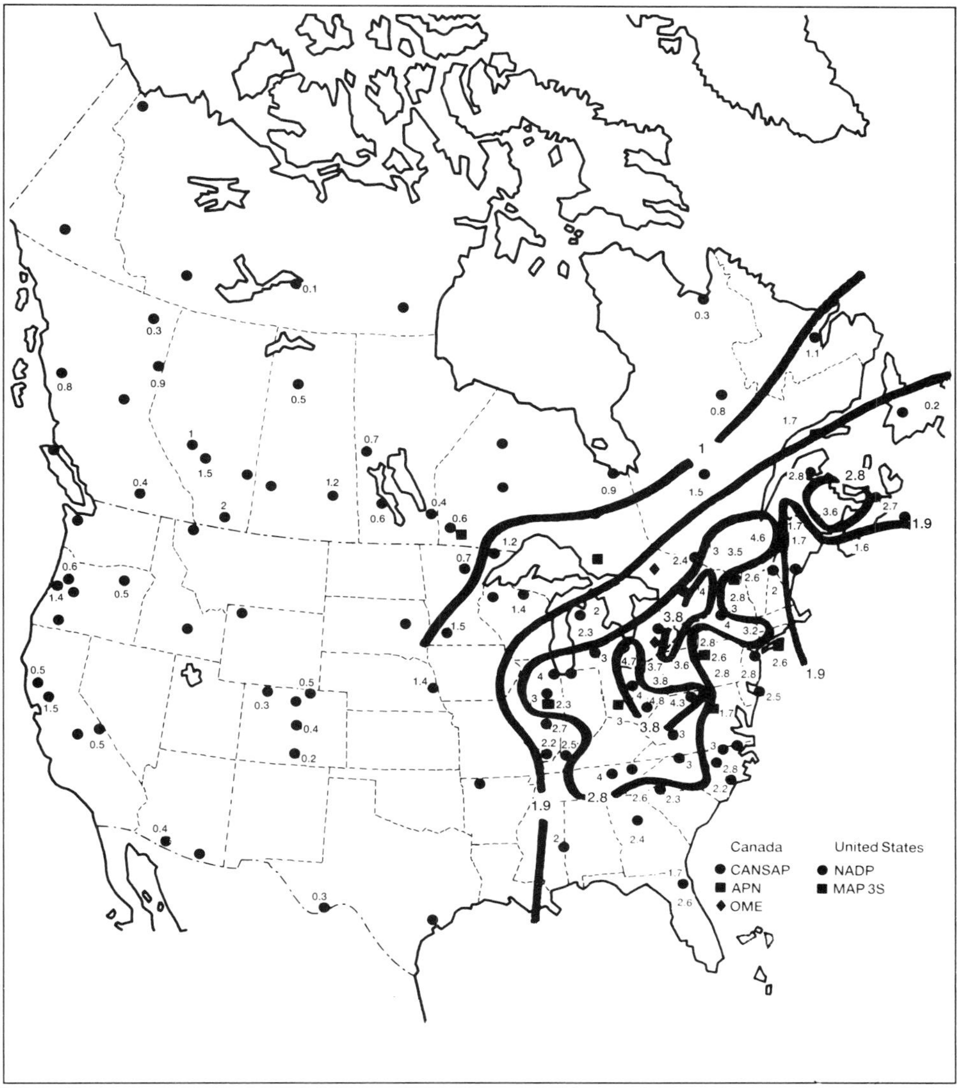

Figure 24.6 Spatial variation of precipitation-amount-weighted mean sulphate ion deposition (g/m^2) in North America in 1980 (after Memorandum of Intent Work Group 2, 1982)

Raynor and Hayes (1981, 1982a, 1982b, 1982d) investigated the variation in composition of precipitation with meteorological conditions at Brookhaven, New York, a station in the MAP3S precipitation chemistry network. Total SO_4^{2-} and H^+ ion concentrations were highly correlated; when only non-marine SO_4^{2-} was considered the correlation with H^+ was even higher. This suggested a common origin and possible prior chemical association of these ions in the atmosphere. Over 85% of SO_4^{2-} on an annual basis was found to be of non-marine, presumably

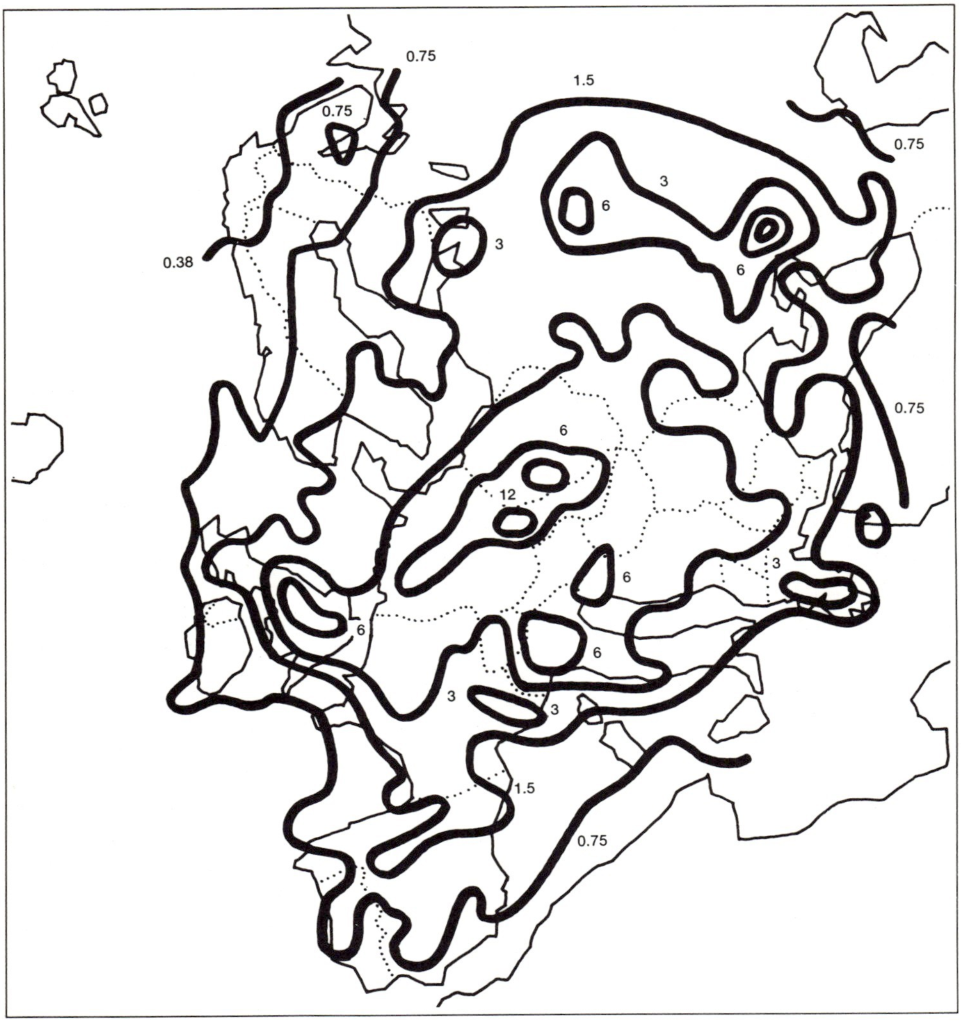

Figure 24.7 Calculated total deposition of sulphur October 1978–September 1980 ($g/m^2/y$) (after UNCE, 1985)

man-made origin. Sulphate concentrations were higher in light precipitation, but annual average concentrations were quite uniform, possibly due to the widespread distribution of emission sources and the amount of SO_4^{2-} aerosol and SO_2 always present in the atmosphere. Both H^+ and SO_4^{2-} ion deposition was greatest with cold front and squall line precipitation, thundershowers and rain showers, westerly wind directions, at wind speeds of 3 to 5 m/s and at high temperatures. The highest rate of SO_4^{2-} deposition was seen with low precipitation rates whereas H+ deposition was greatest with high precipitation rates.

Deposition of SO_4^{2-} and H^+ ions in north-eastern USA was greatest in the summer months when precipitation was lowest and least in winter when precipitation was greatest. Wilson *et al.* (1980) found that 65–75% of the annual deposition of SO_4^{2-} and 55–65% of H^+ was deposited during the summer months (May to October). Concentrations of these ions in precipitation were also highest in summer and lowest in winter, reflecting the frequent presence of stagnant or slow moving air masses heavily loaded with aerosol SO_4^{2-} during the warmer months (Raynor and Hayes, 1982b, 1982d; Semonin *et al.*, 1981; Bowersox and de Pena, 1980).

In Colorado, USA, the most acidic precipitation, with the highest sulphate concentrations, was seen in frontal or pre-frontal precipitation from low pressure centres tracking across to the south of Colorado. Maritime surface air from the Gulf of Mexico, modified by passage across the highly populated, oil-refining areas of south-east Texas, in the regional scale and Denver, in the mesoscale, was thought to be responsible (Nagamoto *et al.*, 1983). Reddy and Claassen (1985) found no systematic deviation from mean concentration values with season at two high-altitude sites in south-western Colorado.

Concentrations in Europe were generally highest in spring and lowest in autumn. Annual cycles of deposition were less marked than those of concentration and since they were related to seasonal variation in the amount of precipitation, they differed markedly in different climatic zones. Maximum deposition was seen in summer in Scandinavia although in maritime areas the minimum could fall in summer. Maximum concentrations of sulphate in precipitation were out of phase by about three months with the seasonal variation in sulphur emissions, which exhibited a maximum in winter and a minimum in summer. The reason for this is unclear but may be related to seasonal variations in meteorological phenomena or in the residence time of sulphur, related to different removal processes or rates which predominate at different times of the year (Barrett *et al.*, 1983; Granat, 1978). This pattern was seen in the European Air Chemistry Network (EACN) data. Rodhe and Granat (1984) found that, in most parts of the network, the annual cycle of concentration of sulphate had a maximum in February to May and a minimum in July to October. The wet deposition however showed a maximum in May to August and a minimum in December to March. Man-made SO_2 emissions were greatest in December to March, indicating that the fraction of sulphur exported out of the region was larger in winter than in summer.

Sulphate ion concentration and acidity increase with elevation. In Colorado, USA, deep convective storms entrain air and aerosols from the upper troposphere which are later rained out. Acidic aerosols are transported across the Pacific by the prevailing winds to the Hawaiian Islands where an orographic effect causes the entrained aerosols to be scavenged and deposited with the very heavy rains (up to 10 000 mm/y). The effect is also seen with orographic precipitation in Europe (Nagamoto *et al.*, 1983; Fowler *et al.*, 1982; Miller and Yoshinaga, 1981; Ottar, 1978).

The use of annual or monthly mean values, for example in Table 24.3, conceals large variations in individual rain events. Like sulphate aerosol concentration, sulphate ion deposition has an episodic character. In some remote areas, up to 30% of annual sulphate deposition can come down in one rain storm (Ottar, 1978). Smith and Hunt (1978) define episode days at a site as those days with the highest wet deposition which, when summed, make up 30% of the total annual wet deposition. At a site in England in 1974 30% of wet sulphate deposition occurred on 5.3% of wet days while 50% occurred on 15% of wet days. Episodicity is defined as the ratio, expressed as a percentage, of the number of episode days to the annual number of wet days. An area is highly episodic if episodicity is $<5\%$ and unepisodic if its episodicity is $>10\%$. Since some precipitation typically falls on 30–70% of days a site is highly episodic if 30% of deposition falls on 9 days. Episodicity is a relative measure and not related to total deposition at a site. Table 24.4 (after OECD, 1979) shows the degree of episodicity at several European sites.

Since precipitation is itself episodic, usually occurring between five to ten per cent of the time but reaching 20% in some mountainous areas, precipitation episodicity can be compared to sulphate episodicity. Figure 24.8 (after OECD, 1979) shows this comparison for the OECD LRTAP sites in 1974 and indicates that sulphate episodes cannot be explained by the amount of precipitation alone. This sulphate episodicity arises through variations in the quantity of pollutant in

Table 24.4 Percentages of the total wet deposition of sulphate for the 5, 10 and 25 days with largest deposition and the amount of precipitation (%) on the same days (after OECD, 1979)

Station	*5 days*		*10 days*		*25 days*		*Total number of days with precipitation*
	SO_4^{2-}	*Precipitation*	SO_4^{2-}	*Precipitation*	SO_4^{2-}	*Precipitation*	
1973							
N1* Birkenes	26	16	42	32	68	55	124
NL1* Wageningen	12	6	20	12	41	35	140
S4 Ryda Kungsgaard	18	8	33	16	60	46	105
UK1* Cottered	29	21	42	37	65	58	110
1974							
CH6 Magadino	28	27	41	41	66	67	100†
N1* Birkenes	19	12	32	20	57	44	129
N28* Fillefjell	42	10	52	15	71	33	139
NL1* Wageningen	13	8	21	14	40	29	158
S4 Ryda Kungsgaard	17	13	29	18	53	40	129
UK1* Cottered	22	11	34	18	50	34	158

*excess sulphate
†February–December 1974

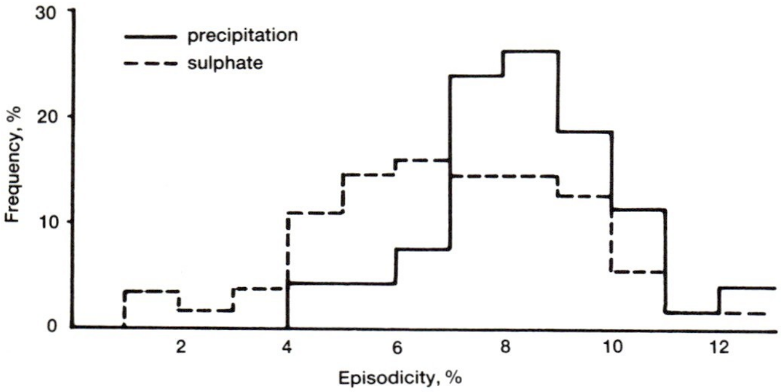

Figure 24.8 Frequency distributions of sulphate and precipitation episodicities at the OECD LRTAP sites for 1974 (after OECD, 1979)

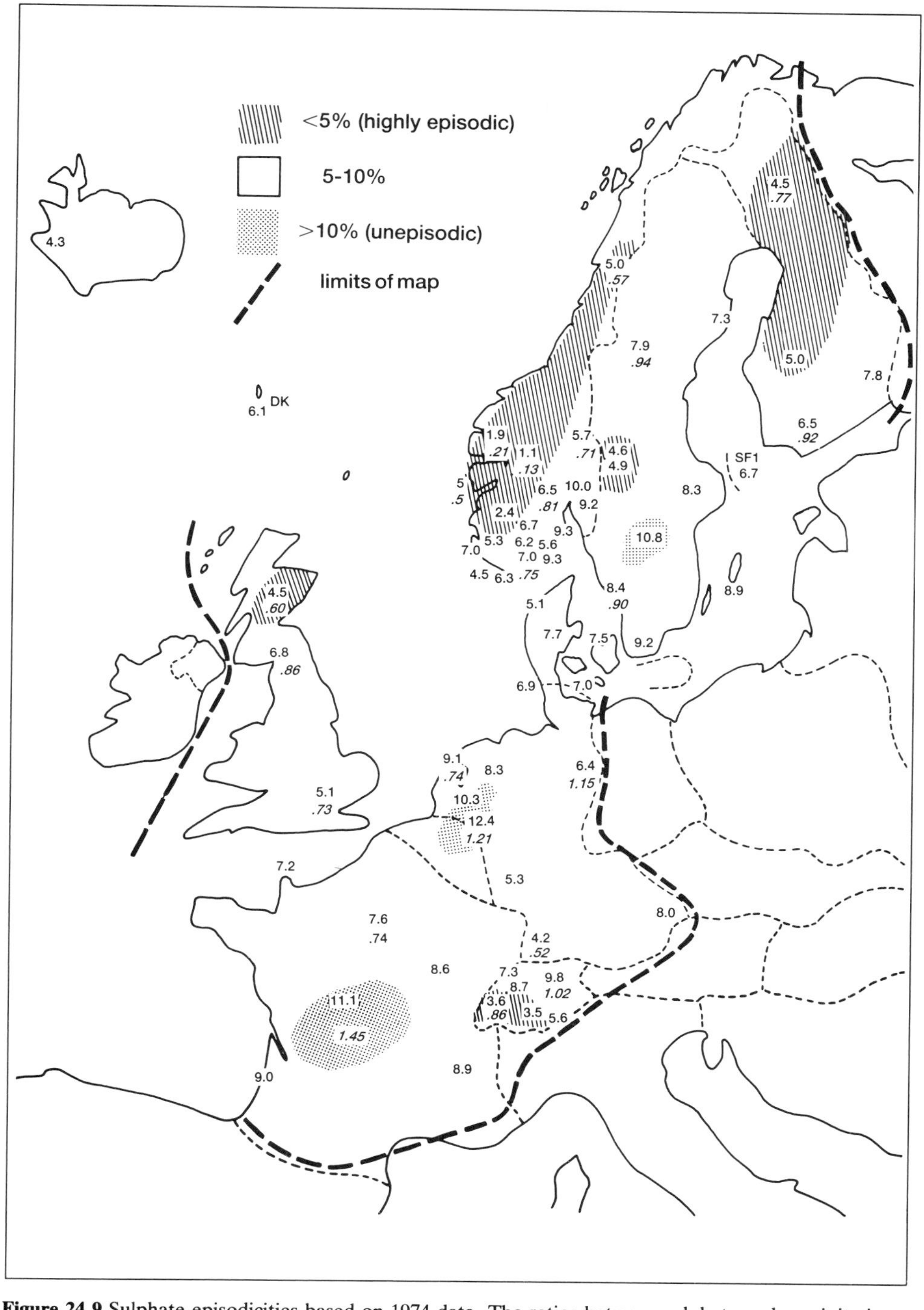

Figure 24.9 Sulphate episodicities based on 1974 data. The ratios between sulphate and precipitation episodicities at some of the sites are shown in italic (after OECD, 1979)

the air associated with rain. The pattern of episodicity in Europe is shown in Figure 24.9 (after OECD, 1979) using data from 1974. In episodic areas, the ratio between sulphate and precipitation episodicities is <1, in Western Norway down to <0.05, indicating that sulphate episodes are not caused by precipitation episodes alone. Analysis of 1973 data shows a similar general picture but larger areas of Norway and Finland would then be classified as highly episodic (Smith and Hunt, 1978; OECD, 1979; Barrett *et al.*, 1983). Dry deposition is much less episodic. The episodicity of dry sulphur deposition at a site in southern Scotland is about a factor of 1.5 less than that for wet deposition of sulphate, and a factor of four less than that for acidity. Wet deposited acidity exceeded the episodicity of sulphate by a factor of two. However, episodes showing the highest acidities (pH <4) were characterised by large concentrations of sulphate in rain, sulphate particles and SO_2 in air, poor visibility and 48 h back-trajectories over industrial regions of the UK or continental Europe (Fowler and Cape, 1984).

Spatial variability can be seen in many studies. Liljestrand and Morgan (1981) in southern California, found that non-marine sulphate decreased from the western coastal measuring stations to the inland and mountain sites. Their calculations indicated that less than 2% of the man-made emissions of SO_2 and NOx were locally scavenged by precipitation, on an annual basis, because of the arid climate. In Florida, northern sites in the Florida Atmospheric Deposition Study showed significantly higher concentrations of excess sulphate than southern sites, corresponding roughly to the distribution of sulphur sources. On a smaller scale, Martin (1982) studied the influence of a valley near the sea in Wales, UK, on the composition of rainwater sampled weekly. In the valley bottom 15% more H^+ was deposited than on the adjacent hillside, often accompanied by its equivalent of SO_4^{2-} ion. This is consistent with polluted air being held below the valley tops until washed out by rain, as occurs with fog, thus affecting the valley bottom but not the hilltops.

The acidity of precipitation is not determined just by the acidic species present. Other components such as Ca^{2+}, Mg^{2+} and NH_4^+, deriving from alkaline soil dust and gaseous NH_3 generated by bacterial action, can neutralise acidic species and thus influence the extent of acid precipitation (Sequeira, 1982; Chan and Kuntz, 1982). NH_3 was the most important neutralising agent on the coast of the Netherlands. Relative neutralisation by NH_3 varied from 15 to 72%; the average neutralisation seen with winds blowing from the Continent was 56% and with winds from the sea 33%. Lack of this neutralisation would result in a decrease in precipitation pH from 4.1 to 3.6 in rain of Continental origin and a drop in pH from 4.1 to 3.9 in maritime precipitation (Asman *et al.*, 1981). Georgii (1981), reviewing WMO precipitation network results for 1972 to 1978, suggested that SO_4^{2-} concentration and its relation to Ca^{2+} largely determines precipitation pH. A stoichiometric relation of SO_4^{2-} to Ca^{2+} gives pH 5 while an excess of SO_4^{2-} leads to a lower pH, for example an excess of 40 μmol of sulphur would give pH 4. If partial neutralisation of H_2SO_4 in rain by Ca^{2+} in particulate matter is taking place, then the correlation between $[H^+]$ and $[SO_4^{2-}–Ca^{2+}]$ should be positive. This is seen in data from Schauinsland, FRG. In India also, alkaline soil dust was mainly responsible for the neutralisation of man-made acidic pollutants (Khemani *et al.*, 1985).

Studies in north-central USA (Munger, 1982; Thornton and Eisenreich, 1982; Gorham, 1980) demonstrate the role of basic ions. Major ions were determined in precipitation-event and snow core samples from several sites along a 600 km transect from the agricultural prairies of North Dakota (on calcareous soils) to the north-eastern Minnesota forest (on thin, non-calcareous soils and underlain with granitic bedrock). The latter area is subject to man-made pollution, including SO_4^{2-} haze, from eastern North America. Acidity increased 4-fold from west to east; Ca^{2+} and Mg^{2+} decreased 2–3-fold across the transect as the effect of alkaline dust and NH_3 decreased with increasing distance from the cultivated prairie. Sulphate had both natural sources from soil in the west and man-made sources in the east; the lowest concentration was found at the central site. Wet deposition of H^+ decreased 8-fold and SO_4^{2-} deposition decreased 1.5–2-fold from east to west of the transect. Acidity increased during winter when snow and freezing temperatures decreased airborne soil dust and the evolution of NH_3. Sulphate concentration was also greatest in winter at the eastern sites. The authors concluded that precipitation chemistry in central north America was controlled by interaction between soil-derived alkaline dust and gaseous NH_3 from the cultivated prairie and man-made acid aerosols from the urban-industrial Lower Great Lakes–Ohio Valley region. Feeley and Liljestrand (1983) in eastern Texas calculated a background contribution to precipitation alkalinity from soil dust sources of 40 μ equivalents/l. In areas already acidified, NH_3 would not provide significant buffering capacity against further acid deposition. The SO_4^{2-} and NO_3^- concentrations in both rain and snow in central Alberta are large enough to acidify precipitation if at least part of the potential acidity is not neutralised. Basic ions, especially Ca^{2+} and NH_4^+, are the determinants of pH in this area (Klemm and Gray, 1982).

In a review of recent precipitation chemistry data for North America Munger and Eisenreich (1983) concluded that precipitation chemistry was dominated by local and regional land use patterns and proximity to man-made emissions. Acidity was highest in precipitation where emissions of precursor gases, derived from human activity, were highest and concentrations of neutralising soil components were lowest.

Historical trends

Sulphate concentration in rain water has been measured for over a century in short term projects. Annual averages from around the 1880s showed sulphate levels in rural rain of 3 mg/l with suburban values up to 20 mg/l and industrial urban values up to 50 mg/l in the UK. An annual free acidity for Manchester in 1870 is reported equivalent to 420 μm/l of H^+ or pH 3.4. Historical measurements of sulphate and acidity of rain are reviewed by Martin (1979).

The European Air Chemistry Network (EACN) has been collecting and analysing precipitation for over 20 years at between 50 and 120 sites in western Europe. The network used bulk sampling techniques to collect both wet and dry deposition from containers continuously exposed to the atmosphere. Bulk samples are subject to contamination from windblown dust, leaves, insects and bird droppings amongst other things, and so are of lower quality than wet deposition only samples collected using later techniques. The difference in chemical composition of precipitation collected in bulk and wet-only collectors was investigated by Soederlund (1982). He found that, on average, bulk collectors gave 4.5% higher SO_4^{2-} concentrations than wet-only collectors. No difference was observed for H^+ (calculated from pH). At sites with large local sources of contamination, such as Ca^{2+} from soil which also affects pH, the difference could be several times higher. Lewin and Torp (1982) showed that biological contamination had no significant influence on SO_4^{2-} and NO_3^- concentrations in precipitation samples so this data can still be used even from heavily polluted samples. However data on the concentration of strong acid and pH were severely affected. Changes which occurred in analytical techniques, sampling techniques and analytical laboratories during the sampling period may also have affected the results. Details of the EACN network and an analysis of early results are given by Granat (1972). Concentration data collected over the period 1955 to 1979 are presented by Rodhe *et al.* (1984).

The long term behaviour of non-marine, or excess, SO_4^{2-} in precipitation was analysed by Granat (1978). Marine SO_4^{2-} was important only within about 50 km from the sea coast. Both concentration and deposition of SO_4^{2-} showed considerable long term fluctuations which were markedly different between areas in different directions and at different distances from major sulphur-emitting areas. Deposition was constant or decreasing over the decade to 1975 in most of the area covered by the network, although some smaller areas showed a relative increase. At stations in Norway and Sweden, where 20 years of data were available, both concentration and deposition increased at about the same rate as sulphur emissions in Europe. This increase could be explained almost completely as a change in the frequency of winds from major sulphur emitting areas associated with precipitation at these stations. The general decrease in deposition over Europe together with the observed meteorological shifts suggests that more European sulphur is now being transported and deposited elsewhere possibly further east.

Kallend *et al.* (1983) analysed data from 120 EACN stations where 5 or more years' data over the period 1955 to 1975 could be selected, although many of the stations operated over different periods. A statistically significant trend of increase in sulphate concentration was seen at 23 stations with one showing a decrease. In Scandinavia, 12 sites with at least 18 years' data showed an average increase in sulphate concentration of about 2.5% per year. Nitrate concentrations showed a significant increase at 55 sites, with no station showing a decrease. An average rate of increase of about 6% per year was seen at the 12 Scandinavian sites. These sites also showed increases in H^+ concentration averaging 7% per year.

Rodhe and Granat (1984) analysed more than 25 years of data on sulphate in precipitation from the EACN at up to 100 monitoring stations. In Scandinavia, the sulphate concentration increased by roughly 50% between the late 1950s and the late 1960s. Most stations in Norway and Sweden showed a decline by an average of 20% since the early 1970s. Data from the UK and continental Europe showed less systematic variations, but it was only in data from the UK, Ireland and Iceland that no upward trend was noticeable from 1955 to 1970. The long term changes in precipitation sulphate in Scandinavia were in fair agreement with changes in man-made SO_2 emissions over the period in countries known to contribute significantly to deposition in Scandinavia.

Barrett *et al.* (1983) found some increase in European rainfall acidity in EACN data and, on sulphate and nitrate deposition, saw a large number of positive trends which, even though not statistically significant individually, gave the impression of an upward trend. Ferguson and Lee (1983) studied past records of rainfall chemistry in a rural area in the southern Pennines, UK (an area of uplands surrounded by industrial towns) and the nearby city of Manchester. Rain in the 1970s was found to be substantially more acidic than it was in 1950 and 1954. The difference in rain composition between Manchester and the rural site today may be smaller than in 1954 but the actual levels of sulphate in rural rainfall have not declined.

Given the limitations associated with both sulphate concentration and SO_2 emissions data it is not possible to conclude that there is a non-linear relationship between wet deposition and emissions in western Europe. In smaller areas with longer deposition records, such as Norway and Sweden, a linear increase in deposition with European SO_2 emissions can be seen. However there remains some doubt about the reliability of the bulk deposition data that these conclusions are based on (Rodhe and Granat, 1984; Kallend *et al.*, 1983).

Data comparable with those of the EACN network do not exist in America. Precipitation chemistry networks operated for only a few years at a time and the only long term continuous measurement program is the Hubbard Brook Ecosystem Study, a single site project dating from 1963 located in the White Mountain National Forest, New Hampshire. Cogbill and Likens (1974) and Likens and Butler (1981) used the fragmentary available data from the past two decades for eastern North America to draw conclusions about general temporal trends in precipitation chemistry in eastern USA. Ion balance methods were used to calculate pH where this was not originally measured. Their analyses indicated that precipitation with pH <5.6 was prevalent over much of eastern USA by 1955–1956. The measurements from 1975–1976 showed a substantial southward and westward extension of the area subjected to acidic precipitation and an intensification of acidity in north-eastern and south-eastern regions of USA since the mid 1950s. The SO_4^{2-} contribution to acidity in central New York State, dropped from 78% to 65% and NO_3^- increased from 22% to 30% from 1955 to 1973. Stensland (1980) compared precipitation data collected in central Illinois in 1977 with data taken from the same site in 1954. The 1977 samples had a median pH of 4.1 compared with a median pH of 6.05 calculated for 1954 by ion balance methods. The SO_4^{2-} concentration was lower in 1954, at 2.88 mg/l, than the 3.36 mg/l observed in 1977. However higher pH in 1954 was considered to be due to elevated Ca^{2+} and Mg^{2+} levels, from soil erosion caused by the dry weather experienced at that time, rather than to lower concentrations of acid-forming anions.

A continuous record of precipitation chemistry starting in 1963 exists at the Hubbard Brook Experimental Forest, based on bulk deposition sampling measurements. From 1964 to 1977 there was a downward trend in SO_4^{2-} concentration together with a marked upward trend in NO_3^- concentration. The mean annual pH of precipitation from 1964 to the present was variable but showed no significant long-term change, although short-term changes occurred (Likens *et al.*, 1980). Over nineteen years, from 1963 to 1982, a 34% decrease in sulphate concentration occurred which was roughly comparable with the 39% decline in SO_2 emissions in the north-eastern US states, while emissions from midwestern states decreased by 18% from 1965 to 1978. The trends in sulphate deposition paralleled the trends in concentration. Although H^+ ion concentrations have varied by a relatively small amount over the whole period, there has been a significant decline of about 28% since 1970 (Likens *et al.*, 1984). A set of precipitation event samples were collected at Hubbard Brook over three years from 1975 to 1978. Over this period a downward trend of 7% was seen in sulphate concentrations as compared with a decrease of 5% in regional emissions of SO_2 over the same period (Munn *et al.*, 1984). Lipfert and Dupuis (1985) investigated implications that the impact of emissions from a small paper-making plant at Lincoln, about 10 km to the north of Hubbard Brook, would tend to mask the effects of regional SO_2 reductions. Both theoretical analysis of the source characteristics and statistical analysis of the precipitation chemistry record indicated that any influence of local pollution sources on Hubbard Brook precipitation chemistry was minimal.

Since 1965 the US Geological Survey (USGS) has operated a network of nine precipitation quality monitoring stations in and near New York State collecting bulk deposition samples. Peters *et al.* (1982) analysed data from 1965 to 1978 and found that during this period SO_4^{2-} concentration in precipitation decreased by an average of 1 to 4% annually at seven of the nine stations, a similar trend to that seen at Hubbard Brook. In the west of New York State H^+ concentration increased by about 3% annually since 1965, equivalent to a total decrease of 0.2 pH units. In the eastern part of the state H^+ concentration decreased by a comparable amount. As a whole these trends balanced out to suggest that precipitation pH was unchanged since 1965 though station-to-station comparisons revealed significant local differences. Bilonick and Nichols (1983) concluded, from a statistical study of the USGS data, that there was no evidence for a long term change in the mean level of acidity and SO_4^{2-} levels were also constant.

Miles and Yost (1982) attacked the USGS data for inadequate analytical quality control procedures and produced a data series containing a pattern of elevated H^+ ion concentrations, primarily in low pH samples, from the later collection years thus making meaningful ion concentration or pH trend analysis impossible. Peters (1984) described the extensive quality control techniques used for USGS data, similar to those of Miles and Yost (1982), and reiterated the finding of Peters and others (1982) of a decreasing trend in sulphate concentration.

Hansen and Hidy (1982) critically reviewed north-eastern USA rain chemistry data relating to acidity to quantify uncertainties and biases associated with both original sampling and analytical methods, and problems of re-interpretation such as merging of inconsistent data and inclusion of partial selected data rather than other available information. They concluded that precipitation in this region was more acidic than expected from natural baseline conditions but that historical data were of insufficient quality and quantity to demonstrate detailed long-term trends. A region of elevated acidity from Alabama eastward through Florida was seen which was not present in data from the 1950s. The relationship between climatology and total ionic composition of precipitation was thought to show a strong link between the droughts of the 1950s and elevated pH, which could obscure any connection with air pollution.

The National Research Council (1983) concluded that the Hubbard Brook and other data demonstrated an overall reduction in sulphate concentration similar to the general reduction in SO_2 emissions. They found 'no evidence for a strong non-linearity in the relationships between long term average emission and deposition' in eastern North America. The only direct evidence of non-linearity came from European data, based on bulk precipitation collection. Any difference between the two regions may be the result of differences in meteorology, latitude, the spatial distribution of sources or other factors.

Oppenheimer *et al.* (1985) compared the variation in SO_2 emissions from smelters in western USA over a four-year period with the variation in sulphate concentration in precipitation in the Rocky Mountain States. The data showed a linear relationship between emissions and sulphate concentration at sites more than 1000 km away from the emissions sources. Emissions varied throughout the test period, and the concentration of sulphate in wet deposition varied in parallel.

Snow and ice cores collected from glaciers provide samples of trapped chemical species which may have remained unaltered since their deposition, thus providing a record of past deposition patterns. Herron (1982) found that SO_4^{2-} was enriched by about a factor of three in Greenland snow and ice in the past 200 years and concluded that this was due to emissions from fossil fuel combustion. Neftel *et al.* (1985) found that sulphate ion deposition in the Greenland ice sheet increased by a factor of ~2 between 1895 and 1978. The sulphate level was already considerably elevated over pre-industrial values by 1904. Emissions of man-made SO_2 already surpassed natural sources in the late 1950s, as shown by comparison of the concentration increase in the ice with source strength. No corresponding increase was seen in Antarctic ice where SO_4^{2-} concentrations are of purely natural origin.

Summary and comments

Combustion-generated sulphate aerosols are predominantly found in the size range 0.1 to 2 μm where atmospheric dry deposition processes are slow and inefficient; dry deposition is not therefore considered a significant removal process in most areas. However, recent field measurements suggest that dry deposition may be more important than theory suggests. Wet deposition is the most important removal process for sulphates, with nucleation contributing an estimated 65% to wet-deposited sulphur and the solution and oxidation of SO_2 contributing 20%, in regions remote from sources. Once the sulphate ion is attached to cloud water elements it is subject to the storm behaviour of the associated water. In the past, deposition of sulphates in fog, dew and frost was overlooked in measurements of wet deposition. This can lead to a significant underestimation of total sulphate deposition, especially since the ion concentrations in these forms of deposition are relatively high. A large proportion of annual sulphate deposition can come from fog and cloud droplet capture even where wet deposition rates are high. Precipitation scavenging occurs in plumes from coal-fired power plants, removing mainly primary sulphates and dissolved SO_2, relatively close to the plants.

On a regional basis the relative importance of wet and dry sulphur deposition processes to total sulphate and hydrogen ion concentrations is dependent on factors such as distance from sulphur emission sources and meteorological conditions. Wet deposition of sulphate aerosol is of most relative significance in areas of high rainfall far from emission sources.

Precipitation contains a complex mixture of acidic and alkaline ions with the overall acidity of the solution governed by the ion balance. Unpolluted rainwater, in equilibrium with dissolved atmospheric CO_2, would theoretically have a pH of 5.6 but natural variability in the sulphur cycle alone could cause pH values to range from 4.5 to 5.6 in areas unaffected by human activity. Only precipitation with a pH < 5.6 is termed acidic. In areas receiving acidic precipitation the H^+ ions are largely associated with SO_4^{2-} and NO_3^- ions. Although the proportions vary from place to place, in remote areas of industrialised regions commonly about 70% of the acidity is due to H_2SO_4 and 30% to HNO_3. In areas of Europe, up to 15% can be associated with HCl. There is evidence of a recent increase in the contribution of HNO_3 to precipitation acidity in some areas, possibly due to mobile NO*x* sources. Basic species, especially Ca^{2+}, Mg^{2+} and NH_4^+, have a great influence on precipitation acidity. The quality of precipitation chemistry is dominated by local and regional land use patterns and proximity to man-made emissions. In general, acidity in precipitation is highest where emissions of precursor gases derived from human activity are highest and concentrations of neutralising components are lowest.

The rate of sulphate and H^+ ion deposition with precipitation is episodic in nature. In some remote areas up to 30% of the annual sulphate deposition can fall in one rainstorm. These sulphate episodes cannot be explained by the episodicity of precipitation alone, but by episodicity of sulphate concentrations in the ambient air associated with the precipitation.

In Europe there is evidence that there have been changes in sulphate ion deposition and rainfall acidity similar to trends in man-made emissions. Over the European area as a whole, it is not possible to conclude that there is a non-linear relationship between deposition and emissions, given the limitations of the data. In Scandinavia, where a longer data record exists, changes in precipitation sulphate were in fair agreement with changes in man-made SO_2 emissions in countries known to contribute to deposition in the area.

In eastern North America, although the evidence is more fragmentary, an increase in precipitation acidity from the 1950s to the 1970s is seen. This has been ascribed to a decrease in basic cations rather than an increase in man-made anions. Trends in precipitation sulphate concentration over a 20-year period are consistent with a general proportionality between SO_2 emissions and sulphate deposition. The National Research Council (1983) found 'no evidence for a strong non-linearity in the relationship between long term average emission and deposition' in this area. A linear relationship between SO_2 emissions and sulphate concentration in wet deposition has been seen in western USA.

Sulphate concentration has increased by up to a factor of 3.7 in Greenland ice over the last 200 years, probably due to emissions from fossil fuel combustion, whereas no change is seen in Antarctic ice where sulphate concentrations are of natural origin.

Chapter 25

Long-range transport and deposition models

Traditional scientific methods have so far not proved adequate to determine the relationship between source emissions and deposition of sulphur, largely because of a lack of suitable field data. Regional scale transport is not directly measured at present for the large number of emissions sources and deposition areas of interest under the variety of meteorological conditions required. Hence mathematical modelling of the processes involved, to the extent that they are understood, can provide a useful framework for putting the diverse field data into perspective.

The relationship between emissions and deposition can be evaluated either as theoretical material balance calculations or by the statistical manipulation of field observations such as air trajectories. The modelling process itself is outside the scope of this review. Detailed overviews of transport and deposition models and the modelling process are presented by Altshuller *et al.* (1983a) and the US National Research Council (1983) together with critical assessments of the applications and performance of these models.

Statistical trajectory analyses

An air mass trajectory is the path that would be followed by the centre of mass of an air mass parcel if it were carried by the wind. Rodhe (1972) first applied trajectory analysis to the origins of long-range transported sulphates and found that highest concentrations at sites in southern Sweden occurred with trajectories from the UK and much of continental Europe. Man-made emissions from sources up to 1000 km distant made significant contributions. The considerable uncertainties involved in using trajectories are described by Pack *et al.* (1978) and Clarke *et al.* (1983). Individual calculated trajectories are subject to a range of errors but the technique can prove valuable in relating air quality data at isolated locations to pollution transported from large area sources (Pack *et al.*, 1978). The use of statistical analyses on an aggregate of many trajectories to reduce random error can give a useful comparison of contributions from different source regions, or sectors, where the source areas are large enough. The method has the advantage of requiring few predetermined assumptions but those made in trajectory calculations, unlike other mathematical models.

The European LRTAP study (OECD, 1979) performed sector analyses of observed ambient air sulphate aerosol concentrations according to their direction of origin as determined by air mass trajectories. Long-range transport of sulphur

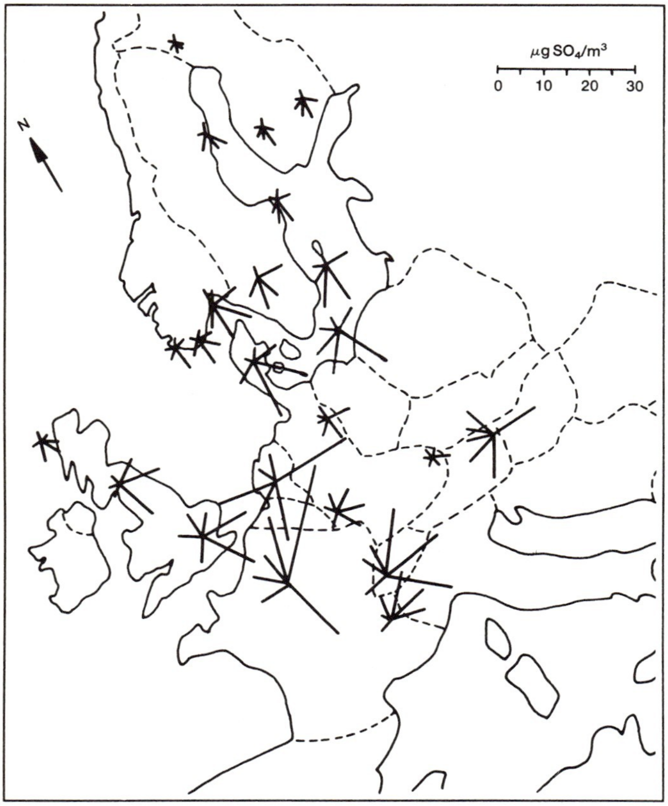

Figure 25.1 Mean concentration of aerosol sulphate for six transport sectors at various European sites (OECD, 1979)

compounds mainly takes place in the lowest 2 km of the atmosphere; so for the air trajectory calculations the observed wind at 850 mb level was chosen, which is typically 1200–1500 m above sea level. The data were allocated to six directional areas and a mean value was calculated. Figure 25.1 shows results from groups of stations for aerosol sulphate. The values are shown by straight lines in the middle of each sector with lengths proportional to the concentrations. Sectors with high mean concentrations are not randomly oriented but are generally directed towards the main sulphur emitting areas. The results show that long range transport of sulphur contributes significantly to the concentration of sulphate aerosol in air but has less effect on SO_2 concentrations (OECD, 1979).

Figures 25.2 and 25.3 show the mean SO_4^{2-} and H^+ ion concentrations respectively deposited in precipitation for six transport sectors. The values are shown by straight lines in the middle of each sector with lengths proportional to the concentrations. The maxima for precipitated sulphate and strong acid fall in westerly or south-westerly sectors for most locations since most precipitation takes place with winds from these sectors. The sectors with high mean concentrations are generally directed towards the major sulphur source areas. These results give

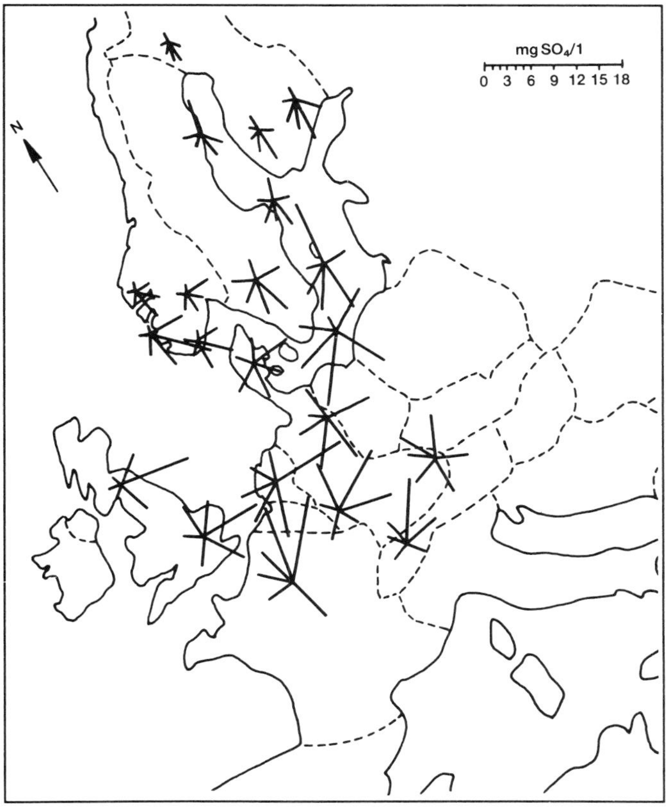

Figure 25.2 Mean concentrations of sulphate in precipitation (mg/l) for six transport sectors at various European sites (after OECD, 1979)

no clear indication of which sources give rise to what fraction of the concentration in a given sector. However, it is possible to conclude from geographical position, that emissions in the UK and Ireland on average give rise to a sulphate concentration in precipitation of about 1 mg/l in south Norway and the Netherlands in addition to the 0.8 mg/l already present as background in Atlantic air. If this calculation is applied to strong acid concentration in precipitation, sites in southern Norway receive an average of 21 μg/l and the Netherlands 20 μg/l from the UK and Ireland. For Norwegian sites the average concentration in clean oceanic air is 25 μg/l. In some areas the concentration of strong acid seems to be affected by regional differences, however in areas such as Scandinavia the distribution of acid is similar to that of sulphate. The evidence indicates that considerable transport of sulphur pollution is taking place over national boundaries in Europe. The total deposition in a given country depends on the size and geographical position of the country in relation to the major emission sources (OECD, 1979).

Pacyna *et al.* (1984) used a simple trajectory model to show that measured concentrations of sulphates and trace elements from long-range transport at

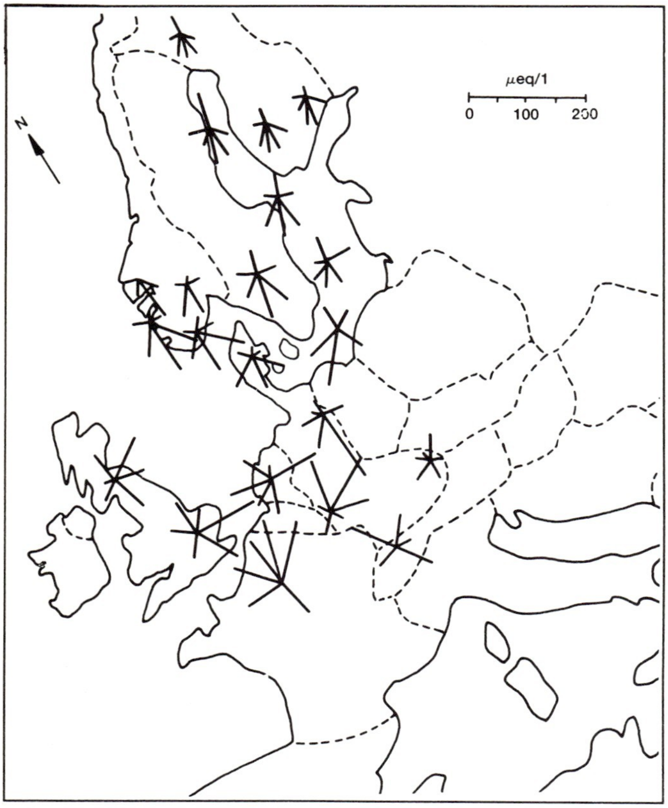

Figure 25.3 Mean concentrations of strong acid in precipitation μeq/l for six transport sectors at various European sites (after OECD, 1979)

Birkenes, Norway, were in good agreement with estimated man-made emissions in European countries, for days with only slowly changing air trajectories.

North American data from the MAP3S network have been used for trajectory analysis. Raynor and Hayes (1982d) found that the evidence from spatial distribution of pollutant concentrations and trajectories to six MAP3S stations for a single storm event strongly suggested the midwest, in particular the Ohio River Valley, as an important source region. Concentrations of man-made air pollutants in precipitation at the sampling stations used were roughly proportional to the number of major point sources in the regions traversed by the air parcels arriving at the stations during the precipitation period. Directional variability in wet deposition at Whiteface Mountain, New York, and Urbana, Illinois, MAP3S sites in 1978 was studied by Wilson *et al.* (1982, 1980). Results showed that 56% of the wet deposition, delivering 64% and 62% of the annual SO_4^{2-} and H^+ ion deposition respectively, at Whiteface Mountain came from air masses that had previously (<48 h) passed over the midwest Ohio Valley sectors. Air masses from the Canadian sectors provided 26% of the annual precipitation delivering 31% each of the annual SO_4^{2-} and H^+ ion deposition. At the Illinois station the south-westerly sector accounted for 71% of the total wet deposition, delivering 67% and 78% of the annual SO_4^{2-} and H^+ ion deposition respectively. No steep gradients in normalised wet deposition of pollutant ions were found in the area of the study. Normalised annual SO_4^{2-} deposition values ranged from 35 mg/m^2 in the Ohio Valley to 27 mg/m^2 at Ithaca, New York, and Penn State University and 18 mg/m^2 at Whiteface Mountain, in 1979. Atmospheric mixing of pollutants on the regional scale and varying precipitation patterns appeared to act together to make the distribution of pollutants deposited per unit amount of precipitation fairly uniform over time scales of a year or more.

Henderson and Weingartner (1982) reached a similar conclusion as to directionality using the MAP3S data covering October 1977 to October 1979 from the Ithaca station. In summer about 63% of the SO_4^{2-} and 60% of the H^+ ions deposited arrived from the octant covering the Ohio River Valley and the midwest. In winter this sector again predominated contributing 47% of SO_4^{2-} and 44% of H^+ ions. However, most of the NH_4^+ ions which act to neutralise SO_4^{2-} also came from this octant in both seasons. The precipitation event producing the largest acidic loading recorded at Ithaca over the period was associated with trajectories which passed through the Ohio River Valley; the air masses spent about 36 h in the area between the south of Ohio and Pittsburgh.

Tanaka *et al* (1980) found that the elemental composition of rain in Florida differed with northerly or southerly air flows. In southern rain the average sulphur concentration was 0.56 mg/l with an average pH of 5.3 while northern rain had an average sulphur concentration of 0.21 mg/l with an average pH of 4.4. In northern rain the existence of a strong relationship between pH and sulphur suggested that the measured form was largely H_2SO_4 transported from northern pollution sources while the weak relationship found in southern rains implied that sulphur occurred as neutralised sulphates. Transport of North American pollutants has been detected at even greater distances. Jickells *et al.* (1982) found that rain in Bermuda originating in the sector covering continental North America had an average pH of 4.4 compared to an average pH of 5 for the Caribbean sector and 4.9 for the North Atlantic sector. The acids present were almost entirely H_2SO_4 with a small amount of HNO_3: the upper limits of their contributions to free acidity were 94% and 26% respectively. This same effect was seen in Hawaii where the more acidic precipitation was associated with air flows from the north (Miller, 1979).

At the Hubbard Brook experimental forest, New Hampshire, five-day back trajectories associated highest values of SO_4^{2-} and H^+ with winds from the south–southeast to the south–southwest and with looping trajectories over New England. These trajectories accounted for about two thirds of the wet deposition of H^+. A trend analysis for SO_4^{2-} concentrations over three years, using trajectory statistics and a simple linear chemistry model, showed a downward trend of 7% after normalising for meteorological variations. This is comparable with a decrease in regional SO_2 emissions of about 5% over the same period (Munn *et al.*, 1984).

Kurtz *et al.* (1984) performed a sector analysis of an eight-day acidic deposition episode in the Muskoka–Haliburton region of Canada during which 12% of the total 1981 precipitation fell, accounting for 28% of the wet deposition of both SO_4^{2-} and free H^+ ions. Both air and precipitation concentrations were found to depend on the sector of origin. Almost every day with a high sulphur concentration ($> 5\ \mu g/m^3$) was associated with a south-west back trajectory. All of the SO_4^{2-} deposited was associated with trajectories from the south-west or south, the direction of industrialised southern Ontario and the Ohio Valley. During the period of this study a depression was observed in the pH in two streams flowing into a nearby lake from pre-episode values of 6.7 and 6.3 to post-episode values of 5.6 and 4.7, a greater decrease than was seen during spring runoff. This indicates that stream pH decreases associated with episodic rainstorms can be of similar magnitude to pH decreases associated with snow melt and spring runoff.

Theoretical models

Regional-scale models are largely based on the extension of air pollution analyses of plumes from single sources or the multiple sources found in urban areas. Trajectory (Lagrangian) models, the most common approach, require calculations of pollutant diffusion, transformation and removal to be performed in a moving frame of reference tied to air parcels which are transported in accordance with an observed or calculated wind field. The second main type (Eulerian) divides the area of interest up into a 2- and 3-dimensional array of grid cells of fixed columns, through which all air parcels flow, to produce calculated deposition patterns. In either case the calculations provide a differential material balance of the emitted constituents and their secondary chemical species. The elements simulated usually include emission injections, air flow, mixing, chemical transformation and wet and dry deposition (Altshuller *et al.*, 1983a; Hidy, 1984). Figure 25.4, after Altshuller

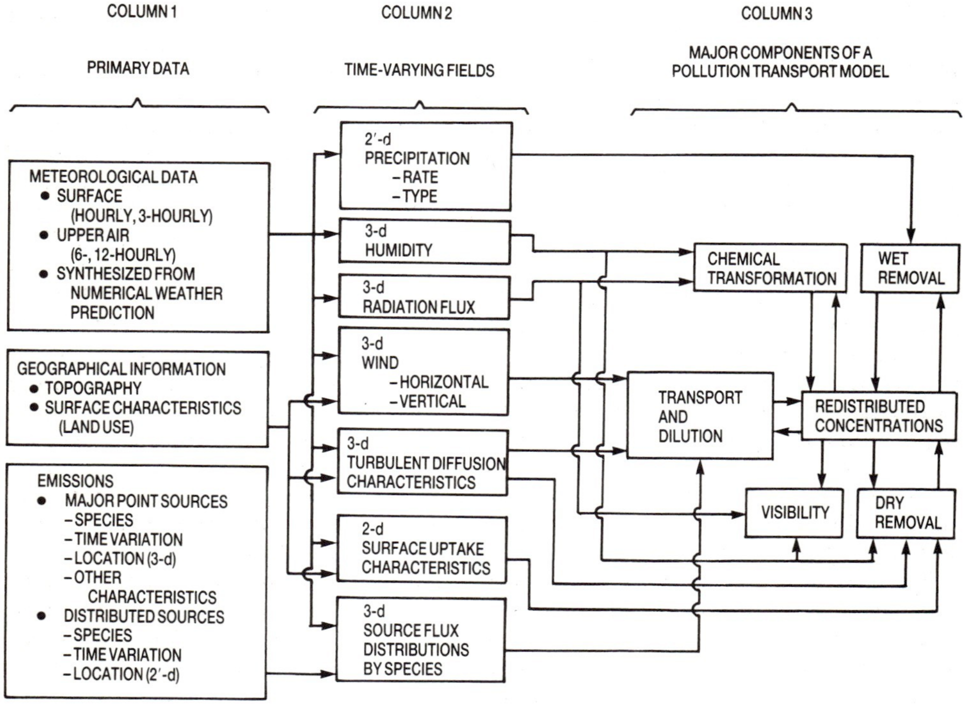

Figure 25.4 Interaction among the data sources and components of a pollution transport model (after Altshuller *et al.*, 1983a)

et al. (1983a), shows diagrammatically how the major components of a general transport model are interconnected and how the primary data interacts with the time-varying fields, shown in column 2, which drive these models. Some processes must be simulated in all models but the choice of formulation influences the character of the model's other components. Useful information can only be gained from a model if the input data is of a quality and quantity consistent with the structure and assumptions of that model. Model performance must be tested quantitatively by comparing model calculations with field measurements. However data for comparison are limited in both spatial and temporal coverage and in the types of parameters which have been measured. Examples of the type of results obtained follow.

A major problem in establishing source–receptor relationships in Europe is the large uncertainty in SO_2 emission inventories for the various European countries, especially those in Eastern Europe (Beilke, 1985). As a result of the OECD LRTAP programme a further study including all European countries was undertaken known as the Cooperative Program for Monitoring and Evaluation of Long-Range Transmission of Air Pollutants in Europe (EMEP) under the auspices of the Economic Commission for Europe of the United Nations (UNECE). During the first phase of EMEP 70 monitoring stations in 20 countries were in operation and sector analysis of the results was carried out. From the linear trajectory model transboundary fluxes of sulphur pollution were calculated as shown in Table 25.1 (after Eliasson and Saltbones, 1983) which shows the calculated average monthly sulphur budget for the two year period 1 October 1978 to 30 September 1980. It can be see that, according to model estimates, 17.3×10^3 t of sulphur is deposited in the Netherlands (No. 17) in an average month. The two largest contributors are the Federal Republic of Germany with 4.5×10^3 t of sulphur and the Netherlands itself with 4×10^3 t of sulphur. Sulphur in the undecided category (UND) is the background concentration of sulphate in precipitation from clean air established by OECD (1979) and amounts to 0.27 mg/l for Norway and Finland and 0.40 mg/l for other countries (EMEP, 1982). Table 25.2 (after Eliasson and Saltbones, 1983) shows the deposition due to foreign and indigenous sources for each country as a percentage of the total sulphur deposition, based on the figures in Table 25.1, together with the average monthly sulphur emissions for each country. The figures given are only correct to within a factor of two for most of Europe but they do give an indication of the extent and magnitude of transboundary transport of sulphur compounds. In most European countries, sulphur deposition due to foreign emissions is an important contribution to total deposition; in 13 countries it greatly outweighs the deposition from indigenous sources. The undecided background deposition contributes less than 20% of the total in all but five countries (Eliasson and Saltbones, 1983). Hoegstroem *et al.* (1981) showed that sulphur deposition in Sweden increased by about 30% between 1971/1972 and 1978, a period when the contribution from local sources decreased by 30%. In 1978 barely 20% of total sulphur deposition originated in Sweden compared with 30% in 1971/1972.

Table 25.3, after the National Research Council (1983), shows the characteristics of several regional-scale models for north-eastern North America used to assess sulphur transport and deposition in the context of acid deposition. The US–Canada Memorandum of Intent on Transboundary Air Pollution Working Group compared eight of these models. They were all run to calculate monthly and annual averages for January and July 1978, months in which air quality data, precipitation chemistry data and emissions inventories were available for comparison. Tables

Table 25.1 Calculated average monthly sulphur budget for the two-year period 1 October 1978 to 30 September 1980, unit - 100 t sulphur (after Eliasson and Saltbones, 1983)

Country*	1	2	3	4	5	6	7	8	9	10	11	12	13	14	15	16	17	18	19	20	21	22	23	24	25	26	27	28	29	UND	Total
1	10	0	0	3	0	0	0	0	0	0	2	2	0	0	8	0	0	0	0	0	2	0	0	0	0	0	0	13	0	12	67
2	0	52	3	0	35	0	0	20	24	32	0	14	0	0	56	0	2	0	13	0	0	2	0	3	0	0	8	35	0	30	341
3	0	0	67	0	0	0	0	28	2	24	0	0	0	0	0	0	4	0	0	0	0	0	0	0	0	0	18	0	0	9	161
4	0	0	0	153	8	0	0	2	6	4	7	12	0	0	8	0	0	0	7	0	34	0	0	0	3	12	2	45	0	29	346
5	0	22	10	2	483	2	0	45	195	108	0	68	0	0	39	0	7	0	95	0	14	4	0	2	0	14	32	53	0	91	1301
6	0	0	0	0	3	39	0	3	12	11	0	0	0	0	0	0	0	0	4	0	0	0	0	0	0	0	11	0	0	11	109
7	0	0	2	0	8	3	77	5	19	15	0	3	0	0	2	0	2	2	15	0	2	0	14	0	0	43	14	3	0	55	293
8	0	0	33	0	7	0	0	629	20	98	0	0	0	2	41	4	12	0	4	0	0	66	0	7	0	0	99	6	0	166	1212
9	0	0	8	0	56	3	0	20	497	85	0	4	0	0	3	0	6	0	23	0	0	0	0	0	0	3	22	6	0	26	778
10	0	8	37	0	48	4	0	108	118	561	0	6	0	0	24	5	23	0	18	0	0	6	0	6	0	3	72	15	0	80	1158
11	3	0	0	34	4	0	0	3	3	3	93	5	0	0	14	0	0	0	2	0	8	2	0	0	4	4	0	27	0	31	253
12	0	9	0	0	46	0	0	7	18	13	0	194	0	0	27	0	0	0	21	0	16	0	0	0	0	4	3	75	0	20	467
13	0	0	0	0	0	0	0	0	0	0	0	0	0	0	0	0	0	0	0	0	0	0	0	0	0	0	0	0	0	18	24
14	0	0	0	0	0	0	0	2	0	0	0	0	0	18	0	0	0	0	0	0	0	0	0	0	0	0	12	0	0	26	65
15	0	6	2	0	12	0	0	53	11	22	0	11	0	0	793	0	0	0	6	0	3	17	0	7	0	2	9	65	5	93	1132
16	0	0	0	0	0	0	0	3	0	2	0	0	0	0	0	3	0	0	0	0	0	0	0	0	0	0	0	0	0	0	11
17	0	0	16	0	2	0	0	15	6	45	0	0	0	0	0	0	40	0	0	0	0	0	0	0	0	0	27	0		11	173
18	0	0	4	0	8	6	2	11	22	20	0	2	0	0	0	0	3	20	10	0	0	0	10	0	0	8	40	2	0	74	255
19	0	7	7	3	136	8	0	26	213	80	0	43	0	0	18	0	7	0	565	0	13	2	4	0	0	34	32	40	0	77	1330
20	0	0	0	0	0	0	0	0	0	0	0	0	0	0	0	0	0	0	0	20	0	17	0	0	0	0	0	0	0		73
21	0	3	0	30	37	0	0	7	26	16	4	71	0	0	27	0	0	0	38	0	287	0	0	0	4	48	6	115	0	60	797
22	0	0	2	0	0	0	0	38	4	20	0	0	0	0	2	0	3	0	0	9	0	367	0	0	0	0	15	0	2	111	583
23	0	0	5	0	18	16	10	15	42	35	0	5	0	0	4	0	5	8	31	0	3	0	83	0	0	24	35	7	0	113	472
24	0	0	0	0	2	0	0	23	3	13	0	0	0	0	47	0	0	0	0	0	0	2	0	14	0	0	5	3	0	17	141
25	0	0	0	28	5	0	0	3	6	5	20	7	0	0	13	0	0	0	5	0	12	3	0	0	175	17	0	22	0	79	416
26	3	14	18	69	189	25	50	67	283	190	22	159	0	2	94	3	23	5	386	0	205	10	38	3	45	3610	103	208	3	106	16901
27	0	0	7	0	4	0	0	27	11	21	0	0	0	7	2	0	5	0	2	0	0	4	0	0	0	0	675	0	0	72	847
28	3	14	2	32	38	0	0	20	25	22	6	72	0	0	131	0	2	0	21	0	26	7	0	0	0	8	7	557	2	83	1093
29	10	13	51	85	106	52	30	295	214	252	79	68	0	20	448	6	53	11	130	9	76	175	45	10	85	226	611	216	21	107	44486

*1, Albania; 2, Austria; 3, Belgium; 4, Bulgaria; 5, Czechoslovakia; 6, Denmark; 7, Finland; 8, France; 9, German Democratic Republic; 10, Germany, Federal Republic of; 11, Greece; 12, Hungary; 13, Iceland; 14, Ireland; 15, Italy; 16, Luxembourg; 17, Netherlands; 18, Norway; 19, Poland; 20, Portugal; 21, Romania; 22, Spain; 23, Sweden; 24, Switzerland; 25, Turkey; 26, USSR; 27, United Kingdom; 28, Yugoslavia; 29, Remaining area inside the region; UND, Undecided

The horizontal axis shows the emitter countries; the vertical axis shows the receiver countries

Table 25.2 Sulphur deposition based on average monthly sulphur emissions, units are 100 t sulphur (after Eliasson and Saltbones, 1983)

Country	*Deposition from foreign sources* (% of total)	*Undecided background deposition* (% of total)	*Deposition from indigenous sources* (% of total)
Albania	67	18	15
Austria	76	9	15
Belgium	53	6	41
Bulgaria	47	8	45
Czechoslovakia	56	7	37
Denmark	54	10	36
Finland	55	19	26
France	34	14	52
German Democratic Republic	32	3	65
German, Federal Republic of	45	7	48
Greece	51	12	37
Hungary	54	4	42
Iceland	25	75	0
Ireland	32	40	28
Italy	22	8	70
Luxembourg	73	0	27
Netherlands	71	6	23
Norway	63	29	8
Poland	52	6	42
Portugal	33	40	27
Romania	56	8	36
Spain	18	19	63
Sweden	58	24	18
Switzerland	78	12	10
Turkey	39	19	42
USSR	32	15	53
UK	12	9	79
Yugoslavia	41	8	51

(transfer matrices) were produced for these periods relating sulphur emissions from specific regions to SO_4^{2-} deposition in precipitation and ambient sulphur concentrations in specific regions. The group concluded that most models were able to reproduce SO_4^{2-} wet deposition patterns better than sulphur concentrations in ground level air and could predict the correct order of magnitude of wet sulphur deposition over large time and space scales. However, it was not possible to choose a 'best model' for regulatory use since the transfer matrices of the different models exhibited variations among the magnitudes of the transfer matrix elements which could lead to substantial differences in the selection of optimum emission reduction scenarios depending on the particular model applied and the level of detail required (Memorandum of Intent Work Group, 1983). When the models were used to rank areas by deposition, that is ignoring differences in the actual numbers produced from the transfer matrices, they all ranked the top five source/receptor areas in the same order (Whelpdale, 1985).

The extent of the agreement between measured and computed values for various models can be seen in individual comparisons. The AES model gives average computed SO_4^{2-} values generally within 60% of measured concentration and

Table 25.3 Characteristics of several regional-scale air pollution models used to assess sulphur and acidic deposition (after National Research Council, 1983)

Characteristic	*AES*	*ASTRAP*	*CAPITA*
Type	Lagrangian-box	Statistical-trajectory	Monte Carlo
Output	Monthly SO_2 and SO_4 concentrations and dry and wet S depositions	Monthly SO_2 and SO_4 concentrations and dry and wet and bulk S wet depositions	Monthly SO_2 and SO_4 concentrations and dry and wet depositions
Input	Annual and seasonal SO_2 emissions; daily precipitation amounts; winds and temperatures twice daily at four heights	Annual and seasonal SO_2 emissions for six-layered grid; stack parameters for major sources; 6 h precipitation amounts; wind profiles twice daily	Annual SO_2 emissions; 6-h precipitation probabilities; twice daily upper-air wind profiles and three times daily surface winds
Number of cells in the grid	52 × 37	User specified	52 × 37
Grid size (km)	127 × 127	Receptor point locations	127 × 127
Analysis of precipitation (by preprocessor)	Objective analysis of daily precipitation amounts	Hourly data are summed to produce 6-h totals across a grid of about a 76-km spacing	No actual precipitation rates used: time averages of precipitation probabilities used for each grid square
Mixed layer	Monthly climatological heights	Diurnal pattern including nocturnal surface-based inversion: maximum: 1000 (w) and 1800 (s)	Day: 800 (winter) 1200 (spring/fall); 1350 (summer); Night: 300
Oxidation rate for SO_2 (%/h)	1.0	Varies diurnally: 0.2–5.5 (summer); avg. of 2.0; 0.1–1.5 (winter avg. of 0.5	0.6 (winter); 1.2 (summer)
Dry deposition velocity (cm/s)			
SO_2	0.5	varies diurnally: 0.45 (summer avg.); 0.25 (winter avg.) (SO_2 and SO_4 similar but not identical)	0.31 (winter); 1.20 (summer)
SO_4	0.1		0.07 (winter); 0.15 (summer)
wet removal rate (%/h)			
SO_2	$3 \times 10^6 P_{24}(t)/H$	–	0.6 PP (winter); 11.7 PP (summer)
SO_4	$85 \times 10^6 P_{24}(t)/H$ where H is the mixed layer (mm)	–	5.0 PP (winter); 29.0 PP (summer) where PP = probability of precipitation (%) each 6-h period
Bulk S	–	minimum of $100(P_6(t)/10)^{0.5}$ and 80	–

Table 25.3 continued

Characteristic	*ENAMAP-1*	*MEP*	*MOE*
Type	Puff-trajectory	Lagrangian	Statistical
Output	Monthly SO_2 and SO_4 concentrations and dry and wet depositions	Monthly concentration and dry and wet depositions of S	Long-term SO_2 and SO_4 concentrations; annual dry and wet sulphur depositions
Input	Annual SO_3 emissions; 3-h precipitation amounts; wind profiles twice daily	Annual and seasonal SO_2 emissions; 3-h precipitation amounts; 6-h surface pressures	Point sources; area sources treated as effective point sources; statistics of durations of wet and dry periods and average precipitation rate during wet periods; statistical treatment of winds
Number of cells in the grid	46 × 41	User specified	User specified
Grid size (km)	70 × 70	Point receptors	Point receptors
Analysis of precipitation (by preprocessor)	Hourly US: and 6-h Canadian data are summed to produce 3-h totals	Objective analysis of 3-h amounts	Climatological lengths of Eulerian and Lagrangian wet and dry periods; rate of 1 mm/h
Mixed layer	1150 (winter), 1300 (spring/fall); 1450 (summer)	Varies diurnally over model domain	1000
Oxidation rate for SO_2 (%/h)	1.0	Seasonal and diurnal variation: mean of 1.0	1.0
Dry deposition velocity (cm/s)		seasonal and diurnal variation: mean of	
SO_2	0.38 (winter); 0.48 (summer)	0.75	0.5
SO_4	0.22 (winter); 0.28 (summer)	0.25	0.05
wet removal rate (%/h)			
SO_2	$28.0\,P_1(t)$*	Dependent on pH and temperature	10.8
SO_4	$7.0\,P_1(t)$*	Precipitation rate dependent: mean of $30\,P_1(t)^{0.5*}$	36.0 (used in stochastic scavenging model with $T_d = 56\,h$ and $T_w = 7\,h$ (the avg. dry- and wet-period durations)
Bulk S	–	–	–

Table 25.3 continued

Characteristic	*RCDM-3*	*UMACID*	*ELSTAR*	*RTM-II*
Type	Analytical	Puff-trajectory	Lagrangian trajectory	Hybrid Lagrangian/Eulerian
Output	Monthly concentration and dry and wet depositions of sulphur	Estimates of source contributions to downwind concentrations and contributions of upwind sources on receptors at 6-h time steps	Hourly concentrations of photochemical products and intermediates; dry deposition; budgets for SO_x, NO_x	3-h, daily, monthly concentrations of SO_2, SO_4; wet and dry sulphur deposition
Input	Emissions and centroids of emissions for area sources; spatially averaged annual and seasonal wet and dry periods; monthly, season, and annual resultant winds and persistence factors	Annual SO_2 emission rates; 3-h precipitation amount; wind profiles twice daily	Emissions of SO_x, NO_x, HC; winds extrapolated hourly in all layers from surface conditions, temperature profiles	Daily or hourly emissions; meteorological data
Number of cells in the grid	70 × 70	41 × 32	Variable: 100 × 100	52 × 46 × 2 (episode); 26 × 23 × 2 (long-term)
Grid size (km)	80 × 80	80 × 80	5 × 5	40 × 40 (episodic); 80 × 80 (long-term)
Analysis of precipitation (by preprocessor)	Spatially averaged precipitation amounts and the average durations of wet and dry periods	Hourly data summed for 3 h periods for each 80-km grid square	Not included	Objective analysis on hourly data to generate grid-by-grid rates
Mixed layer	600 (winter); 1200 (summer); 1000 (annual)	Varies only with month	4 or 5 layers as function of maximum mixing height on test day	Linear interpolation of daily minimum and maximum; spatial interpolation by spline fit

Oxidation rate for SO_2 (%/h)	Chemical conversion time scale 2.4×10^5 s (1%/h)	Winter: 1.4 (day) 0.1 (night); Summer 2.8 (day); 0.2 (night)	Variable; tied to photochemical model	Function of solar zenith angle and geographic location; northeastern US in summer, varies from 0.1 to 2.0 with 0.78 avg.; northern Great Plains, 0.3 avg.
Dry deposition velocity (cm/s)	Weighted by the percent of dry time	Varies according to time after sunrise and land use:		
SO_2	0.50	0.10–0.55 (winter); 0.10–0.82 (summer)	Varies with aerodynamic resistance	One-dimension diffusional model, depends on land use and time of day
SO_4	0.05	0.05–0.28 (winter); 0.03–0.43 (summer)	0.55 times SO_2 rate	
Wet removal rate (%/h)				
SO_2	–	$5.0 \times 10^4 P_1(t)/H$	Not included	Depends on precipitation rate and cloud type
SO_4	–	$2.32 \times 10^5 P_1(t)^{-0.625}/H$ where H = mixing height (mm)		
Bulk S	$\Delta P_1 \; T_d T_w / T_w$ where T_d and T_w are the average duration of the dry and wet periods (h), respectively, and Δ is the scavenging coefficient equal to 0.34	–		

Characteristic	STEM-I	STEM-II	SURADS
Type	Time-dependent Eulerian	Time-dependent Eulerian	Eulerian
Output	Hourly and daily SO_2 and SO_4 concentrations; instantaneous and accumulated SO_2 and SO_4 deposition; daily regional sulfur budget	Hourly and daily species concentrations; instantaneous and accumulated deposition rates	Hourly and daily SO_2, SO_4 concentrations; dry deposition rates; daily SO_2 budget

Table 25.3 continued

Characteristic	*STEM-I*	*STEM-II*	*SURADS*
Input	Surface and upper-air winds, cloud cover and ceiling, surface roughness, evaporation rates, vertical temperature and humidity profiles	Surface and upper-air winds, cloud cover and ceiling, rainfall rate, surface roughness, evaporation rates, vertical temperature and humidity profiles	Initial SO_2 and SO_4 updated hourly; hourly SO_2 emissions; estimates of NO_x and HC emissions; radar-based precipitation; 12-h synoptic wind field aloft at five levels; hourly surface winds; temperature profile sunlight, cloud cover
Number of cells in the grid	User specified	User specified	Variable e.g. 20–23
Grid size (km)	80 × 80	Regional scale	80 × 80
Analysis of precipitation (by preprocessor)	–	–	Cloud chemistry and scavenging algorithm in five layers
Mixed layer	Calculated, varies diurnally	Calculated, varies diurnally	Five layers to 1500 m; preprocessors calculate mass conservative winds by layer and estimate diurnally variable mixing in each layer
Oxidation rate for SO_2 (%/h)	–	–	Constant (0.5–3.0) or diurnally variable as calculated
Dry deposition velocity (cm/s)			
SO_2	Calculated based on surface type, evaporation, and meteorological conditions (applies to SO_2 and SO_4)	Calculated based on surface type, evaporation, and meteorological conditions (applies to SO_2 and SO_4)	Diurnally variable, calculated
SO_4			0.1–1.0 times SO_2 rate
Wet removal rate (%/h)			
SO_2	User-specified subroutine to calculate zeroth and first-order removal rate constants (applies to SO_2, SO_4 and Bulk S)	Calculated based on liquid–water content of clouds, temperature, pH, rainfall intensity, drop size and number distribution, and chemical and physical properties of the absorbed species	Not included
SO_4			
Bulk S			

*$P_x(t)$ = liquid precipitation rate (mm/xh)

deposition fields but tends to underpredict wet SO_4^{2-} deposition in areas of Canada remote from major sources (Voldner *et al.*, 1981). The AIRSOX model underpredicts spatially averaged ambient SO_4^{2-} concentration by about 13% but the wet sulphur deposition prediction is more accurate. The western and southern parts of the modelling region are subject to underprediction. The accuracy of ASTRAP simulations over the whole modelling area for periods of a month or longer is estimated to be within a factor of two. Simulations of the contributions of US and Canadian emissions to deposition in the two countries by the ASTRAP model for 1974–1975 and 1978 (Shannon, 1981b; Shannon and Voldner, 1982) indicate that the US contributes about half of the total sulphur deposition in Canada. Less than 5% of total sulphur deposition in the US can be attributed to Canadian emissions. About four to five times as much sulphur is transported northwards across the border as is transported southwards. The Ohio Valley and the upper midwest are the major US source regions which cause an impact on Canadian sulphur deposition. Bhumralker *et al.* (1980) find reasonable agreement between observations and ENAMAP-1 model results for 1977 data, although the agreement varies with season and patches of over- and underprediction exist. An example of one of the types of output from these models is shown in Table 25.4 after Bhumralker *et al.* (1980), a tabulation of interregional exchanges of sulphur pollutants between individual source and receptor regions. For example according to this model run, the emitter region 11 (EPA region II) receives 204 kt of sulphur, or 28%, of its total sulphur deposition from its own emissions. Region 11 sends 65 kt of sulphur to region 12 (EPA region I), which is 14% of the total deposition received by the latter from all sources, and 53 kt to region 3 (Southern Ontario) which is 4% of its total deposition. Region 11 receives 179 kt (24%) from region 10 (EPA region III), 87 kt (12%) from region 3 (Southern Ontario), nothing from regions 1, 4, 5 and 6, and the remaining 35% from the other regions.

The models considered previously are all designed to predict deposition over a long time scale, generating monthly, seasonal or annual averages under the assumption that the effects of acidic deposition are due to long-term build-up of ions. In order to model SO_4^{2-} episodes (to predict 3 h to 24 h averages) more meteorological detail is needed in the model and correspondingly more input data on mixing height, wind, temperature, moisture fields, etc, resulting in greater running costs. Until recently most short-term models dealt with ambient concentrations only so there are few comparative data on the performance of short-term deposition models. A comparison between EURMAP-2 predictions and measured SO_4^{2-} concentrations during two SO_4^{2-} episodes in 1974 showed a reasonably good agreement over large parts of the modelled area, although the model underpredicted in the eastern portion, possibly due to lack of adequate emissions data. The wet deposition patterns corresponded closely to the natural inhomogeneities in precipitation with SO_4^{2-} deposition generally displaced downwind of SO_2 source areas (Bhumralker *et al.*, 1981). The transport and deposition of sulphur compounds from north-western Europe to southern Norway over short time scales was calculated by a time-dependent model and the results compared with OECD daily average measurements (Maul, 1982). Agreement was good with the mean calculated value for all episode days of 22 mg/m^2 of SO_4^{2-} being slightly below the mean measured value of 26 mg/m^2 of SO_4^{2-}. The correlation coefficient between the calculated and measured values was 0.79 but both measured and calculated values were strongly correlated with rainfall. The measured and calculated values of ambient SO_4^{2-} concentration were also in good agreement, with a correlation

Table 25.4 Interregional exchange of airborne sulphur: (1977) (after Bhumralker *et al.*, 1980)

	Total contributions of sulphur deposition within receptor regions												
Emittor region state (EPA region)	1	2	3	4	5	6	7	8	9	10	11	12	13
1 ND,SD (VIII-North)	10	1	0	2	0	0	0	0	0	0	0	0	0
2 MN,WI,MI (V-North)	3	655	290	46	0	3	229	6	24	78	50	18	23
3 S Ontario	0	66	820	2	0	1	49	2	7	74	87	40	87
4 NE,KS,IA,MO (VII)	1	43	10	367	0	26	137	22	41	12	3	2	2
5 CO east (VIII-South)	0	0	0	0	0	0	0	0	0	0	0	0	0
6 AR,LA,OK,TX,NM east (VI-East)	1	4	1	40	1	401	7	35	6	1	0	0	0
7 IL,IN,OH (V-South)	2	186	145	135	0	14	1566	59	425	520	92	30	26
8 MS,AL,GA,FL (IV-South)	0	8	7	16	0	44	31	949	279	25	2	1	2
9 KY,TN,NC,SC (IV-North)	0	19	24	11	0	13	221	108	929	159	15	7	6
10 PA,WV,VA,MD,DE (III)	0	11	57	3	0	1	178	14	141	1363	179	56	21
11 NY,NJ (II)	0	1	53	0	0	0	1	1	4	37	204	65	14
12 ME,NH,VT,MA,CT,RI (I)	0	0	1	0	0	0	1	0	2	9	91	207	22
13 S Quebec	0	2	105	0	0	0	1	0	0	2	8	41	204
Total (kt S)	18	997	1514	621	1	503	2422	1197	1856	2280	732	467	407

	Percentage contributions to sulphur deposition within receptor regions												
Emittor region	1	2	3	4	5	6	7	8	9	10	11	12	13
1 ND,SD (VIII-North)	55	0	0	0	6	0	0	0	0	0	0	0	0
2 MN,WI,MI (V-North)	19	66	19	7	0	1	9	0	1	3	7	4	6
3 S Ontario	3	7	54	0	0	0	2	0	0	3	12	9	21
4 NE,KS,IA,MO (VII)	3	4	1	59	0	5	6	2	2	1	0	0	0
5 CO east (VIII-South)	0	0	0	0	0	0	0	0	0	0	0	0	0
6 AR,LA,OK,TX,NM east (VI-East)	7	0	0	6	92	80	0	3	0	0	0	0	0
7 IL,IN,OH (V-South)	9	19	10	22	1	3	65	5	23	23	13	6	6
8 MS,AL,GA,FL (IV-South)	1	1	0	3	0	9	1	79	15	1	0	0	1
9 KY,TN,NC,SC, (IV-North)	0	2	2	2	0	3	9	9	50	7	2	1	1
10 PA,WV,VA,MD,DE (III)	2	1	4	1	0	0	7	1	8	60	24	12	5
11 NY,NJ (II)	1	0	4	0	0	0	0	0	0	2	28	14	3
12 ME,NH,VT,MA,CT,RI (I)	0	0	0	0	0	0	0	0	0	0	12	44	5
13 S Quebec	0	0	7	0	0	0	0	0	0	0	1	9	50

coefficient of 0.75. A comparison of this short-term model with three long-term models including the OECD (1979) model and EURMAP-1 showed that while the short-term model performed better for individual daily measurements long-term average calculations were much the same, suggesting that simpler and more economical models are adequate for long time scales. Since the sophisticated models are hampered by inadequate data for input and verification and inadequate characterisation of the physical and chemical processes taking place, a simple climatological model could be an efficient way of obtaining long-term average distribution patterns (Fisher, 1978; Venkatram *et al.*, 1982). These models are based on statistical distributions of windspeed, wind direction and dispersion, and rainfall, together with detailed emission inventories. The influence of fluctuations in meteorological and physical conditions is small on average so the use of a statistical model is an efficient way of obtaining average distribution patterns. The comparison of OECD network measurements with sulphur deposition over Europe calculated by Fisher (1978) showed that the calculated annual average values for wet and dry sulphur deposition were generally accurate to within a factor of two. This was comparable with the results from the more complex OECD (1979) model. A similar model for eastern North America (Venkatram *et al.*, 1982) produced predictions which were generally lower than observed annual average wet deposition values but the model explained 76% of the measured variance. The accuracy of the model is good since 77% of the observations are within a factor of two of the model predictions.

The models so far considered assume linear chemistry for the SO_2 oxidation reaction and usually also assume that the physical processes affecting the chemical species, such as dispersion and deposition to the surface, are also proportional to their concentrations. These assumptions mean that linear models will predict a change in sulphur concentration and deposition proportional to any change in SO_2 emissions. The models also assume that total deposition at any given point can be found by adding together all sources, implying that the pollutants from different sources do not interact with one another. This is known as superposition. Rodhe *et al.* (1981) formulated a simple photochemical model with nonlinear chemistry including sulphur, nitrogen and hydrocarbon species to simulate the formation of H_2SO_4 and HNO_3 during long-range transport. Model results suggested that in Europe ambient SO_4^{2-} concentrations rose proportionally less than sulphur emissions while ambient SO_2 concentrations rose proportionally more. This is in line with the observation of Granat (1978) that deposition of SO_4^{2-} in precipitation over northern Europe during the last 20 years has not risen at the same rate as SO_2 emissions. An increase in hydrocarbon emissions enhanced the production of H_2SO_4. Shaw and Young (1983) adapted this non-linear chemistry model to assess the effect of assumptions of proportionality and superposition on downwind receptors in eastern North America with control scenarios of 50% reductions in sulphur, nitrogen and hydrocarbon emissions. The non-linear chemistry altered the partitioning of airborne sulphur into SO_4^{2-} and SO_2 because of the different deposition velocities of the two species. If concern is focused on deposition of total sulphur to sensitive surfaces, as is increasingly the case, then non-linearity of the chemistry becomes less important. The model predicts that a 50% reduction in sulphur emissions alone would result in about a 35% reduction in H_2SO_4 concentrations in ambient air and a 60% reduction in SO_2 concentrations. A 50% reduction in sulphur, NO_x and hydrocarbon emissions would result in a 40% reduction in H_2SO_4 concentration and a 60% reduction in SO_2 concentration. The reduction in total airborne sulphur concentration would be almost the same as the sulphur emission reductions. Superposing model runs for several sources introduced errors for both physical and chemical reasons. Use of the model with North American data to estimate the effects of non-linear chemistry on historical trends in airborne sulphur concentrations gave a similar result to that seen by Rodhe *et al.* (1981) although the trend was less pronounced.

Fisher and Clark (1983) applied the simple statistical long-range transport model, described by Fisher (1978), to both European and eastern North American conditions. The same choice of parameter values for the main sulphur removal processes gave acceptable agreement, despite eastern North America having SO_2 emissions about half of those in an equivalent area of Europe, a lower density of emission sources and a higher oxidising potential from photochemical oxidants. This strongly suggested that, while numerous non-linear processes may have contributed to the scatter between measurements and calculations, the sulphur system as a whole was approximately linear. This may be because the rate of removal of SO_2 was largely determined by dry deposition velocity and rainfall frequency, neither of which were strongly coupled to emissions. However, in areas of high wet deposition of sulphate, a large 'background' contribution of unknown origin was required, which introduced a non-linear element into the sulphur system.

This was consistent with the conclusions from the empirical model of Venkatram and Pleim (1985), applied to eastern North America, that the model parameters which determined the estimates of wet deposition were not sensitive to sulphur concentration. The emission distribution of SO_2 was a major contributor to the variance of the sulphur wet deposition pattern. Theoretical calculations in the EMEP programme (UNECE, 1985) also indicated that the non-linear effect was small. They estimated that a reduction of sulphur emissions by 30% in Europe would lead to reductions in long-term sulphur deposition by 26–34%, depending on the distance from the source, indicating a nearly linear relationship.

Understanding of the complex processes that govern the transport, transformation and deposition of pollutants is incomplete and computer capacity is limited so all the models used in the acidification debate contain simplifying assumptions to such an extent that the larger and more complex models have not been proved to outperform simpler ones. Examples of these assumptions are the use of linear approximations of non-linear processes; the use of simplified sulphur chemistry only, ignoring nitrogen and other species; the use of only one deposition velocity for each gas or particle in dry deposition modules and the treatment of rain and snow in the same way in wet deposition modules. The effects of such simplifications are not fully understood and will not be until the models are rigorously evaluated. Unfortunately there is a lack of field data, especially meteorological data, both to act as input to the models and to validate the findings, so confidence in the results is limited to the extent that the results are only indicators of general trends. Models and the data bases they are built from are constantly being improved and where the level of uncertainty is estimated model results can be used to expand on our present limited data. Current air quality models can give guidance on sulphur transport and deposition on the regional scale, although they have not been demonstrated to predict accurately the locations where high deposition levels would occur under given circumstances. At present models are not sufficiently developed to give reliable predictions of the effects of alternative control strategies on patterns of sulphur deposition in the environment (Altshuller *et al.*, 1983a; National Research Council, 1983).

Contribution of local sources

Sector analysis of air trajectories carried out by Kurtz and Scheider (1981) based on precipitation measurements taken in the ecologically sensitive Muskoka–Haliburton region of Ontario, Canada, showed that in summer, local events contributed 16% and 20% of the SO_4^{2-} and H^+ ion deposition respectively whereas in winter, no local contributions could be detected. Events which originated in the highly populated, industrialised south and south westerly octants contributed 70% of the SO_4^{2-} and 75% of the H^+ ions in summer compared with 87% and 85% respectively in winter. Since most of the SO_4^{2-} and H^+ ion deposition was associated with southerly airflows insufficient data were gathered from the north-western sector to enable reliable estimates of deposition from this sector to be made although average SO_4^{2-} and H^+ ion concentrations from the north-west octant were high.

About 225 km north west of the Muskoka–Haliburton region lies Sudbury where the nickel smelting industry has emitted between 1.5 and 2.7×10^6 Mt/y of SO_2 for over 25 years. The effect of the Sudbury emissions on bulk deposition in this region was studied by Scheider *et al.* (1981) before and for seven months during the shutdown of the Inco Ltd smelter. While deposition of sulphate decreased significantly (5 to 50% decrease) at the stations within 12 km of Sudbury there were no significant decreases in SO_4^{2-} or H^+ ion deposition up to 50 km further afield or in the Muskoka-Haliburton region during the shutdown. Chan *et al.* (1982) carried out an air sector analysis on wet-only precipitation data collected within a 50 km radius of the Inco smelter and influenced by its plume. Precipitation quality was highly dependent on the origin of the air mass associated with precipitation. During the passage of warm fronts, predominantly with southerly and south westerly air flows, the local contribution to SO_4^{2-} and H^+ ion deposition was 12% and 9% respectively of the total deposition. With precipitation from cold fronts, where the predominant wind directions were north westerly and northerly, 31% of SO_4^{2-} and 21% of H^+ deposition was of local origin. Chan *et al.* (1984) found that the Inco smelter contributed 20% or less of total sulphur deposition within a radius of 40 km of the plant.

Three years of meteorological data were used by Domaracki *et al.* (1983) to produce forward trajectories from Albany, Buffalo and New York City to estimate how often air pollutants emitted at or near these cities might be expected to drift over specific downwind regions. From the New York City air circulation region about 64% of trajectories were over the Atlantic Ocean after 24 h of transport, rising from 40% after 3 h. Over inland areas about 85% of trajectories were outside New York State after 6 h travel time, rising to 95% after 24 h. A maximum of 2.8% of trajectories traversed the Adirondack region after 12 h. Most emissions from New York City had a minimal impact on New York State. On the other hand 19% of trajectories from Albany after 3 h and 12 per cent of trajectories from Buffalo after 9 h traversed the area including the Adirondacks. Buffalo and Albany had a similar potential for trajectories to reach the Adirondacks but New York City, although about as far distant as Buffalo, had only one fourth the likelihood. A similar method was used by Sistla *et al.* (1982) to provide an estimate of the frequency of occurrence of trajectory end points terminating over or near areas protected by regulations against visibility degradation in the north-eastern US. Less than 1% of trajectories leaving the New York City, Albany and Buffalo air circulation regions crossed over one of these areas. No seasonal trends were discernable.

The chemistry of precipitation in urban areas should throw some light on local influences but there are few urban data since rain-sampling networks site samplers in non-urban areas specifically to avoid local impacts. The relative contribution of local and distant sources was addressed by Shaw (1982) based on work at a rural site in Nova Scotia, Canada. During a one year sampling period approximately 50% of the wet deposition of SO_4^{2-} and H^+ ions was due to SO_2 emissions from a city 25 km from the sampling site; a moderate source which produces 1800 g/s of SO_2. This source had an important effect on wet deposition up to at least 40 km away although at 100 km the local effect was small compared to emissions from source regions in the eastern USA and central Canada. Back trajectory analyses indicated that about 65% of the deposition from distant source regions came from the eastern USA south of the Great Lakes, 31% from the Lower Great Lakes–St Lawrence Valley region and 4% from the rest of Canada. This implied that 92% of the imported wet deposition at this site originated in the USA and 8% in Canada.

Gatz (1980a, 1980b) sited 85 collectors within 75 km of St Louis, Missouri, and the nearby coal-fired power plant and performed factor analysis on the results to evaluate the impact of urban areas on deposition patterns. Ion deposition patterns were found to group consistently into four main types: soluble pollutants, SO_4^{2-}, Zn, Cd and Pb; insoluble pollutants; soluble soil elements; and insoluble soil elements. Deposition peaks for SO_4^{2-} were associated with both heavy rain and urban–industrial source areas. On seasonal time scales, deposition maxima associated with rainfall maxima in single events tended to average out away from the city, as did rainfall, while the local maxima near the sources persisted. A rural site in south-east Michigan was compared with a nearby urban site about 7 km north of Detroit, where SO_2 emissions were 300 times higher (Dasch and Cadle, 1984). The urban site had 76% higher atmospheric particle concentrations of SO_4^{2-} and correspondingly higher dry deposition but the SO_4^{2-} wet deposition concentrations, whether considered in terms of average concentrations, weighted averages, total wet flux or on an individual storm basis, showed no significant difference between the two sites. The high local emissions near the urban site showed no apparent effect on precipitation chemistry. It seems that the proportion of emitted material deposited in the local and mesoscale is highly variable.

In Scotland, UK, Fowler *et al.* (1982) established a regional pattern of precipitation chemistry from measurements made in rural areas. To provide an estimate of the influence of a city, rainfall chemistry was also measured in Glasgow. The regional pattern predicted a rainfall pH of 4.45 (equivalent to 39 μg/l of H^+) whereas a pH of 4.26 (55 μg/l of H^+) was actually recorded, indicating that about 30% of acidity at the site was of local origin. The excess SO_4^{2-} concentration was also large in Glasgow, being about twice the regional pattern estimate (4.42 compared with 2.06 mg/l). The most acidic rain episodes in rural areas ($pH < 4$) were seen when air mass trajectories had passed over industrial SO_2 source regions in the UK or continental Europe.

The US–Canada Work Group 2 (1982a) on local and mesoscale effects concluded that up to about 20% of sulphur emitted from a source was converted, in the long term, to sulphate and deposited within 50 or 100 km of its source, leaving most of the emitted sulphur available for long range transport. Deposition of sulphur during precipitation events was more problematical. The long-range transport models tested indicated that, at the nine sensitive receptor points, from 10 to 40% of the annual wet sulphur deposition came from sources within 300 km while the remaining 60 to 90% came from sources further away.

Summary and comments

The use of mathematical models has established the long term source and receptor areas of man-made pollutants on a large scale but it is still not possible to account for deposition on a day-to-day basis. Model results suggest that considerable transport of sulphur pollution is taking place over national boundaries in both Europe and North America. In most European countries sulphur deposition due to foreign emissions is an important contributor to total deposition and in 13 countries it greatly outweighs indigenous sources. In eastern North America, US emissions provide about half of the total sulphur deposited in Canada while less than 5% of total sulphur deposition in the US can be attributed to Canadian emissions. In the long term, up to about 20% of emitted sulphur is deposited within 50 to 100 km of its source, leaving most of the emitted sulphur available for long-range transport. However, some investigators find a significant influence of local sources on precipitation chemistry (up to 50%) while others can detect no contribution.

Understanding of the complex processes governing the transport, transformation and deposition of pollutants is incomplete and there is a lack of field data for both input and validation purposes. Current models have not been demonstrated to predict accurately the locations where high deposition levels would occur under given circumstances. They have not been shown to predict accurately the effects of alternative control strategies on patterns of sulphur deposition, although it is clear that reducing present SO_2 emissions would reduce total deposition of sulphur.

Chapter 26

Human health

Concern over the health effects of sulphates centres on their effects on the respiratory tract, especially their possible contribution to the initiation or worsening of respiratory diseases, bronchitis in particular. One problem in studying bronchitis is that the term is applied to a variety of symptoms ranging from persistent cough or excessive sputum production to irreversible obstruction of lung ventilatory function. It is not known whether these symptoms are the progressive stages of a single disease or are related to different conditions. In addition smoking habits in humans may mask the effects of other pollutants. Eye and skin irritation are also effects of sulphate aerosols (EPA, 1982; IERE, 1981; Okita and Ohta, 1979).

The population groups at special risk from air pollution include young children, the elderly and individuals predisposed by particular diseases including asthma, bronchitis, cystic fibrosis, emphysema and cardiovascular disease. In the normal population there are also non-diseased but hypersensitive individuals. Such individuals have been found among normal, chronic bronchitis and asthmatic groups who have been studied under controlled exposure conditions (EPA, 1982).

In an atmosphere containing sulphates formed from combustion sources, nitrate aerosols will also be present together with trace metal particles and polycyclic organic matter. Elevated concentrations of the primary gaseous pollutants SO_2 and NOx may also be present. NOx levels are related to oxidant levels, especially O_3, and the presence of elevated hydrocarbon levels, for example from traffic, increases oxidant levels. Because of the number of substances present, it is difficult to estimate the effects of sulphates from exposure to ambient air. Experiments using sulphate aerosols alone do not allow for the possible additive or even synergistic effects of other pollutants. Prior to or during inhalation, SO_2 may be oxidised to sulphuric acid which can react with the ammonia present in the surrounding air or in the breath. Thus some of the observed effects of SO_2 in experimental studies may be due to acidic or neutralised sulphates but the extent to which this has affected test results cannot be quantified (EPA, 1982; National Research Council, 1979).

Adverse effects of air pollutants are investigated by combining the results of toxicological studies on animals and clinical exposure of human beings with epidemiology. Figure 26.1, after Hackney *et al.* (1984), illustrates the type of information that can be obtained from each type of study. All these types of study have disadvantages. Although animal toxicology can be well controlled and independently replicated extrapolation to normal human health depends on how

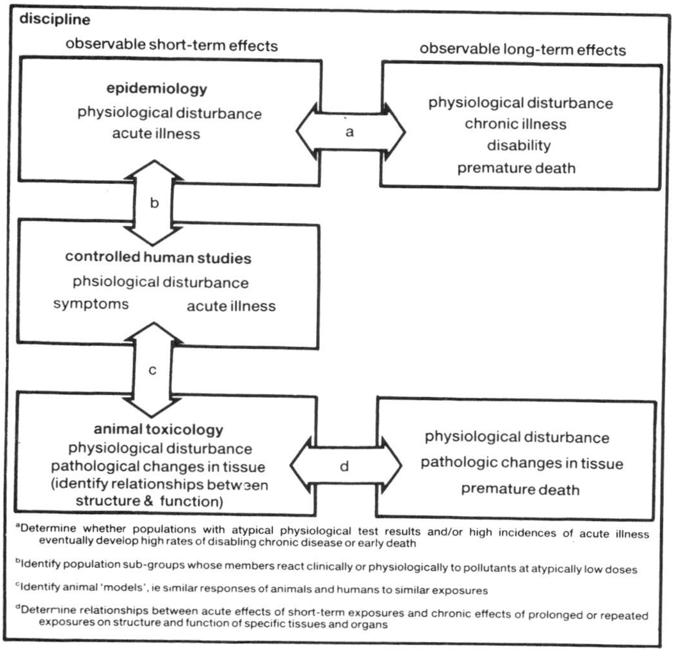

Figure 26.1 Types of study used to assess health risks from air pollution (after Hackney *et al.*, 1984)

well the various animal models represent human characteristics and this is often in doubt. Controlled human exposure studied can deal with only relatively small groups of people and with relatively brief exposures which produce only mild and reversible effects. Study groups may not adequately represent the sensitivity of the population since healthy adult subjects are usually used. However, it is possible to extrapolate from the results and the interfering variables encountered in epidemiology can be controlled. Epidemiology suffers from many drawbacks. Evidence is based mainly on human disease that has already occurred, available methods are insensitive in detecting anything other than very gross and marked effects, studies cannot be done on small specific populations for statistical reasons, there is frequently a lack of adequate exposure data both at the time covered by the study and in the practice of using a few years' data to substitute for a lifetime of fluctuating exposure. It is difficult to correct for biases and variability in the human population (Hackney *et al.*, 1984; IERE, 1981).

Particle deposition in the respiratory tract

Knowledge of the mechanism of particle deposition in the respiratory tract can help in understanding findings from human and animal exposure studies. Aerosol penetration to the lungs is a complex process but is largely determined by particle size. Only the 0.1 to 1 μm size range will be considered here since sulphate aerosols generated by coal combustion lie in this range. The airways of the human

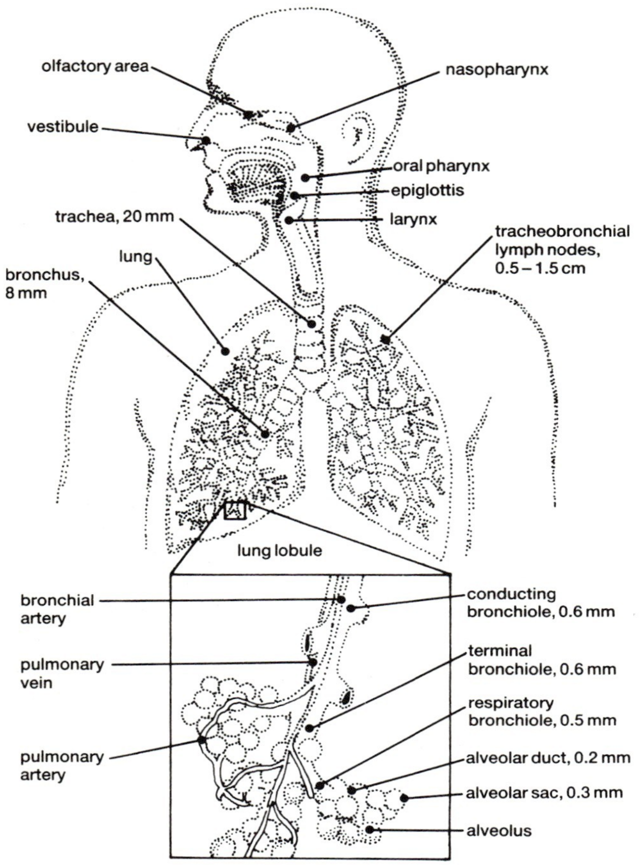

Figure 26.2 Conducting airways in the human respiratory tract (after Josephson, 1983)

respiratory tract are represented schematically in Figure 26.2, after Josephson (1983). The pulmonary or gas-exchange region of the lung begins with the respiratory bronchioles. Deposition in different regions of the respiratory tract depends on breathing patterns; inhalation through the nose results in different deposition patterns from oronasal breathing or breathing through the mouth only. In dogs, rats and hamsters the relative distribution of particles $>3\,\mu m$ along the respiratory tract during nasal breathing follows a regional deposition pattern similar to that found in humans during nasal breathing. Therefore it is possible to extrapolate from these animals to humans, given appropriate corrections. Sulphate particles are so small that they can be carried deep into the airways and deposited by diffusion, mainly in the alveoli, with deposition efficiencies of about 30%. The highly soluble nature of sulphates means that contact with body fluids, such as mucus, in any part of the airways can result in immediate absorption into the bloodstream (EPA, 1982; IERE, 1981; National Research Council, 1979).

All sulphates are hygroscopic and take up water at the high relative humidities found in the respiratory tract thus increasing the size of the particle. Cocks and

Fernando (1982) estimated growth rates for inhaled sulphuric acid, ammonium bisulphate and ammonium sulphate aerosols with initial radii of 0.1 and 1 μm. In this size range the aerosols doubled or trebled their radii in 0.1 seconds for particles ~0.1 μm and ~10 seconds for particles of 1 μm. The deposition characteristics of human airways are such that this growth would have little effect on deposition efficiency for particles of ~0.1 μm and most particles would be deposited in the alveolar region. Growth of the initially larger aerosols would significantly increase the extent of deposition in the upper airways. Similar growth curves were predicted for all three solutes so neutralisation of sulphuric acid aerosols by endogenous ammonia would have a negligible effect on growth rate.

Studies on physiological responses to air pollutants have largely been concerned with changes in respiratory mechanical function such as airway resistance or conductance. The effective diameters of conductive airways are defined by the depth of the mucus layer. In healthy individuals the reduction in air path cross-section by mucus is negligible. In people with bronchitis the mucus layer can be much thicker, narrowing or even blocking airways. Air flowing through partially restricted airways forms jets which may increase small airway particle deposition. Airway resistance provides a rapid and easily measured indication of reversible broncho-constriction produced by short term exposure to pollutants. The other change measured in experimental studies is in mucociliary particle clearance. The tracheal and bronchial airways and part of the nose are lined with ciliated cells which propel the overlying mucus, together with any deposited pollutant matter, by the coordinated beating of the cilia towards the pharynx where the mucus is swallowed and passes out of the body. Particle clearance from these zones in healthy individuals is completed in less than one day although a wide range of rates is found in apparently healthy subjects. Air pollutants can cause either an increase or decrease in clearance rates under different circumstances (IERE, 1981; Lippmann, 1980a; National Research Council, 1979).

Animal studies

Animal studies can be used to reveal the pathological mechanisms of pollutant damage and the factors governing individual susceptibility. More detailed investigation of the progress of the effects can be gained by killing animals for autopsy during the course of the exposure. Tests using high exposure levels or gross invasive investigatory techniques can be carried out. Large numbers of animals can be exposed to a variety of pollutant combinations to permit adequate statistical evaluation of the data. Chronic effects of long-term exposures can be studied. However, it cannot be assumed that animal results can be directly extrapolated to humans. The intrinsic biological activity of the pollutant may be different, geometrical and morphological differences exist between the airways of small rodents and humans and inappropriate respiratory diseases may be present (IERE, 1981).

There is a considerable difference in sensitivity to sulphates among small laboratory animals with the guinea pig being most responsive; rats, mice and rabbits are less sensitive, as are humans. The guinea pig appears to be a useful model for asthmatic humans. Sulphuric acid aerosol produces an increase in pulmonary flow resistance with the irritant effect increasing as particle size decreases. The greatest effect is seen with sub-micron particles in the accumulation size range, the most common size range for sulphates found in the ambient air. Particles in this size range penetrate to the lung and produce a rapidly occurring response similar to that seen from exposure to irritant gases such as SO_2. However, alterations produced by sulphuric acid are not rapidly reversible as is the case with the effects of irritant gases. Sulphuric acid aerosol is a more potent irritant than SO_2 but the sulphate ion does not appear to be an irritant and none of the sulphate salts studied is as potent as sulphuric acid. These species can be ranked in terms of their short-term irritant potency (measured as increased airway resistance) in guinea pigs to one hour exposures to 0.3 μm aerosols. If sulphuric acid is assigned the value of 100 then the ranking is zinc ammonium sulphate 33; ferric sulphate 26; zinc sulphate 19; ammonium sulphate 10; ammonium bisulphate 3; cupric sulphate 2; ferrous sulphate 0.7; sodium sulphate 0.7 and manganous sulphate −0.9

Table 26.1 Animal responses to sulphuric acid exposure (after EPA, 1982)

*Concentration and particle size**	*Duration*	*Species*	*Results*
Chronic exposure			
100 μg/m^3 (0.3–0.6 μm)	1 h/d 5 d/week 6 months	Donkey	Within the first few weeks all four animals developed erratic bronchial mucociliary clearance rates, either slower than or faster than those before exposure; animals not exposed to H_2SO_4 before this study had slowed clearance during the second 3 months of exposure and for 3–4 months after the end of the exposure
80 μg/m^3 (0.84 μm); 100 μg/m^3 (2.78 μm)	52 weeks cont.	Guinea pig	No significant blood effect; no lung alterations; no effect on pulmonary function
380 μg/m^3 (1.15 μm); 480 μg/m^3 (0.54 μm)	78 weeks cont.	Monkey	No significant blood effect; 380 μg m^{-3} increased respiratory rate; 480 μg m^{-3} had no effect on respiratory rate but altered distribution of ventilation early in exposure period but not later
900 μg/m^3 (2.78 μm)	52 weeks cont.	Guinea pig	No significant effects on haematology, pulmonary function, or morphology
Acute exposure			
100 μg/m^3	1 h	Guinea pig	Pulmonary resistance increased 47%; pulmonary compliance decreased 27%
500 μg/m^3	1 h	Dog	Slight increases in tracheal mucociliary transport velocities immediately and 1 day after exposure. One week later clearance was significantly decreased
510 μg/m^3	1 h	Guinea pig	Pulmonary resistance increased 60%; pulmonary compliance decreased 33%
1000 μg/m^3	1 h	Guinea pig	Pulmonary resistance increased 78%; pulmonary compliance decreased 40%
190–1400 μg/m^3	1 h	Donkey	Bronchial mucociliary clearance was slowed
1000 μg/m^3	1 h	Dog	Depression in tracheal mucociliary transport rate persisted at 1 week after exposure
1400 μg/m^3	1 h	Donkey	No effect on tracheal transport

*All particle sizes are mass median diameters (geometric median size of a distribution of particles, based on weight) (the negative indicates a decrease in airway resistance). These rankings can give no information on the effects of long-term exposure (EPA, 1982; Lippmann, 1980a; Amdur, 1978).

Individual studies are reviewed extensively by IERE (1981) and EPA (1982). Table 26.1 shows the results of some recent animal studies after EPA (1982). For short term exposure, 100 μg/m^3 is the lowest concentration of sulphuric acid aerosol found to increase airway resistance. More effect on respiratory function is seen with particles <2 μm than with larger ones. Duration of exposure, concentration, particle size and chemical composition all contribute to changes in pulmonary function. Chronic effects are less certain than acute ones but in a sulphuric acid exposure study on donkeys a sustained impairment of mucociliary clearance occurred in two previously unexposed individuals out of four test subjects which persisted for at least three months after the exposure had ended (Schlesinger, 1979).

In studies at sulphate concentrations found in ambient air, toxicological responses are rarely observed but exposure to a second stress such as infectious bacteria, can provide a more sensitive animal model. The estimated concentrations of sulphates which induced 20% excess mortality in mice exposed for three hours and infected with Streptococcal pneumonia were (in mg/m^3) 0.2 cadmium sulphate; 0.6 cupric sulphate; 0.5 zinc sulphate; 2.2 aluminium sulphate; 2.5 zinc ammonium sulphate; 3.6 magnesium sulphate. Exposure to ammonium sulphate at concentrations up to 5.3 mg/m^3 had no effect. The response of the mice suggested that the basic defence mechanisms of the lungs were impaired by exposure to sulphate aerosols, possibly by a reduction in the animal's ability to clear the inhaled micro-organisms from the lungs. A decrease in resistance to respiratory infections after exposure to air pollutants was also seen in hamsters and monkeys so, under the proper circumstances, a similar response to these stresses could be expected in humans (Ehrlich, 1980). Gardner *et al.* (1977) found an increase in mortality in mice infected with Streptococcal pneumonia when the animals were exposed to 0.196 mg/m^3 of ozone for one hour followed by a two hour exposure to 0.9 mg/m^3 of sulphuric acid. Neither pollutant alone caused a significant increase in mortality and a statistically significant additive effect of the two pollutants was seen only when exposure to the oxidant preceded that of the acid. Synergistic as well as additive effects were also observed by some investigators using sulphuric acid and ozone in hamsters and rats without bacterial infections (EPA, 1982).

Wolff *et al.* (1979b) found that the concentration of sulphuric acid required to produce 50% mortality (LC50) in young guinea pigs was 30 mg/m^3 for 0.8 μm diameter aerosol. For 0.4 μm aerosol the LC50 was above 109 mg/m^3. An LC50 of 18 mg/m^3 was earlier reported in young guinea pigs exposed to 1.2 μm sulphuric acid aerosol. During these tests the animals either tended to exhibit symptoms and die within minutes (0.4 μm diameter) or hours (0.8 μm) after exposure, or appeared nearly unaffected.

Human clinical studies

The studies considered here all deal with short-term exposure (<4 h) thus the potential for chronic health effects cannot be fully evaluated from this data. No clear evidence of respiratory changes was found by Hackney *et al.* (1980) after 2 h exposures to 100 μg/m^3 ammonium sulphate, 85 μg/m^3 ammonium bisulphate or 75 μg/m^3 sulphuric acid. A 2 h exposure of 20 normal and 18 asthmatic subjects to 75 μg/m^3 ferric sulphate did not cause significant changes in total respiratory system resistance or forced expiratory flow/volume performance. Five individuals showed small decremental trends in pulmonary function and nine subjects showed some improvement but there was no significant change in the group as a whole (Kleinman *et al.*, 1981). The same group exposed 21 normal and 19 asthmatic individuals for 2 h to 20 μg/m^3 zinc ammonium sulphate, in a simulation of conditions typical of pollution episodes occurring in autumn in north-eastern USA or in winter in southern California. Small but statistically significant changes were found in some lung function measurements but these were not consistent and could not necessarily be considered adverse effects. Reports of symptoms showed no significant variation which could be ascribed to sulphate aerosol exposure (Linn *et al.*, 1981). Kerr *et al.* (1981) found no effect on 28 subjects after exposure for 4 h to 100 μg/m^3 sulphuric acid aerosol.

Exposure of normal and asthmatic subjects to 1000 μg/m^3 of ammonium sulphate, sodium bisulphate, ammonium bisulphate or sulphuric acid aerosol for 16 minutes produced significant reductions in airway conductance and flow rates in asthmatics. Subsequent inhalation of a known broncho-constrictor significantly enhanced the effect compared to exposure after sodium chloride aerosol or alone in both normal and asthmatic subjects. Exposure to 450 μg/m^3 of the two acidic sulphates altered the effect of the broncho-constrictor on airway conductance in asthmatics but not in normal subjects. None of the sulphates altered the effect of the broncho-constrictor on any pulmonary function at a concentration of 100 μg/m^3 in either normal or most asthmatic subjects. However, effects on airway conductance were seen in the two asthmatics most responsive to high doses of sulphate after 100 μg/m^3 of sulphuric acid. The broncho-constrictor action was potentiated by sulphates more or less in relation to their acidity (Utell *et al.*, 1984, 1982).

Lippmann (1980a, 1980b) found that mucociliary clearance was markedly altered in 12 healthy individuals after exposure to 100 μg/m^3 sulphuric acid for 1 h, either in terms of bronchial clearance times or tracheal transport velocities, although no consistent changes were seen in respiratory mechanics. This transient alteration was consistent with results in similar inhalation tests on donkeys. When donkeys were repeatedly exposed to sulphuric acid at comparable concentrations the majority of the animals developed persistently slowed clearance which remained abnormal for several months after the tests. This suggests that under chronic exposure conditions at these concentrations, persistent changes could occur in mucociliary clearance in previously healthy individuals or exacerbate pre-existing respiratory disease (Lippmann, 1980a, 1980b; EPA, 1982). Short-term human sulphuric acid exposure studies and both short-term and chronic sulphuric acid exposures in donkeys were compared by Lippmann *et al.* (1982). Since most human bronchitis is attributable to cigarette smoke the effects of inhaled sub-micron sulphuric acid were compared with those of tobacco smoke. The effects of sulphuric acid aerosol and fresh cigarette smoke on mucociliary bronchial clearance of tagged ferric oxide particles were shown to be essentially the same in terms of (a) transient acceleration of clearance in low-dose exposures (1 h at 100 μg/m^3 sulphuric acid or 2–6 cigarettes); (b) transient slowing of clearance following high dose exposures (one h at 1000 μg/m^3 sulphuric acid or ~15 cigarettes); and (c) alterations in clearance rates persisting for several months following multiple exposures (>6 1-h exposures at 200–1000 μg/m^3 sulphuric acid, six months of daily exposure to

100 μg/m^3 sulphuric acid for 1 h a day or 4 to 8 months of 30 cigarettes three times a week).

Exposure of 20 subjects to 100 μg/m^3 sulphuric acid aerosol for 4 h a day for two consecutive days led to no significant response in any of the seven biochemical blood parameters measured by Chaney *et al.* (1980). At this concentration the body's defence mechanisms seemed to limit any effects to the upper airways.

Although studies of exposure to sulphates alone at 100 μg/m^3 have not shown ill effects, exposure to sulphates with other pollutants have shown some changes. Exposure of 19 healthy subjects for 2 h to a mixture of sulphuric acid at 100 μg/m^3, O_3 at 0.37 ppm and SO_2 at 0.37 ppm showed significant decrements in forced expiratory performance. There was also an increase in reported symptoms during exposure, relative to a control, which persisted after completion of the study (Hackney *et al.*, 1980). Stacy *et al.* (1983) found that exposure of 231 healthy subjects for 4 h to 100 μg/m^3 sulphuric acid, 133 μg/m^3 ammonium sulphate or 116 μg/m^3 ammonium bisulphate (a sulphate concentration equivalent to 100 μg/m^3 of sulphuric acid) produced no significant effects on respiratory function.

Mixtures of each aerosol with 0.75 ppm SO_2 or 0.5 ppm NO_2 also showed no effect. However, significant differences were noted in specific airway resistance, forced vital capacity, forced expiratory flow and related measurements when any of these aerosols were present with 0.4 ppm of O_3.

Epidemiology

Present day studies usually explicitly include sulphates, but many previous studies did not have access to sulphate monitoring data. The only air pollution parameters measured in earlier years were SO_2 and some indication of particulate matter. However, many episodes occurred in the presence of high humidity, for example the London episode of December 1952, which caused about 4000 excess deaths, occurred in thick fog. High concentrations of sulphuric acid were found in later fogs. Sulphates form the largest proportion of fine atmospheric particulate matter today and are likely to have been present in the coal smoke of earlier years. Sulphuric acid aerosols were certainly present in the London episodic peaks although not measured in earlier episodes. A maximum 24 h average value of about 350 μg/m^3 was recorded in 1962. 'Net particulate acid', which was measured, showed very sharp increases during episodes of high pollution associated with increased morbidity and mortality. It is impossible now to determine which pollutants actually contributed to particulate matter but adverse effects may have depended on the whole range of pollutants present (Waller, 1985, 1979; EPA, 1982; IERE, 1981).

Measurement, and intercomparison of different measurements, of particulate matter poses considerable problems. Variations of the British Smokeshade light reflectance method were used in the UK and Europe to give a measurement called Smoke. Particles were collected on a filter and the reflectance of light from the stain measured. Neither the mass nor chemical composition of the particles can be determined by this measurement since the light reflectance measurement depends on both the density of the stain and the optical properties of the collected particles. Since the proportion of carbon and non-carbon particles can be highly variable from site to site and from time to time, the same reflectance measurement can be associated with different concentrations of particulate matter. Two standard curves have been derived from site specific calibrations to relate reflectance to particle concentration (in $\mu g/m^3$). However the uncertainties of the method are such that actual measurements at a given site can vary by a factor of 2 or more from the values associated with a given reflectance reading on either of the standard curves. Therefore Smoke measurements cannot give accurate quantitative estimates of particulate matter.

Two types of particle measurement are found in the USA. The coefficient of haze (CoH), like the Smoke method, measures the light transmittance of a filter stain giving a rough index of the soiling capacity of mostly fine atmospheric particles. This reading is also site specific. More commonly used in the US is the high volume (hi-vol) sampler method to measure total suspended particles (TSP). The hi-vol sampler collects both fine- and coarse-mode particles and the mass is directly measured by gravimetric means. No consistent relationship typically exists for Smoke and TSP measurements at the same site. Waller (1980) determined that Smoke measurements provide a closer approximation to the fine particle fraction from a dichotomous sampler than to the total of the two fractions from that sampler or to TSP measured by a hi-vol sampler. Problems also exist with earlier analytical methods for SO_2 so aerometric data cited in published epidemiological studies often provide only approximate estimates of atmospheric levels. Since the actual mix of pollutants is often essentially unknown, the aerometric data and associated health effects may be seen as only relatively non-specific indicators of the effects of pollutant mixtures containing SO_x and particulate matter (EPA, 1982; IERE, 1981).

Problems also exist in characterising human populations to enable the effects of a relatively small pollution-related factor to be separated out from other environmental factors. Age, sex and race are important determinants of death rates as are the economic, social and welfare factors affecting individuals. Smoking habits are an extremely important factor in respiratory morbidity and mortality and smoking may be synergistic with air pollutants. Compared with non-smokers, age-specific death rates from all causes in smokers are greater by a factor of 1.5, and by a factor of 3 to 10 for deaths from bronchitis. Indoor air pollution, including involuntary exposure to tobacco smoke, is also associated with health effects and is difficult to determine from outdoor pollution levels. Occupational exposures can also contribute to observed health effects.

Other variations also exist which cannot definitely be associated with any specific cause. For example, in the UK, death rates from bronchitis in the north and north-west areas, where coal consumption has generally been higher, are about twice those in the south and south-east. A variety of environmental stresses also affect daily mortality including seasonal trends, heat waves, cold spells, epidemics and day to day temperature changes (EPA, 1982; IERE, 1981).

The earliest reliably documented case of mortality associated with air pollution took place in Belgium in 1930 when 60 deaths were associated with the episode. This and other episodes with high mortality levels are listed in Table 26.2. Up to the mid-1960s sudden increases in deaths in London occurred quite commonly in association with periods of high pollution lasting only one or two days. Day to day hospital admissions also showed clear associations with peaks in pollution. The episodes in the Meuse Valley, Belgium; Donora, Pennsylvania, and London were all associated with extremely dense fog which persisted for several days and in which pollutants accumulated. The largest increases in mortality tended to occur on the later days of each episode. In the London episodes, substantial increases in

Table 26.2 Air pollution episodes involving excess mortality (after Lipfert, 1980)

Location	*Date*	*Excess deaths %‡*	*Pollutant concentrations* $\mu g/m^3$ (24-hour)	
			SO_2	*Smoke**
Meuse Valley, Belgium	Dec 1930	950	25 000†	12 500†
Donora, PA	Oct 1948	800	1600§	4500§
London	Dec 1952	72;200	3830	4460
	Jan 1955	12.5;25	1200	1750
	Jan 1956	30;50	1500	3250
	Dec 1956	25;40	1100	1200
	Dec 1957	27;70	1600	2300
	Jan 1959	10;20	800	1200
	Dec 1962	21;70	3300	2000
New York	Nov 1953	9;16	2200	1000
	Nov 1962	8;16	1800	800(6.3)
	Jan 1963	19;21	1300	800(6.3)
	Feb 1963	23.5	1260	900(7.0)
	Apr 1963	17	470	180(2.3)
	Mar 1964	5.3	1730	520(4.8)
Pittsburgh	Nov 1975	8.5	200	900(7.0)

*The numbers in parentheses represent coefficient-of-haze values, a US unit of measurement for Smoke. The London data are British Smoke measurements, which give generally lower values than corresponding US figures
‡Where there are two figures, the first is the average percentage over the duration of the episode and the second is the percentage on the peak excess mortality day
†Estimated from calculations and assumed emission factors
§Based on retrospective measurements

excess mortality occurred when the population was exposed for several days to air containing SO_2 concentrations $>1000\,\mu g/m^3$ with particulate matter levels $>1000\,\mu g/m^3$. However short-term peak values above these levels occurred with no detectable increases in mortality at other times (EPA, 1982; Waller, 1979).

Studies of the relationship between mortality and air pollution in the UK during periods with no unusual pollution episodes are reported by EPA (1982) and IERE (1981). Mazumder *et al.* (1981, 1982) re-analysed the relationship between daily mortality, SO_2 and Smoke concentrations in London for 14 winters from 1958 to 1972 and concluded that there was an association with Smoke but not with SO_2 and that there was no synergism between SO_2 and Smoke. However, the Smoke measurement was only a surrogate for an unidentified variable, possibly some component of particulate matter. A dose–response relationship was developed which indicated that small but significant increases in mortality were associated with London Smoke levels in the range of 150 to 500 $\mu g/m^3$ and more marked mortality increases occurred as Smoke levels rose above 500 to 1000 $\mu g/m^3$. Deficiencies in the data and weaknesses in the statistical analyses indicated that these results should be treated with caution. However, taken overall, increased mortality seemed to be most clearly associated with increases in fine- and small coarse-mode particles to levels above 500 to 1000 $\mu g/m^3$. Temperature change did not appear to be a key factor.

Acidic sulphur air pollutants have caused health problems in the past. Kitagawa (1984) concluded that the prevalence of respiratory diseases in Yokkaichi, Japan, from 1960 to 1969 was caused by the emission of large quantities (100 to 300 t) of highly concentrated primary sulphuric acid from an industrial plant located to the windward of residential areas. Mean concentrations of 270 μg/m^3 SO_2 and 130 μg/m^3 SO_3 were found in the Isozu area in February 1965 but emission figures show that sulphuric acid aerosol concentrations could have been as high as 800 to 12 800 μg/m^3 when the wind was blowing from the plant.

Cross-sectional studies, in which comparisons of mortality or morbidity are made between different areas at one time, show contradictory results. This may be due to deficiencies in either the recorded data or statistical analyses of the data, or to the presence of confounding variables which could not be distinguished from co-varying or uncontrolled factors. At the current levels of air pollution it is very difficult to isolate any small additional effect of pollution from the dominant factor of socio-economic circumstances. This is all the more difficult in that pollution levels can be linked to socio-economic factors, such as residence in sub-standard housing, so that a correlation between mortality or morbidity and pollution may be seen even if there is no causal connection. The quantitative findings from several reliable epidemiological studies of respiratory effects of sulphur pollution are summarised in Table 26.3, after EPA (1982). Evaluations of recent studies can be found in EPA (1982) and IERE (1981).

Longitudinal studies, where a sample population is followed through time with personal exposure monitoring and representative outdoor exposure estimates, offer better prospects of relating health effects to pollutant exposure. The Six City Study is a longitudinal study of the respiratory health effects of air pollutants, including sulphates, in six US towns. Cross-sectional analyses of the data can also be made. The six cities were selected on the basis of their historical levels of pollutants to include two clean cities (Portage, Wisconsin, and Topeka, Kansas), two cities close to the present US SO_2 and particulate matter standards (Watertown, Massachusetts and Kingston/Harriman, Tennessee) and two cities that often exceed the standards (St Louis, Missouri and Steubenville, Ohio). About 1500 adults, 25 to 74 years old, and 2000 children per city were included in the sample. Both indoor and outdoor pollutant concentrations were measured. Outdoor mean sulphate concentrations in Portage and Topeka were approximately 4 μg/m^3, levels in Watertown, St Louis and Kingston were between 5.5 to 7.5 μg/m^3 and Steubenville had a substantially higher mean sulphate concentration of 14 μg/m^3. All sites showed seasonal patterns in monthly mean sulphate levels. Monthly means in July were two to three times the mean concentration in January in all cities except Steubenville, which did not show a distinct seasonal trend. In all cities, indoor respirable sulphate concentrations were consistently lower than the outdoor concentrations. Significant differences occurred in indoor mean sulphate concentrations measured at sites in each of the cities, indicating that personal exposure varied significantly among people in the sample population in each city. From model estimates of indoor concentrations the mean infiltration rate of outdoor respirable sulphates was estimated to be approximately 70%, with a value of 30% for fully air-conditioned buildings. Indoor sources of sulphates were the use of gas for cooking, the mean impact of which was to raise indoor sulphate concentrations approximately 1 μg/m^3, and the use of matches when smoking, which was not significant. In Watertown and Steubenville outdoor sulphate levels gave fairly good predictions of personal sulphate exposure (Spengler and Thurston, 1983; Spengler *et al.*, 1981; Dockery and Spengler, 1981a, 1981b; Ferris *et al.*, 1979).

Analysis of the initial cross-sectional data showed that both adults and children had higher disease rates in Steubenville than in the other cities but the effect was

Table 26.3 Summary of quantitative conclusions from studies relating health effects to exposure to ambient air PM and SO_2 (after EPA, 1982)

Acute exposure

Type of study	*Effects observed*	*24-hour average pollutant level* ($\mu g/m^3$)	
		Smoke	SO_2
Mortality	Clear increases in daily total mortality or excess mortality above a 15-day moving average among the elderly and persons with pre-existing respiratory or cardiac disease during the London winter of 1958–1959	⩾1000	⩾1000
	Analogous increases in daily mortality in London during 1958–1959 to 1971–1972 winters		
	Some indications of likely increases in daily total mortality during the 1958–1959 London winter, with greatest certainty (95% confidence) of increases occurring at Smoke and SO_2 levels above 750 $\mu g/m^3$	500–1000	500–1000
	Analogous indications of increased mortality during 1958–1959 to 1971–1972 London winters, again with greatest certainty at Smoke and SO_2 levels above 750 $\mu g/m^3$ but indications of small increases at Smoke levels <500 $\mu g/m^3$ and possibly as low as 150–200 $\mu g/m^3$		
Morbidity	Worsening of health status among a group of chronic bronchitis patients in London during winters from 1955 to 1960	⩽250–500*	⩾500–600
	No detectable effects in most bronchitics; but positive associations between worsening of health status among a selected group of highly sensitive chronic bronchitis patients and London Smoke and SO_2 levels during 1967–1968 winter	<250*	<500

Chronic exposure

Type of study	*Effects observed*	*Annual average pollutant levels* ($\mu g/m^3$)		
		Particulate matter		
		Smoke	*TSP*	SO_2
Cross-sectional (4 areas)	Likely increased frequency of lower respiratory symptoms and decreased lung function in children in Sheffield, UK	230–301‡	–	181–275
Longitudinal and cross-sectional	Apparent improvement in lung function of adults in association with decreased PM pollution in Berlin, NH, USA	–	180	†
Longitudinal and cross-sectional	Apparent lack of effects and symptoms, and no apparent decrease in lung function in adults in Berlin, NH, USA	–	80–131	†

*The 250–500 $\mu g/m^3$ Smoke levels stated here may represent somewhat higher PM concentrations than those actually associated with the observed effects. This is because estimated Smoke mass value cannot be clearly determined

‡Smoke levels stated here in $\mu g/m^3$ must be viewed as only crude estimates of the approximate PM (BS) mass levels associated with the observed health effects, given ambiguities regarding the use or non-use of site-specific calibration in Sheffield to derive the reported Smoke levels in $\mu g/m^3$

†Sulfation rate methods indicated low atmospheric sulfur levels in Berlin, NH during the time of these studies. Crude estimation of SO_2 levels from that data suggest that SO_2 levels were generally <25–50 $\mu g/m^3$ and did not contribute to observed health effects

small. For all respiratory diseases in children, except asthma in females, the particular city was a significant factor. Steubenville and Kingston/Harriman had the highest illness rates and Watertown the lowest. Small decreases in pulmonary function were seen in children exposed to sudden increases in TSP and SO_2 concentrations near or above the current 24 h standards during air pollution episodes in Steubenville. The changes were small relative to sampling variability and were only marginally significant. Although the Six Cities Study was well designed the initial results were ambiguous and showed that the analyses were sensitive to sample composition and to the assumptions and variable definitions used. Resolution of effects from the data was hampered by the many confounding factors and the small signal-to-noise ratio (Ferris *et al.*, 1983; Ferris *et al.*, 1980; Ferris *et al.*, 1979).

An earlier longitudinal study in the UK took samples of people born in some areas of London in 1952 or 1953 and in 1957 or 1958 and control samples of people born in a relatively clean town in 1957. They were all seen on reaching the age of eighteen and a small sub-sample of Londoners was followed through to the ages of 23 and 28. The major factors affecting the prevalence of respiratory symptoms and ventilatory capacity were smoking habits and past respiratory illness. Contrasts between London and the cleaner town only showed up when non-smokers were considered separately. Some respiratory symptoms were more prevalent in London subjects, but there were no significant differences in ventilatory capacity. Overall, growing up in the more polluted area made some contribution to the development of symptoms in the early adult years, but by then smoking was the dominant factor (Waller, 1983).

Aggregations of analogous data for entire populations across large areas have been manipulated to evaluate chronic air pollution effects on mortality. Although these studies are subject to all the errors and limitations of both epidemiological studies and statistical manipulations, the results have been reported without qualification and used as a basis for cost-benefit calculations, in studies such as OECD (1981). The most widely cited of this work was done by the economist Lave and his co-workers, culminating in an extended study (Lave and Seskin, 1977) employing linear regression analyses to evaluate relationships between mortality rates and indices of air pollution (sulphates and particulate matter) in standard metropolitan statistical areas (SMSAs) in the United States. They concluded that the levels of certain pollutants in the air of US cities during the 1960s caused an increase in mortality, and presumably also in morbidity. They estimated that a 58% reduction in particulate matter and an 88% reduction in SO_x would lead to a 7% decrease in total mortality. This study has been widely criticised, with perhaps the most fundamental defects being the use of aerometric data known to be of questionable accuracy, use of data from one monitoring station to characterise air quality for whole SMSAs (areas $> 1000\,km^2$) with widely varying pollution levels, inadequate data on tobacco smoking and the use of a sulphate measure which was not uniformly defined over the set of SMSAs (EPA, 1982; IERE, 1981).

Hamilton and Cooper (1979) re-examined the Lave and Seskin data base using a technique which did not depend on initial arbitrary selection of significant variables to extract from all the possible combinations the set of explanatory variables which best fit the regression equation. Lave and Seskin's significant pollution terms, minimum sulphate and mean particulate matter, were both found not to be statistically significant using this method. Crocker *et al.* (1979) analysed 1970 US urban mortality data and found an effect of air pollution on mortality about an order of magnitude lower than Lave and Seskin with smoking, dietary variables and medical care the most significant variables. Sulphates were statistically insignificant across all diseases.

Recently workers at the Brookhaven National Laboratory (Office of Technology Assessment, 1984; Hamilton, 1984) used a long-range transport model (RCDM-2, see Table 25.2) to generate sulphate concentration data. They attempted a probabilistic approach by generating a 'subjective' representation of the likelihood for any size of health effect, from the range of estimates for the relationship between sulphate pollution and mortality quoted in the literature. The range reflected the current controversy between those who believed there was a negligible effect at current sulphate concentrations and those who saw a significant association. Sulphate was used as an index of the sulphur–particulate matter mix but health effects found were the result of exposure to the total mixture including SO_2, NOx and particles. They estimated that about 50 000 deaths or 2% (a range of 0–50%) of total deaths per year in the USA plus a comparable level of morbidity might be attributable to current levels of sulphate and other particulate matter pollution. The risk varied with region from increases of $<1\%$ in western USA to about 4% (a range from about 0–10%) in regions of highest sulphate concentration. The usual caveats about epidemiological studies apply and in addition the use of a long-range transport model reduces the uncertainty of representativeness of exposure only to the extent of the accuracy of the model. Most models are less reliable at producing sulphate aerosol concentration patterns than they are at reproducing wet deposition patterns.

Morgan *et al.* (1985) compared mortality models provided by seven health experts for chronic steady-state exposure of the general public to specified sulphate levels. It was possible to derive a fairly high upper boundary, of a few thousand excess deaths per year, on the level of mortality due to exposure to sulphates. However, there was no agreement among the experts on the likely mortality that would result from exposure to sulphates beneath this upper boundary. Uncertainty arising from differences between the models was generally larger than the uncertainty associated with the output of any single model.

Non-respiratory health effects

Numerous complaints of eye, throat and skin irritation were reported in the mid-1970s in the Kanto region of Japan, which includes Tokyo. Major episodes occurred during the rainy season; for example 32 000 complaints of eye and skin irritation from fog and drizzle were reported in July 1974. The episodes occurred over a wide area as a result of exposure to fog or drizzle droplets. No complaints were received after periods of heavy rain. The meteorological conditions were similar in each case including high humidity, calm air, an inversion capping the mixing layer which trapped pollutants and a warm front to the south.

The pH of atmospheric moisture was at or below three and the droplets contained a wide range of substances including sulphates, nitrates, oxidants and organic irritants such as formaldehyde and acrolein. The irritant effect was caused by the pollutant mix rather than by one particular species. The role of sulphur compounds may have been only in lowering the pH of the fog. This health effect has only rarely been seen in Japan since the 1970s (Okita, 1977; Okita and Ohta, 1979).

Gaseous SO_2 produces lung tumours in mice exposed to massive doses (much higher concentrations than are generally found in the atmosphere) and acts as a co-factor in benzo(a)pyrene (BaP) carcinogenesis but few studies have been done on sulphates. BaP is a product of inefficient coal combustion and a known carcinogen. Dimethyl sulphate and its hydrolysis product monomethyl sulphate are known to have mutagenic and carcinogenic properties and both have been found in fly ash from and in the plumes of coal-fired power plants. Inhalation studies of sulphuric acid with BaP have not shown an increased cancer yield. BaP is known to react at room temperature with sulphuric acid to produce sulphonic acids and the reaction is thought most likely to occur under plume conditions. In general the sulpho group is seen to be strongly detoxifying to polyaromatic hydrocarbons such as BaP, and BaP-monosulphonic acid has been reported to be non-carcinogenic to mice (Nielsen *et al.*, 1983; Lee *et al.*, 1980; Falk and Jurgelski, 1979).

The results of a case control epidemiological study of occupational exposure to high concentrations of sulphuric acid showed a positive association between the development of upper respiratory cancers and exposure to sulphuric acid. The relative risks for all upper respiratory cancers in highly exposed subjects were higher by a factor of four in comparison with low or no exposure subjects. The relative risk for laryngeal cancer, the most common site, was up to 13 times as great, after allowing for the other main risk factors of tobacco use, alcoholism and history of ear, nose or throat disease. The concentrations of sulphuric acid were much higher than those generally encountered in the atmosphere (Saskolne *et al.*, 1984).

Although aerosol sulphates are not thought to be primary carcinogens, sulphuric acid, or other acidic air pollutants, may enhance the risk of cancer in those also exposed to carcinogens such as tobacco smoke. In a statistical study a significant correlation was found between sulphate levels and age-standardised death rates from lung cancer in 50 SMSAs in the USA (Falk and Jurgelski, 1979). In south-eastern USA lung cancer mortality rates are very high by US standards while the mortality rates for most other forms of cancer are generally at or below the national average and not rising. On the Atlantic seaboard the death rate from lung cancer among white males has risen steadily from 1950 to 1975 in a pattern similar to that of air quality, while smoking frequency has remained constant or declined. Lung cancer death rates for white females have quadrupled in the same period, an increase much greater than their increase in smoking frequency. Since no direct measurements of sulphuric acid aerosol were taken during this period a simple model was used to reconstruct the pattern of geographic variability of sulphate aerosol concentration in this region from 1950. The average annual relative aerosol sulphur concentration with latitude was compared with smoothed lung cancer mortality rates by county for white males, broken down into successive five year intervals from 1950 to 1975. Generally higher values were seen in the north than in the south in both sulphate and mortality trends, with a broad maximum in the northern region, the same pattern seen in lung cancer deaths. The rise in death rate from lung cancer from 1950 to 1970 was about 4% per year on average, about the same as the probable rise in SO_2 emissions from air pollution sources in eastern USA. The trends of both aerosol sulphur concentration and lung cancer mortality pertain largely to non-urban areas. Irritation of the respiratory tract by chronic inhalation of acid droplets may allow carcinogens in cigarette smoke or polluted air to more readily enter the body. Another possible mechanism is the reaction of sulphuric acid with other pollutants to form new carcinogenic compounds which are then inhaled. Hydrochloric acid, released from sea salt particles by sulphuric acid, may react with formaldehyde in tobacco smoke or polluted air to form the potent carcinogen bis(chloromethyl)ether which is then inhaled. This reaction is known to occur but its yield has not been measured under these conditions (Winchester, 1984, 1984; Winchester *et al.*, 1981).

Summary and contents

Atmospheric sulphates are suspected of having direct detrimental effects on human health. Toxicological studies in animals suggest that there is a causal link between high concentrations of acidic airborne sulphates and health damage. Human exposure studies have shown comparable pulmonary reactions to sub-micron sulphuric acid as to fresh cigarette smoke. Sulphuric acid is more reactive than neutral sulphates so it may be acidity, or hydrogen ion concentration, which is the active component. Epidemiological studies can only give an indication of the pollution levels at which exposure effects are likely to be seen, under some circumstances, among certain human populations. Effects from acute exposure are seen in the elderly and those with cardio-respiratory diseases. Increased risk from chronic exposure has been convincingly demonstrated only in children. It must be stressed that no respiratory effects have been demonstrated at the current ambient sulphate levels. Modelling studies give a fairly high upper boundary to the level of mortality connected with exposure to sulphates. However there is no agreement on the likely level of mortality that would result below this upper limit. Whilst eye and skin irritation from atmospheric sulphates has been seen it is thought to occur only in specific meteorological conditions and in the presence of organic air contaminants. A study of occupational exposure showed a positive association between upper respiratory tract cancers and exposures to high concentrations of sulphuric acid. At the much lower concentrations found in the atmosphere the combination of acidic air pollution and smoking, or exposure to other carcinogens, may enhance the risk of lung cancer. Statistical associations of airborne sulphates with lung cancer have been proposed, but there is no experimental evidence for a connection at present-day atmospheric sulphate concentrations.

Chapter 27

Visibility

Perception of visual air quality involves general atmospheric clarity or haziness, the total distance over which objects can be seen, their apparent colour and contrast with the sky and discerned details of line, texture and form. The optical properties of the non-cloudy atmosphere are determined by the fixed components, the molecular gases oxygen, nitrogen, etc., and the variable components, primarily aerosol particles. Visibility is a term often used to denote visual range, which is defined as the farthest distance at which a large, black object can be seen against the horizon sky in the daytime. Pollution affects visibility either as coherent plumes or haze layers, visible because of their contrast with the background or as a widespread, relatively homogeneous haze that reduces overall contrast and visual range. In the US visual range is often 100 km or more in the absence of air pollution. In polluted atmospheres visual range can be reduced to <20 km and can even go as low as 1 km during severe smog episodes. Visibility reduction may be limited to a local air circulation region or emissions source area but over western Europe and eastern USA it is often a regional effect covering scales of 100 to 1000 km and resulting from the combined emissions of many sources which cannot easily be distinguished (EPA, 1982; Covert *et al.*, 1980).

Visual range is inversely related to total extinction. Extinction refers to the attenuation of radiation passing through a medium caused by absorption and scattering. The extinction coefficient is a measure of the reduction in light intensity due to scattering and absorption of light by particles and gases and it varies with particle and gas concentration, particle size distribution and composition, and wavelength of light. The sum of the scattering coefficients of particles and gases with the absorption coefficients of particles and gases constitutes the extinction coefficient of the medium. For typical atmospheric aerosol size distributions, 80–90% of extinction is due to particles in the accumulation mode, which is dominated by sulphates. In urban industrial areas, particle scattering accounts for 50–65% of extinction, in urban residential areas 70–85% and in remote areas 90–95%. Absorption of light by fine carbonaceous particles can be important in urban areas and absorption by NO_2 can be significant in power plant plumes but on the regional scale, extinction due to scattering is dominant. Scattering extinction is closely proportional to the fine particle mass concentration, typical extinction/fine mass concentration ratios of about $3\,m^2g$ (for $<70\%$ relative humidity). In terms of contribution to scattering extinction, sulphate dominates over other chemical species with a ratio of extinction to mass of the order of $10\,m^2/g$. An increase in relative humidity from 30% to 80% will generally result in an increase in particle radius by a factor of 1.2 and a mass increase by a factor of 1.7 due to uptake of water. By 80% relative humidity scattering will generally have increased by a factor of 1.6 to 1.9 so a hygroscopic aerosol will cause a greater reduction in visibility in a humid atmosphere than it would in a dry one (Hegg *et al.*, 1985; Horvath and Pirich, 1983; EPA, 1982; Waggoner *et al.*, 1981; Covert *et al.*, 1980).

Visibility reduction on the regional scale was strongly related to elevated fine sulphate concentrations in south-eastern England, UK; Toronto, Canada; rural eastern USA, mid-western USA; California, USA; and in the deserts of south-western USA (Malm and Johnson, 1984; Trijonis, 1982; Weiss *et al.*, 1982; Waggoner *et al.*, 1981; Pitchford *et al.*, 1981; Ouimette *et al.*, 1981; Ferman *et al.*, 1981; Cahill *et al.*, 1981; Anlauf *et al.*, 1980; Covert *et al.*, 1980; Barnes and Lee, 1978). The general visibility levels experienced in the USA are shown in Figure 27.1, after Trijonis (1982). This illustrates the median midday visual range determined largely from national airport observations. Visibility is best in the mountainous south-west where median visual range exceeds 112 km. Visual range exceeds 72 km to the north and south of this region but to the east and west fairly sharp gradients occur. In a narrow band along the Pacific Coast median visual range falls to less than 40 km and to less than 24 km in parts of California. Visibility is poorer in eastern USA where median visual range is generally less than 24 km.

Ambient sulphate aerosol concentrations in eastern USA exhibit strong seasonal variations and this variability is reflected in visibility measurements. The summertime maximum in sulphate aerosol and light extinction can be seen not only in the average of data over many sites but also in data for almost any rural eastern location. Figure 27.2 (after Trijonis, 1982) illustrates the median midday visual range using summer data only. Comparing this with the annual patterns seen in Figure 27.1, it can be seen that although visibility over much of the country is

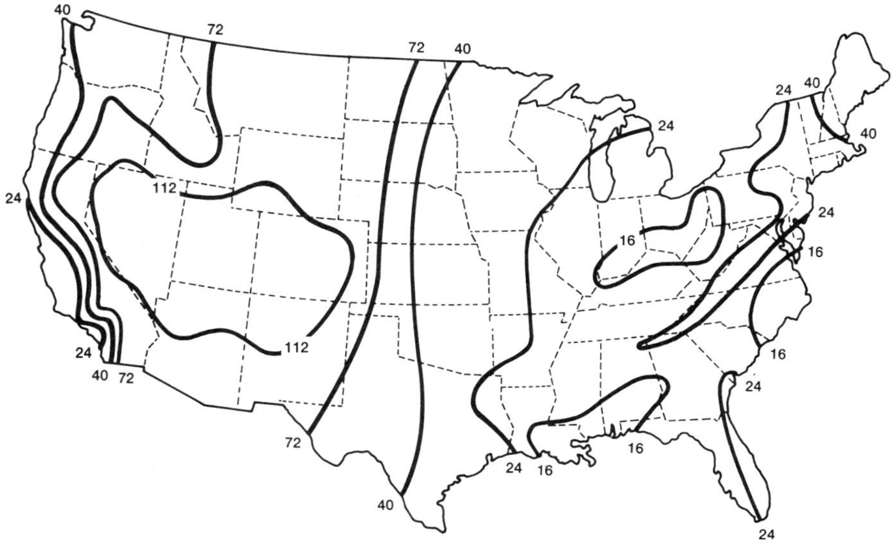

Figure 27.1 Median annual visual range in km at suburban/non-urban locations in the United States, 1974–1976 (after Trijonis, 1982)

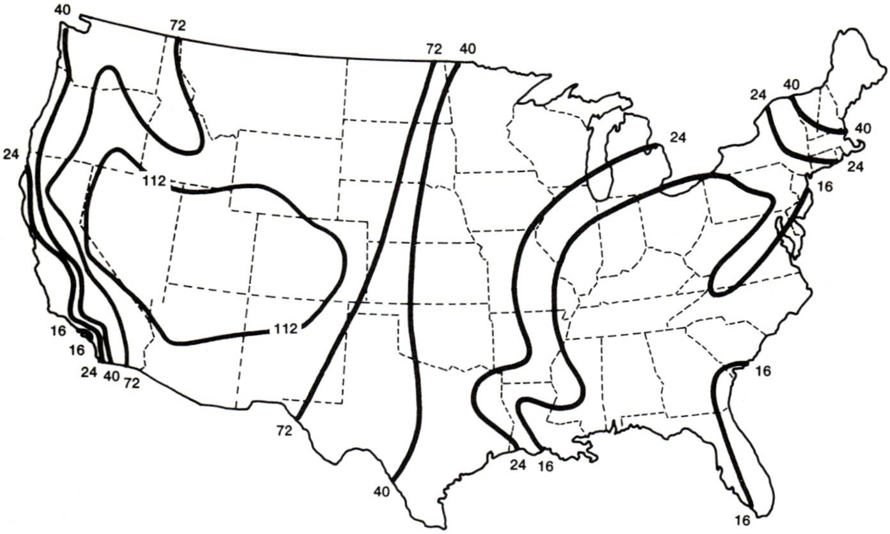

Figure 27.2 Median summer visual range in km at suburban/non-urban locations in the United States, 1974–1976 (after Trijonis, 1982)

relatively unchanged a large area of eastern USA which has a 16 to 24 km median visual range annually exhibits a 13 to 16 km median visual range during the summer.

Satellite observations of the regional movement of sulphate aerosol hazes have been made and visibility reduction associated with sulphate aerosol has been related to long range transport of pollutants (Lyons, 1980; Lyons *et al.*, 1978). Barnes and Lee (1978) found that the worst mean visibility (6.4 km) and highest sulphate levels in an area representative of south-east England, UK, occurred with south-easterly winds while south-westerlies were associated with the best mean visibility (13.7 km) and lowest sulphate levels. The most likely source of the sulphate was SO_2 emissions on the mainland of Europe. High sulphate concentrations and impaired visibility on a regional scale observed at Toronto, Canada, were associated with long-range air mass transport from the south and south-west (Anlauf *et al.*, 1980). Analysis of haze in the Shenandoah Valley on the western slopes of the Blue Ridge Mountains, Virginia, showed that on all days when average sulphate exceeded 10 $\mu g/m^3$, and indeed on 78% of all days in the study, the air parcels responsible for haze had passed through some of the major SO_2 source regions in the midwest before arriving at the monitoring site (Ferman *et al.*, 1981). Factor analysis of fine aerosol composition measured in the Shenandoah Valley grouped together SO_4^{2-}, NO_3^-, H^+, NH_4^+, Se, sulphur and total mass implying that they had a common source and similar dispersion behaviour. The link with Se suggested that this source was coal combustion since 62% of Se released into the atmosphere by human activities was estimated to come from coal burning and field measurements have linked ambient concentrations of sulphur and Se to coal combustion emissions (Stephens *et al.*, 1984). Episodes of haze observed on the Louisiana Coast were primarily due to aged sulphate aerosols which had formed over and downwind of one of two major SO_2 emission areas; the eastern midwest or a north-east corridor running from north of Boston south-westward into North Carolina (Wolff *et al.*, 1982).

Data from the VISTTA study showed that in the clean air of the deserts of south-western USA, episodes of elevated sulphate concentrations, and corresponding deterioration in visual range, were associated with transport of air masses from southern California. Measurements of sulphate particle formation and SO_2 oxidation rates in the plumes of two modern coal-fired power plants in this arid area showed that most of the mass of sulphate particles was not contained in the optically critical size range of 0.3 to 1.5 μm diameter and therefore had little effect on visibility. Plume colouration leading to the appearance of a visible plume was due to NO*x* and fly ash. Transported air masses had significantly greater impact on regional visibility than the emissions from well controlled, coal-fired power plants in the deserts (Hegg and Hobbs, 1983; Hobbs and Hegg, 1982; Hering *et al.*, 1981; Macias *et al.*, 1981b; Blumenthal *et al.*, 1981; Richards *et al.*, 1981; Vanderpol and Humbert, 1981).

Visibility is also linked to ambient sulphate loading in urban areas. Leaderer *et al.* (1981) found that sulphate aerosol was strongly correlated with light scattering in New York City during both summer and winter measurement periods but the apparent origin of the aerosol varied with the season. The lack of a diurnal pattern for light scattering in summer together with a strong association with ozone levels and wind direction dependency suggested that the transport of an aged aerosol in photochemically rich air masses rather than locally generated photochemical aerosol was responsible for increased levels of light scattering. In the winter however levels of light scattering, although about 50% lower than summer levels, were much less dependent on wind direction and highly correlated with SO_2, vanadium and lead as well as sulphates. The main source of light scattering aerosol in winter was thought to be primary and locally-generated secondary aerosol from oil combustion with a contribution from automotive emissions.

Visibility showed an approximately linear decline with increased sulphate concentration in several industrial towns in the UK. Increasing relative humidity also decreased visibility, probably due to the hygroscopic nature of sulphate aerosols, while visibility increased with wind speed. Highest sulphate concentrations and below average visibility occurred with winds from the east, which tended to bring relatively polluted, hot, dry air from northern Europe at below average wind speed (McInnes, 1982, 1980). At a suburban site in the Netherlands, 83% of total extinction was due to particle scattering. Sulphate contributed 35%, nitrates 35% and the remainder of the fine particle mass, mainly elemental carbon and organics, 13%. Over 70% of the sulphates found resulted from transboundary transport (Diederen *et al.*, 1985).

The Denver winter 'brown haze' was studied by Waggoner *et al.* (1983) and Groblicki *et al.* (1981). They found that fine aerosol particles were more effective than NO_2 in determining haze colour and reduced visual range. At the peak of visual response, NO_2 caused only seven per cent of optical extinction, 20% was caused by particle absorption and 64% by particle scattering. However, ammonium sulphate and its associated water accounted for only 20% of the visibility reduction. Carbon was the most important visibility-reducing species. Elemental carbon caused 38% of total extinction, mainly by light absorption although it did also contribute to scattering, while organic carbon caused 13% of extinction. Wolff *et al.* (1981) estimated that coal combustion contributed more than 25% of visual range reduction and more than 20% of fine particle mass (including 75% of sulphate, which was present as secondary ammonium sulphate) during this period in Denver. Carbon was the predominant species in Portland, Oregon (Shah *et al.*, 1984) where it accounted for 62% of optical extinction while sulphate was responsible for only 12%. Carbonaceous aerosol was also of importance in Houston, Texas (Dzubay *et al.*, 1982) where it contributed 17–24% of average daytime light extinction coefficient but sulphate and its associated cations (NH_4^+ and H^+) accounted for 32% and water on particles for 16%. During the summer in Detroit, sulphate and its associated water contributed 65% of the observed light extinction while carbon contributed 20% (Wolff *et al.*, 1982b).

Visibility reduction or haziness on the regional scale is related to sulphate aerosol concentration which in turn is associated with SO_2 emissions from fossil fuel combustion. Changes in visibility should therefore be comparable with historical trends in man-made SO_2 emissions. Trijonis (1982) examined visual range observations for three-year periods early in each decade from the 1930s to the 1970s from eight airports in suburban or rural areas in eastern USA. At seven of the airports, visibility increased from the 1940s to the 1950s and decreased from the 1950s to 1970s in apparent agreement with national coal consumption and SO_2 emission charges. However, the visibility trends at Albany, New York, showed exactly the opposite pattern to trends at the other seven locations.

Regional trends in haziness, expressed as light extinction coefficients, over the last 80 years were studied by Husar and Patterson (1980) and Husar *et al.* (1981) using visual range, turbidity (or vertical optical depth) and solar radiation data bases. The spatial distributions of five year seasonally averaged extinction coefficients from routine visual range data for the periods 1948 to 1952, 1960 to 1964 and 1970 to 1974 showed drastic increases in summer extinction coefficients with time in the Carolinas, Ohio River Valley and Tennessee/Kentucky area. The area in and around the Smoky Mountains showed the most pronounced change, a decline in characteristic visibility from 24 km to 10 km. Winter visibility showed improvements in the north-east and worsening in the southern regions. Both spring and autumn quarters showed a moderate decline in visibility over the entire eastern USA. The authors concluded, from use of all three data bases, that the trend in haziness showed a sharp drop around 1930 and a strong peak during the 1940s. Another increase in haziness occurred from 1960 to the mid-1970s. The lowest visibility occurred in the region of the Ohio River in the 1970s. These trends in haziness were compared to patterns of 'dirty' fuel consumption since 1850. Before 1958, the trends in haziness resembled the pattern of coal consumption more than the use of crude oil and its products. The spatial distribution of coal use also showed a similar pattern to that of haze. The largest increase in coal consumption was in the states of Kentucky, West Virginia, North and South Carolina and Tennessee where the greatest increase in haze was seen, while in the Ohio Valley region a high level of consumption in the late 1950s and early 1960s has further increased. The north-eastern states reduced their coal consumption from about 1965 when fuel use shifted from coal to oil. However a cause–effect relationship cannot be established from trend analysis alone.

The episodic nature of conditions of poor visibility suggests that meteorological fluctuations could have played a significant role in determining visibility trends. Using statistical methods and a coarse meteorological screening method, Sloane (1982) found a peak in optical air quality in the mid-1950s and a trough in the mid-1970s followed by a slight upturn since 1974 over the mid-eastern region of the

Table 27.1 Magnitude and direction of percentage change in characteristic visibility 1948–1981 (after Sloane, 1984)

Site	*QTR 1*	*QTR 2*	*QTR 3*	*QTR 4*
Metropolitan				
Cleveland	f	f	−m	f
Washington	+m	f‡	−m	m‡
Philadelphia	+l	f	f	m
Newark	+m	+m	f	+l
Urban				
Charlotte	f*	−l‡	−l‡	f*
Columbia	+m*	−m‡	−m	f‡
Greensboro	−m*	−l§	−l	f
Richmond	f	−m	−l	f
Columbus	+s	−s	−m	+m
Dayton	f	−s‡	−l	f
Louisville	−m*	−l*	−m	−s*
Knoxville	f	−m	−l	+l
Rural				
Roanoke	+l*	+l	−m	+l*
Lynchburg	+l	f	−m	+l
Williamsport	+s‡	+s‡	f	+l

Percentage changes evaluated from weighted linear least-squares fit to the 60th percentile visibility trend lines having valid estimates for at least 29 years of the 32 tri-year periods, unless marked
f - flat trend; s - small (<10% change); m - moderate (10–30% change); l - large (>30% change)
*60th percentile visibilities valid for 26–28 tri-year periods
‡75th percentile visibilities valid for at least 30 tri-years
(60th percentile valid for less than 27 years)
§75th percentile visibilities valid for 26–29 tri-year periods used
(60th percentile valid for less than 26 tri-years)

USA. The deterioration of visibility in the 1960s was most pronounced at medium-sized cities which experienced high growth. Metropolitan and rural sites have since shown a slight upturn. Sloane (1983) concluded that meteorological parameters affecting summertime visual air quality in the north-eastern USA differed significantly between the mid-1950s, a period of exceptional summertime visibility, and the early 1970s, a period of greatly reduced visibility, in a direction conducive to better visibility in the mid-1950s. The weather patterns associated with the 1950s high visibility period were not representative of general weather conditions in the area so the change in visibility between the mid-1950s and the early 1970s could not be attributed solely to changes in SO_2 emissions.

In order to present visual air quality trends which would be more closely linked to changes in emissions, Sloane (1984) used only observed changes in mid-range visibility under meteorological conditions typical for each location and season in mid-eastern USA over the period 1948 to 1981, thus minimising the influence of local atypical meteorological factors. The trends revealed by this method were more optimistic than earlier studies which did not consider meteorological factors and error analysis. The overall percentage changes in characteristic visibility are shown in Table 27.1 (after Sloane, 1984). Declines in summertime visibility at metropolitan and rural sites appeared moderate and improvements were seen in the first and fourth quarters.

However, a decline in both spring and summer visibility levels was apparent in the fast growing medium-sized urban areas and no significant improvement was seen in autumn or winter. A parallel between coal combustion and deterioration in visibility was seen, although this was not exact, probably due to meteorological factors. The large decline in visibility in non-metropolitan urban areas may be attributed to rapid local growth and the dispersal of new emissions sources from metropolitan areas. In the 1960s, 80% of new coal consumption by utilities in Ohio, Indiana, Kentucky, Virginia and West Virginia occurred at new or enlarged plants located between urban areas and well away from metropolitan centres (Sloane, 1984, 1982).

Visibility data was also related to fuel use in Europe. Lee (1983) analysed visual range data from five sites in London and southern England, from 1962 to 1979. A steady deterioration in summer visibility was seen at all sites up to 1973, especially in the long-range categories. After 1973 a sharp improvement occurred which was largely maintained to the end of the sampling period and was unrelated to meteorological factors. This improvement was coincident with and might have been related to a decrease in oil consumption by power stations and industry resulting from the 1973 oil crisis.

Summary and comments

On the regional scale, sulphates are the largest single contributors to observed light extinction in polluted atmospheres but in urban areas the contribution of sulphates is more variable. In some urban areas carbonaceous aerosol is the most important species in visibility reduction because of its ability both to absorb and to scatter light while in others sulphate aerosol is the predominant species but a seasonal change in light scattering efficiency may be seen. The visibility reducing effect of sulphate aerosols is at its greatest in the warmer months of the year. Some correspondence has been seen in the eastern USA between patterns of coal use and meteorologically adjusted visibility trends, though this does not prove that a cause and effect relationship exists.

Chapter 28

Climate

The sulphate aerosol burden in the troposphere, largely man-made may influence the atmospheric heat balance both directly by scattering and absorbing solar light and infrared radiation and indirectly by affecting the physical processes in clouds that lead to rain and snowfall. Absorption of radiation generally produces a warming of the aerosol body while scatter reflects part of the energy back upwards, although a portion of the radiation is scattered in the forward direction and this reaches the surface along with the direct beam. The reflection of incoming radiation is known as albedo. The extent to which aerosols affect incident radiation depends on the optical properties of the bulk material, the size of the particles and their concentration in the atmosphere. In spite of the lack of detailed data, it is evident that particle sizes between 0.1 and 5 μm are the most important for climatology both because of their abundance and their relatively long residence times in the atmosphere, provided they are not removed by precipitation. Combustion generated sulphate aerosols fall into this size range (Jaenicke, 1980b; Kellogg, 1980).

Models of atmospheric radiation balance have been developed to determine whether more or less heat is being introduced into the lower atmosphere when aerosols are present. There is no agreement at present on whether the net effect of increased airborne aerosol concentrations would be warming or cooling of the planet. Most models can predict either an increase or a decrease in the effective albedo of the Earth under cloudless skies depending upon which combination of surface albedo, solar angle, particle size distribution and refractive index is assumed. In addition, the effects of clouds must be considered in order to obtain a complete energy budget (EPA, 1982; Lodge *et al.*, 1981).

Sulphates are important as cloud condensation nuclei (CCN). An increase in CCN concentration usually results in increased optical depth, flattening of the scattering function and longer lifetimes of clouds. It is not known how changes in the lifetimes of clouds would affect planetary albedo or the dynamics of the atmosphere. Increased optical thickness leads to an increase in cloud albedo at first but then to a decrease. Thus the direction of effect of increasing sulphate concentrations is still uncertain. The limited residence time of CCN suggests that their effect on cloud forming processes will be seen on a regional rather than a global scale. Urban areas have been shown to alter the pre-urban levels of wind, humidity, air temperature, cloud and rainfall but it is not clear whether these alterations can be related to atmospheric aerosol concentrations (Lodge *et al.*, 1981; Jaenicke, 1980b; Oke, 1980).

There is a layer of particles (the Junge layer) composed mainly of impure sulphuric acid in the lower stratosphere with peaks in concentration at around 18–25 km and 35 km altitude. The sulphuric acid particles consist of variable mixtures of Aitkin nuclei ($<0.1\ \mu m$ diameter) found below 20 km, probably of tropospheric origin and unrelated to volcanic activity, and larger droplets (0.1–1 μm) probably formed *in situ* by gas to particle conversion of sulphur gases via the OH radical. The particle load is primarily of volcanic origin and concentration varies by as much as a factor of 30 over time, reflecting volcanic activity. A residence time of one year is suggested for the stratospheric aerosol, but in the absence of direct injection of SO_2 by volcanic eruptions, a background aerosol concentration level is maintained, possibly resulting from the diffusion of COS into the stratosphere where photochemical oxidation to sulphuric acid can occur (Inn *et al.*, 1982; Keesee and Castleman, 1982; Viggiano and Arnold, 1981; McCormick *et al.*, 1981; Jaenicke, 1980b; Turco *et al.*, 1980; Sze and Ko, 1980a).

There is evidence that after a volcanic eruption which injects SO_2 into the stratosphere the number of stratospheric aerosols is greatly enhanced and their optical depth increases to the level of the tropospheric aerosol. This has the effect of increasing stratospheric temperatures by several degrees and cooling tropospheric temperatures by several tenths of a degree for a year or two after the eruption. These small average temperature changes have in the past been accompanied by severe local weather changes. In periods of intense volcanic activity, lower tropospheric temperatures were found than in non-volcanic epochs. The stratospheric sulphate aerosol could have a definite effect on global climate by influencing both the radiation balance and atmospheric turbidity (Toon and Pollack, 1982; Jaenicke, 1980b).

Low level emissions of SO_2 cannot pass through the tropopause but COS can. Human industrial activities, especially fuel processing and combustion, could produce 50% or more of annual global emissions. Thus man-made COS could have some effect on the sulphate concentration in the troposphere (Turco, 1982; Turco *et al.*, 1980). Khalil and Rasmussen (1984) however suggested that less than 25% of global emissions were produced by human activity.

Mathematical modelling was used to estimate the possible impact of industrial COS emissions on stratospheric aerosols and climate, assuming an eventual tenfold increase in atmospheric COS concentrations from increasing coal and oil consumption by the middle of the next century. This concentration of COS caused a fourfold increase in mass and optical depth of the aerosol layer. The additional opacity could lead to a global surface cooling of about 0.1 K, which would be climatically significant. However COS, like CO_2, can create a 'greenhouse' effect in which far infrared radiation emitted from the earth is trapped in the lower atmosphere. This effect would cause some surface warming which would partially offset the cooling due to sulphate aerosol enhancement. If this increase in COS emissions from fuel consumption were to take place, the environmental and climatic problems caused by the enormous increase in emissions of other combustion products such as SO_2, CO_2, NO_x and CO would probably render insignificant the effects of COS (Turco, 1982; Turco *et al.*, 1980).

Summary and comments

Although theory suggests that tropospheric atmospheric aerosols would have an influence on climatic change this has not yet been proved. Climate modellers are not yet able to agree on whether there will be an increase or decrease in temperature as a result of changes in aerosol concentrations on a global scale. While there is some evidence for sulphates as cloud condensation nuclei affecting cloud dynamics and the probability and amount of rainfall this is not significant on the global scale. The stratospheric aerosol has a greater potential for climatic change on the global scale but at present only volcanic eruptions are known to affect aerosol concentrations. Fossil fuel consumption would affect stratospheric aerosol levels only if massive increases in uncontrolled carbonyl sulphide emissions were to take place and this seems unlikely at present.

Chapter 29

Materials

Increased concentration of air pollutants increases the corrosion and weathering rate of metals, including steel, zinc, copper and nickel; building materials, including stone and concrete; paints; paper; textiles; photographic materials and leather.

Damage to culturally significant materials is widespread. Stone buildings and monuments have deteriorated in much of Europe, most notably the Acropolis in Greece, and damage is coming to light in other areas, such as on the Taj Mahal in India. The detail on bronze monuments has been partially destroyed and medieval stained glass windows throughout Europeare fading. Damage to historical monuments and works of art is especially distressing since these are irreplaceable. Loss of our cultural heritage is difficult to evaluate in material or monetary terms.

Only the effects of SO_2 are widely studied although other pollutants such as NO*x*, ozone and other acid gases are known to affect some materials. Water in the form of atmospheric humidity and surface wetness is an extremely important factor in the degradation of building materials and metals indicating that the transformation of SO_2 to an aqueous acidic form such as sulphuric acid plays a critical role. Pollutants damage materials in ways which are not qualitatively different from the weathering effects caused by natural environmental processes so it can be difficult to distinguish how much of observed damage is due to pollutant sources. There are few reliable quantitative data on materials degradation but those that do exist have been reviewed by Altshuller (1983b) and EPA (1982).

Atmospheric corrosion of metals is an electrochemical process governed by diffusion of moisture, oxygen and acidic pollutants to the surface. Moisture is always required. Each metal has a particular critical humidity, in the range of 60–80% relative humidity, above which corrosion tends to accelerate. Hygroscopic aerosols, such as sulphates (including sulphate corrosion products) cause metal surfaces to remain wet at a lower relative humidity than would otherwise be the case and to deactivate protective oxide films. The relative length of time a metal surface is wet is the most important variable affecting metal corrosion; however, SO_2 levels correlate well with corrosion rates. Steel has been much studied but is usually only used with a protective coating such as paintwork or zinc galvanising. Under urban conditions SO_2-induced corrosion of zinc proceeds at a rate approximately a factor of two faster than that for the equivalent amount of deposited sulphuric acid aerosol (Altshuller *et al.*, 1983b; Harker *et al.*, 1982; Nriagu, 1978).

Atmospheric sulphur compounds react with the carbonates in building materials such as limestone, marble and dolomites, calcareous sandstone and Portland cement mortar and concrete to form more soluble sulphur compounds such as gypsum (calcium sulphate) and ettringite (calcium sulfoaluminate hydrate). Deterioration of concrete by the formation of sulphates is particularly marked in subaqueous environments (dams, bridge piers, sewer pipes, quays, etc.) and in moist soils. The end products of these reactions are relatively soluble salts but chemical weathering associated with the removal of soluble sulphates in solution is generally a minor part of the damage. Where surfaces are protected from rain, black crusts of gypsum and soot form which cause much of the deterioration to stonework. These crusts, aided by alternate freezing and thawing, cause blistering, scaling, exfoliation and loss of cohesion of the surface, and may induce similar effects in neighbouring materials not in themselves susceptible to direct attack. The stone is also damaged by the conversion of carbonates to sulphates leading to an increase in volume of 1.5 to 4.2 times. This process, known as spalling, causes stress within the stone and may cause the surface to peel away. Stone and concrete may also be damaged by corrosion of metallic reinforcing bars, joints and clamps. The presence of atmospheric SO_2 accelerates this deterioration. A clear correlation was seen in the FRG between weight loss of freely exposed specimens of sandstone and limestone and the ambient SO_2 concentration (Kucera, 1983; EPA, 1982; Gauri and Holdren, 1981; Gauri, 1980; Nriagu, 1978).

Skoulikidis (1983) suggested that the degradation of calcareous stone is an electrochemical process, similar to corrosion of metals. A galvanic cell is formed between calcium carbonate acting as the negative pole and the corrosive environment (SO_2 + air + water vapour) acting as the positive pole, with the corrosion product ($CaSO_4{\cdot}2H_2O$) acting as the electrolyte. The cell provides a high enough potential to displace Ca^{2+} ions and force them to diffuse in the solid state. The rate determining step up to a gypsum thickness of about 30 nm is the desorption and diffusion in solution of CO_3^{2-}, Ca^{2+} and CO_2 through the pores of the gypsum film towards the corrosive environment. Water vapour and SO_2 diffuse in the opposite direction through the porous gypsum. New layers of porous gypsum are formed where the ions meet at the $CaCO_3/CaSO_4{\cdot}2H_2O$ interface. As the thickness of gypsum increases, the number and length of pores decrease, due to the larger molecular volume of gypsum than calcite, until at about 30 nm no more pores exist. At this stage, the rate determining step is solid state diffusion of Ca^{2+} to the top of the gypsum film where new layers are formed in contact with the outside environment. This mechanism can continue to operate on stone with polymeric or inorganic coatings and accounts for the cracking of protective coatings and gypsum found in and on them. This mechanism explains the reproduction of original surface details in the gypsum layer on stone monuments attacked by air pollutants. There is experimental evidence for this theory but it has not been proved that this mechanism occurs in the field. The crystallisation of gypsum in the presence of impurities in the laboratory occurs in the form of needles or protruding plates rather than laminar layers. Gypsum crusts on stone usually incorporate significant amounts of impurities such as soot so the formation of non-porous layers may not occur under field conditions (Livingston and Baer, 1983).

Gypsum crusts are unable to form in areas of stone exposed to rain. Acidic solutions freely migrate around the grain boundaries leaving the grains unsupported and resulting in grain-by-grain dissociation of stone. This is a very slow process (Camuffo *et al.*, 1982; Gauri and Holdren, 1981). Stone deterioration patterns in Venice showed a good correlation between weathering and exposure of the surfaces to winds from the relatively unpolluted north-east. Crustal analysis indicated that surface weathering was mainly due to the joint action of dry deposited pollutants (particularly large carbonaceous particles associated with acidic pollutants, arising from combustion of oil) which were later activated by rainfall. Stone deterioration was closely related to the way precipitation, even when chemically almost neutral, wet the stone and activated the dry deposition which had occurred between precipitation events. Heavy deterioration often occurred after relatively small volumes of precipitation (Camuffo *et al.*, 1984).

Microbial action (an indirect result of sulphur deposition) by sulphate-reducing and sulphate-oxidising bacteria on deposited sulphur compounds also contributes to the deterioration of stone surfaces. Large numbers of bacteria of the genus *Thiobacillus* have been found on Italian buildings of historical interest. Limestone buildings and the mortar used in stone and brick built buildings are particularly susceptible to deterioration by the conversion of deposited sulphur compounds to sulphuric acid by microorganisms (EPA, 1982).

Most materials deteriorate more rapidly in urban or industrial areas than in rural areas. Greater damage is seen on stonework in urban areas than in less polluted locations. The useful lifetime of zinc galvanisation is >25 years in rural areas, about 10 years in industrial/urban areas and <5 years in severely polluted districts. Paint durability is generally about 12 years in rural areas and <6 years in urban and industrial locations. The corrosion rate of copper is 0.9–2.2 μm/y in industrial/urban atmospheres and 0.1–0.6 μm/y in rural areas. The SO_2 oxidation rate is usually $<4\%$/h so in urban areas most of the sulphur would still be present as SO_2. The dry deposition rate for sulphate particles is an order of magnitude lower than that of SO_2. This suggests that damage is usually due to local, low-level emissions of SO_2 rather than to long-range transport of sulphates (Altshuller *et al.*, 1983b; EPA, 1982; Nriagu, 1978). Sramek and Buzek (1983) measured the isotopic composition of gypsum crusts created on stone monuments of historical interest in Prague. The results confirmed that dry deposition of SO_2 was the means of delivery of atmospheric sulphur to the stone. The deposition of sulphate aerosols was not a significant factor.

Summary and comments

Materials damage is caused largely by locally emitted, dry deposited, SO_2 rather than airborne sulphates although the effects of NOx, oxidants and hydrocarbons have not yet been fully evaluated. Large carbonaceous particles, originating from the combustion of oil, may catalyse the oxidation of deposited SO_2 on the surface of stone. Ambient pollution levels have decreased over the past few years in most developed countries and SO_2 has been one of the main targets for reduction. It therefore seems possible that much of the damage to building materials seen at present was largely initiated during periods of higher pollution earlier in the century.

Chapter 30

Conclusions

Tne transport, transformation and deposition of sulphur compounds is a very complex chain of events in which many of the links are not yet adequately understood. Most of the sulphur in the atmosphere, in industrialised regions more than 90%, is present as a waste product of human activities. Man-made sulphur is even present in remote regions, such as the Arctic, due to long-range transport of pollutants. The largest proportion of the man-made sulphur, about 90%, is emitted as sulphur dioxide. Most sulphur dioxide derives from the combustion of fossil fuels, 82% on the global scale, with 56% coming from coal. In industrialised regions coal combustion produces about 60% of the total sulphur dioxide emitted. Because sulphur dioxide has a relatively short residence time in the atmosphere, its direct effects occur in the areas surrounding the emission source. The average transport distance for sulphur dioxide is about 100 km. However oxidation of sulphur dioxide takes place in the atmosphere by reaction with a variety of photochemically-produced oxidants to form sulphate aerosols which can travel up to 10 000 km. The rate and level of sulphate aerosol formed does not depend on the fuel burned but is related to external conditions, such as the intensity of sunlight, relative humidity and temperature. Because photochemical reactions produce the oxidants, the intensity of sunlight influences the rate of reaction of sulphur dioxide. This in turn contributes to the seasonal variation in concentration of sulphate aerosols and the episodicity of sulphate in wet deposition. Some of the contributors to oxidant production are themselves pollutants formed by human activities, for example nitrogen oxides and hydrocarbons. The average sulphate formation rate seen in point source plumes is usually less than 4%/h, under conditions conducive to gaseous phase reactions. However the rates seen in cloud or fog, when aqueous phase reactions become significant, are up to an order of magnitude higher.

Depending on the other substances present in the atmosphere, secondary sulphate ions form compounds ranging from the acidic, such as sulphuric acid, to the neutral, such as ammonium sulphate. Atmospheric aerosols commonly contain mixtures of two or more compounds. Although ammonium sulphate is chemically neutral and raises the pH of precipitation the acidifying effect on biologically active soils and surface waters will be just as strong as that of sulphuric acid. This is because hydrogen ions are released when the ammonium ion enters into the natural soil nitrification process. This process may limit the attraction of sulphur dioxide emission control technologies which produce ammonium sulphate as a byproduct, if it is intended for use as a fertiliser. At high humidity an aerosol droplet is likely to consist of many hydrated anions and cations which exhibit the behaviour of the ions present rather than the compounds.

Secondary sulphate aerosols from fossil fuel combustion interact together to form particles with a diameter $< 2\,\mu m$. Because of their size, these are removed from the atmosphere very slowly and so may be transported long distances – up to 10 000 km to the Arctic. This can also cause sulphate concentrations to build up to regional scale episodes, lasting up to a week, which are also related to synoptic scale weather systems, especially regional air mass stagnation. During these high ambient sulphate episodes any precipitation would scavenge higher than normal concentrations of sulphate, leading to an episode of high sulphate concentration in wet deposition. Dry deposition of sulphate aerosols is slow and inefficient because of their particle size, but it can be significant in some areas. Deposition of sulphates in fog, dew, frost and from interception of cloud water can be a significant removal process in some areas, even where rates from wet deposition are high, since the sulphate concentration in those forms of deposition is relatively high. Close to emission sources, precipitation scavenging in the plume of coal-fired power plants removes mainly primary sulphates and dissolved sulphur dioxide. Wet deposition of sulphate aerosol is of greatest relative significance in areas of high rainfall far from emission sources.

Deposition of sulphates is highest over the emissions source areas, where it can be higher by a factor of two than the deposition seen in remote areas. Sulphate deposition declines in a relatively uniform gradient from the source areas, except where conditions cause a relatively high rate of deposition, such as over mountains, where ion deposition tends to be higher than expected from the distance from emissions sources alone. Deposition of sulphate may also be highly episodic, with up to 30% of the annual sulphate deposition falling in one rainstorm. This episodicity cannot be explained by the episodicity of precipitation alone but is also related to episodicity of aerosol particles in the air masses contributing to precipitation.

The acidity of precipitation is governed by the balance of acidic and alkaline ions present. Precipitation with a pH less than 5.6 is termed acidic, by reference to the theoretical value of unpolluted atmospheric water, but variability in the sulphur cycle alone could cause pH values to drop as low as 4.5. In remote areas of industrialised regions receiving acidic precipitation, about 70% of the acidity is associated with sulphates and 30% with nitrates. In parts of Europe, up to 15% can be associated with chlorides. The proportion of acidity associated with nitrates has risen in recent years in some areas, possibly related to mobile sources of NO*x*. However, the overall acidity of precipitation is not determined by acidic species alone and there is evidence that changes in the concentration of alkaline species in precipitation quality are regional land use patterns which affect alkaline species, and a proximity to man-made emissions which produce acidic species.

In Europe, changes in sulphate ion deposition similar to trends in man-made emissions have been seen. This implies an approximately linear relationship between emissions and deposition, that is the changes in emissions were roughly proportional to changes in deposition. In eastern North America, although the evidence is fragmentary, trends in precipitation sulphate concentration are consistent with a general proportionality between sulphur dioxide emissions and sulphate deposition. A linear relationship between sulphur dioxide emissions and sulphate concentration in wet deposition has been found in the western USA. In more remote regions, sulphate concentration has increased by up to a factor of 2.7

over the past 200 years in Greenland ice, which is subject to pollution by man-made emissions, whereas no change is seen in Antarctic ice, where sulphates are of natural origin.

Traditional scientific methods have not been able to pinpoint relationships between source emissions and deposition in areas where emissions sources are dense and widespread so mathematical models have been developed to simulate the process. These models have established the long-term source and receptor areas of man-made pollutants over regions the size of Europe or eastern North America. However, they are not effective on smaller scales and cannot account for deposition on a day-to-day basis. What is clear is that transport of sulphur pollution is taking place over political boundaries in both Europe and eastern North America. In many European countries, deposition from foreign sources is greater than from indigenous sources. Although present day models have not been shown to be able to predict accurately the effects of alternative control strategies on sulphur deposition patterns it is clear that reducing sulphur dioxide emissions would reduce the total deposition of sulphur.

No effects on human health have been demonstrated at present day ambient sulphate concentrations. Sulphuric acid has been shown to be more reactive than other sulphates in both animal toxicological and human clinical studies so it may be the hydrogen ion that is the active component. Acidic sulphates were thought to be present in the London episodic peaks of high mortality in the 1950s.

Sulphates are the largest single contributors to observed light extinction on the regional scale but may play a lesser role in urban areas and areas of low humidity. The reduction of visibility by sulphates is greatest in the summer months, since sulphate concentrations in the atmosphere are highest then. Some correspondence has been seen between coal use and meteorologically-adjusted visibility trends.

At present, there is no evidence that sulphates affect climate on the global scale. Materials damage is caused mainly by locally emitted, dry deposited sulphur dioxide rather than secondary sulphate aerosols. However, the effects of nitrogen oxides, ozone and hydrocarbons have not yet been evaluated. Since sulphur dioxide concentrations in the atmosphere have been reduced in industrialised countries, it seems possible that much of the damage to building materials was initiated during earlier periods of higher pollution.

The only direct effect which can be attributed to atmospheric sulphates at present day ambient concentrations is reduction of visibility in some areas. The association of secondary sulphates with visibility reduction could possibly widen the geographical spread of combustion plants affected by the pressure for reduction of SO_2 emissions.

References

Adams, DF and Farwell, SO (1981) Sulfur gas emissions from stored flue gas desulfurization sludges. *Journal of the Air Pollution Control Association*; **31** (5); 557–564

Adams, DF, Farwell, SO, Pack, MR and Robinson, E (1980b) Estimates of natural sulfur source strengths. In: *Atmospheric sulfur deposition: environmental impact and health effects. Proceedings of the second life sciences symposium. Potential environmental and health consequences of atmospheric sulfur deposition, Gatlinberg, TN, 14–18 Oct 1979*; Ann Arbor, MI, Ann Arbor Science Publishers, pp. 35–44

Adams, DF, Farwell, SO, Robinson, E and Pack, MR (1980) *Biogenic sulfur emissions in the SURE region*; EPRI-EA-1516, Pullman, WA, USA, Washington State University, 170 pp.

Adams, DF, Farwell, SO, Pack, MR and Robinson, E (1981) Biogenic sulfur gas emissions from soils in Eastern and Southeastern United States. *Journal of the Air Pollution Control Association*; **31** (10); 1083–1089

Adams, DF, Farwell, SO, Robinson, E., Pack, MR and Bamesberger, WL (1981) Biogenic sulfur source strengths. *Environmental Science and Technology*; **15** (12); 1493–1498

Ahlberg, MS, Leslie, ACD and Winchester, JW (1978) Characteristics of sulfur aerosol in Florida as determined by PIXE analysis. *Atmospheric Environment*; **12** (1–3); 773–777

Alkezweeny, AJ and Laulainen, NS (1981) Comparison between polluted and clean air masses over Lake Michigan. *Journal of Applied Meteorology*; **20** (2); 209–212

Alkezweeny, AJ and Powell, DC (1977) Estimation of transformation rate of SO_2 to SO_4 from atmospheric concentration data. *Atmospheric Environment*; **11** (2); 179–182

Alkezweeny, AJ, Laulainen, NS and Thorp, JM (1982) Physical, chemical and optical characteristics of a clean air mass over Northern Michigan. *Atmospheric Environment*; **16** (10); 2421–2430

Altshuller, AP (1979) Model predictions of the rates of homogeneous oxidation of sulfur dioxide to sulfate in the troposphere. *Atmospheric Environment*; **13** (12); 1653–1661

Altshuller, AP (1980) Seasonal and episodic trends in sulfate concentrations (1963–1978) in the eastern United States. *Environmental Science and Technology*; **14** (11); 1337–1348

Altshuller, AP (1982) Relationships involving particle mass and sulfur content at sites in and around St Louis, Missouri. *Atmospheric Environment*; **16** (4); 837–843

Altshuller, AP (1984) Atmospheric particle sulfur and sulfur dioxide relationships at urban and nonurban locations. *Atmospheric Environment*; **18** (7); 1421–1431

Altshuller, AP (1985) Relationships involving fine particle mass, fine particle sulfur and ozone during episodic periods at sites in and around St Louis, MO. *Atmospheric Environment*; **19** (2); 265–276

Altshuller, AP, Johnson, WB, Nader, JS, Niemann, BL, Turner, DB, Wilson, WE and D'Alessio, G. (1980) Transport and fate of gaseous pollutants associated with the national energy program. *Environmental Health Perspectives*; **36**; 155–179

Atlshuller, AP, Lindhorst, RA, Nader, JS, Niemeyer, LE and McFee, WW (1983a) *Acidic deposition phenomenon and its effects: critical assessment review papers. Volume 1. Atmospheric sciences (Review draft)*. EPA-600/8-83-016a, PB84-171644, Raleigh, NC, USA, North Carolina State University, 781 pp.

Altshuller, AP, Lindhorst, RA, Nader, JS, Niemeyer, LE and McFee, WW (1983b) *Acidic deposition phenomenon and its effects: critical assessment review papers. Volume 2. Effects sciences*; (Review draft). EPA-600/8-83-016b, PB84-171651, Raleigh, NC, USA, North Carolina State University, 690 pp.

Altwicker, ER, Johannes, AH and Galloway, JN (1981) Wet and dry deposition into Adirondack watersheds. In: *Proceedings of Governor's conference on expanding the use of coal in New York State; problems and issues, 21–22 May 1981*; New York, NY, USA, Research Foundation of State University of New York, pp. 37–41

Amdur, MO (1978) Effects of sulfur oxides on animals. In: *Sulfur in the environment. Part II: ecological impacts*. New York, NY, USA, John Wiley, pp. 61–74

Anderson, RJ, Pilie, RJ and Durham, JL (1979) A preliminary laboratory investigation of the catalytic oxidation of SO_2 in hazes, fogs, and clouds. *American Chemical Society, Division of Environmental Chemistry, Preprints*; **19** (1); 608–610

Andreae, MO, Barnard, WR and Ammons, JM (1983) The biological production of dimethylsulfide in the ocean and its role in the global atmospheric sulfur budget. *Ecological Bulletin*; **35**; 167–177

Aneja, VP (1980) Direct measurements of emission rates of some atmospheric biogenic sulfur compounds and their possible importance to the stratospheric aerosol layer, In: *Atmospheric sulfur deposition: environmental impact and health effects. Proceedings of the second life sciences symposium. Potential environmental and health consequences of atmospheric sulfur deposition, Gatlinberg, TN, 14–18 Oct 1979. Ann Arbor, MI*; Ann Arbor Science Publishers, pp. 47–54 (1980)

Aneja, VP, Overton, JH and Aneja, AP (1981) Emission survey of biogenic sulfur flux from terrestrial surfaces. *Journal of the Air Pollution Control Association*; **31** (3); 256–258

Aneja, VP, Aneja, AP and Adams, DF (1982) Biogenic sulfur compounds and the global sulfur cycle. *Journal of the Air Pollution Control Association*; **32** (8); 803–807

Anlauf, KG, Olson, M and Wiebe, HA (1980) Atmospheric transport of particulate sulphate and ozone into the Toronto region and its correlation with visibility. In: *Atmospheric Pollution 1980, Proceedings of the 14th international colloquium, Paris, France, 5–8 May 1980*; Amsterdam, Netherlands, Elsevier Scientific Publishing Company, pp. 153–158

Anlauf, KG, Fellin, P, Wiebe, HA and Melo, OT (1982) The Nanticoke shoreline diffusion experiment. June 1978 - IV A Oxidation of sulphur dioxide in a powe plant plume. B Ambient concentrations and transport of sulphur dioxide, particulate sulphate and nitrate, and ozone. *Atmospheric Environment*; **16** (3); 455–466

Annegarn, HJ, Leslie, ACD, Winchester, JW and Young, GS (1980) Long range aerosol sulfur transport in Florida inferred from time sequence measurements. In: *Long range aerosol transport of metals and sulphur. Proceedings of a workshop in Lund, Sweden, 13 June 1980*; Report SNV PM-1337, Solna, Sweden, National Swedish Environment Protection Board, pp. 59–65

Anson, D (1981) *Summary of special discussion group on acid plumes and fallout. EPRI-WS-80-127*, From Workshop on applications of fireside additives to utility boilers, Boston, MA, USA, 3 Aug 1980, pp. 3.52–3.55

Appel, BR, Kothny, EL, Hoffer, EM and Wesolowski, JJ (1980) Sulfate and nitrate data from the California aerosol characterization experiment (ACHEX). In: *The character and origin of smog aerosols: a digest of results from the California Aerosol Characterization Experiment (ACHEX)*; New York, NY, USA, John Wiley, pp. 315–335

Ashbaugh, LL, Myrup, LO and Flocchini, RG (1984) A principal component analysis of sulfur concentrations in the western United States. *Atmospheric Environment*; **18** (4); 783–791

Asman, WAH, Slanina, J and Baard, JH (1981) Meteorological interpretation of the chemical composition of rain-water at one measuring site. *Water, Air, and Soil Pollution*; **16** (2); 159–175

Atkinson, R, Lloyd, AC and Winges, L (1982) An updated chemical mechanism for hydrocarbon/ NO_x/SO_2 photooxidations suitable for inclusion in atmospheric simulation models. *Atmospheric Environment*; **16** (6); 1341–1355

Bache, BW and Scott, NM (1979) Sulphur emissions in relation to sulphur in soils and crops. In: *Proceedings of the International symposium on sulphur emissions and the environment, London, UK, 8–10 May 1979*; London, UK, Society of the Chemical Industry, pp. 242–254

Bamber, DJ, Clark, PA, Glover, GM, Healey, PGW, Kallend, AS, Marsh, ARW, Tuck, AF and Vaughan, G (1984) Air sampling flights round the British Isles at low altitudes: SO_2 oxidation and removal rates. *Atmospheric Environment*; **18** (9); 1777–1790

Bandy, AR and Maroulis, PJ (1980) Impact of recent measurements of OCS, CS_2 and SO_2 in background air on the global sulfur cycle. In: *Atmospheric sulfur deposition: environmental impact and health effects. Proceedings of the second life sciences symposium, Potential environmental and health consequences of atmospheric sulfur deposition, Gatlinberg, TN, 14–18 Oct 1979*; Ann Arbor, MI, Ann Arbor Science Publishers, pp. 55–62

Barnes, RA and Lee, DO (1978) Visibility in London and the long distance transport of atmospheric sulphur. *Atmospheric Environment*; **12** (1–3); 791–794

Barrett, CF, Atkins, DHF, Cape, JN, Fowler, D, Irwin, JG, Kallend, AS, Martin, A, Pitman, JI, Scriven, RA and Tuck, AF (1983) *Acid deposition in the United Kingdom*; Stevenage, UK, Department of Trade and Industry, Warren Spring Laboratory, 80 pp.

Barrie, LA (1981) The prediction of rain acidity and SO_2 scavenging in Eastern North America. *Atmospheric Environment*; **15** (1); 31–41

Barrie, LA and Hoff, RM (1984) The oxidation rate and residence time of sulphur dioxide in the Arctic atmosphere. *Atmospheric Environment*; **18** (12); 2711–2722

Barrie, LA, Hoff, RM and Daggupaty, SM (1981) The influence of mid-latitudinal pollution sources on haze in the Canadian Arctic. *Atmospheric Environment*; **15** (8); 1407–1419

Beilke, S (1985) Offenbach am Main, FRG, Umweltbundesamt, Pilotstation Frankfurt, Private communication

Beilke, S and Gravenhorst, G (1978) Heterogeneous SO_2-oxidation in the droplet phase. *Atmospheric Environment*; **12** (1–3); 231–239

Benner, W (1980) Sulfate formation by combustion particles in a laboratory fog chamber. In: *Atmospheric aerosol research FY-1979, LBL-10735.* Berkeley, CA, USA, California University, Lawrence Berkeley University, pp. 8.7–8.10

Benner, WH, Brodzinsky, R and Novakov, T (1982) Oxidation of SO_2 in droplets which contain soot particles. *Atmospheric Environment*; **16** (6); 1333–1339

Benner, WH, McKinney, PM and Novakov, T (1985) Oxidation of SO_2 in fog droplets by primary oxidants. *Atmospheric Environment*; **19** (8); 1377–1383

Bhumralkar, CM, Mancuso, RL, Wolf, DE, Thuillier, RA, Nitz, KC and Johnson, WB (1980) *ENAMAP-1 long-term air pollution model: adaption and application to Eastern North Ameica*; EPA-600/4-80-039, PB80-199003, Menlo Park, CA, SRI International, 103 pp

Bhumralkar, CM, Mancuso, RL, Wolf, DE, Johnson, WB and Pankrath, J (1981) Regional air pollution model for calculating short-term (daily) patterns and transfrontier exchanges of airborne sulfur in Europe. *Tellus*; **33** (2); 142–161

Bilonick, RA and Nichols, DG (1983) Temporal variations in acid precipitation over New York State—what the 1965–1979 USGS data reveal. *Atmospheric Environment*; **17** (6); 1063–1072

Blokker, PC (1978) *Atmospheric sulphates; occurrence, formation, fate and measurement ˇ a critical review*. CONCAWE 7/78, Den Haag, Netherlands, CONCAWE (The oil companies' international study group for conservation of clean air and water ˇ Europe) 67 pp

Blumenthal, DL, Richards, LW, Macias, ES, Bergstrom, RW, Wilson, WE and Bhardwaja, PS (1981) Effects of a coal-fired power plant and other sources on Southwestern visibility. *Atmospheric Environment*; **15** (10/11); 1955–1969

Bonsang, B, Nguyen, BC, Gaudry, A and Lambert, G (1980) Sulfate enrichment in marine aerosols owing to biogenic gaseous sulfur compounds. *Journal of Geophysical Research*; **85** (C12); 7410–7416

Bornstein, RD and Thompson, WT (1981) Analysis of SURE aircraft sulfate data. In: *Second joint conference on applications of air pollution meteorology*, New Orleans, LA, USA, 24–27 March 1980. Boston, MA, USA, American Meteorological Society, 313–319

Borys, RD and Rahn, KA (1981) Long range atmospheric transport of cloud-active aerosol to Iceland. *Atmospheric Environment*; **15** (8); 1491–1501

Bottenheim, JW and Strausz, OP (1982) Modelling study of a chemically reactive power plant plume. *Atmospheric Environment*; **16** (1); 85–97

Boulaud, D, Bricard, J and Madelaine, G (1978) Aerosol growth kinetics during SO_2 oxidation. *Atmospheric Environment*; **12** (1–3); 171–178

Bower, JS and Sullivan, EJ (1981) Polar isopleth diagrams: a new way of presenting wind and pollution data. *Atmospheric Environment*; **15** (4); 537–540

Bowersox, VC and de Pena, RG (1980) Analysis of precipitation chemistry at a central Pennsylvanian site. *Journal of Geophysical Research*; **85** (C10); 5614–5620

Brezonik, PL, Edgerton, ES and Hendry, CD (1980) Acid precipitation and sulfate deposition in Florida. *Science*; **208** (4447); 1028–1029

Briggs, GA (1984) Plume rise and buoyancy effects. In: *Atmospheric science and power production*; Oak Ridge, TN, USA, Department of Energy, Office of Scientific and Technical Information, Technical Information Center, pp. 327–366

Brimblecombe, P and Stedman, DH (1982) Historical evidence for a dramatic increase in the nitrate component of acid rain. *Nature*; **298** (5873); 460–462

Britton, LG and Clarke, AG (1980) Heterogeneous reactions of sulphur dioxide and SO_2/NO_2 mixtures with a carbon soot aerosol. *Atmospheric Environment*; **14** (7); 829–839

Brodzinsky, R (1980) Carbon-catalyzed reactions in aqueous suspensions: oxidation of SO_2 at low concentrations. In: *Atmospheric aerosol research FY-1979*. LBL-10735, Berkeley, CA, USA, California University, Lawrence Berkeley Laboratory, pp. 8.2–8.4

Brooks, L and Salop, J (1983) Chemical and meteorological characteristics of atmospheric particulates in southeastern Virginia utilizing multi-variant analysis. *Journal of the Air Pollution Control Association*; **33** (3); 222–224

Brosset, C (1976) *Black and white episodes*; IVL B 295, Gothenburg, Sweden, Institutet foer Vatten- och Luftvaardsforskning, 28 pp.

Brosset, C (1978) Water-soluble sulphur compounds in aerosols. *Atmospheric Environment*; **12** (1–3); 25–38

Brosset, C (1980) Types of transport episodes in northern Europe. *Annals of the New York Academy of Sciences*; **338**; 389–398

Cahill, TA and Eldred, RA (1984) Elemental composition of Arctic particulate matter. *Geophysical Research Letters*; **11** (5); 413–416

Cahill, TA, Kusko, BH, Ashbaugh, LL, Barone, JB, Eldred, RA and Walther, FG (1981) Regional and local determinations of particulate matter and visibility in the southwestern United States during June and July 1979. *Atmospheric Environment*; **15** (10/11); 2011–2016

Calvert, JG and Stockwell, WR (1983) Acid generation in the troposphere by gas-phase chemistry. *Environmental Science and Technology*; **17** (9); 428A–443A

Calvert, JG and Stockwell, WR (1984) Mechanism and rates of the gas-phase oxidation of sulfur dioxide and nitrogen oxides in the atmosphere. In: *SO_2 NO and NO_2 oxidation mechanisms: atmospheric considerations*. Butterworth, Boston, MA, USA, pp. 1–62

Calvert, JG, Su, F, Bottenheim, JW and Strausz, OP (1978) Mechanism of the homogeneous oxidation of sulfur dioxide in the troposphere. *Atmospheric Environment*; **12** (1–3); 197–226

Calvert, JG, Lazrus, A, Kok, GL, Heikes, BG, Walega, JG, Lind, J and Cantrell, CA (1985) Chemical mechanisms of acid generation in the troposphere. *Nature*; **317** (6032); 27–35

Camuffo, D, Del Monte, M, Sabbioni, C and Vittori, O (1982) Wetting, deterioration and visual features of stone surfaces in an urban area. *Atmospheric Environment*; **16** (9); 2253–2259

Camuffo, D, Del Monte, M and Ongaro, A (1984) The pH of the atmospheric precipitation in Venice, related to both the dynamics of precipitation events and the weathering of monuments. *Science of the Total Environment*; **40**; 125–139

Canosa, C and Penzhorn, RD (1978) Second derivative UV spectroscopy study of the thermal and photochemical reaction of NO_2 with SO_2 and SO_3. Presented at 13th International Colloquium, Paris, France, 25–28 Apr 1978, 22 pp

Cantrell, BK and Whitby, KT (1978) Aerosol size distributions and aerosol volume formation for a coal-fired power plant plume. *Atmospheric Environment*; **12** (1–3); 323–333

Carruthers, DJ and Choularton, TW (1984) Acid deposition in rain over hills. *Atmospheric Environment*; **18** (9); 1905–1908

Cass, GR and Shair, FH (1981) Transport of sulfur oxides within the Los Angeles sea breeze/land breeze circulation system. In: *Second joint conference on applications of air pollution meteorology. New Orleans, LA, USA, 24–27 March 1980*. American Meteorological Society, Boston, MA, USA, pp. 320–327

Chameides, WL and Davis, DD (1982) The free radical chemistry of cloud droplets and its impact upon the composition of rain. *Journal of Geophysical Research*; **87** (C7); 4863–4877

Chan, CH and Kuntz, KW (1982) Lake Ontario atmospheric deposition 1969–1978. *Water, Air, and Soil Pollution*; **18** (1/2/3); 83–99

Chan, WH, Ro, CU, Lusis, MA and Vet, RJ (1982) Impact of the INKO nickel smelter emissions on precipitation quality in the Sudbury area. *Atmospheric Environment*; **26** (4); 801–814

Chan, WH, Vet, RJ, Ro, CU, Tang, AJS and Lusis, MA (1984) Impact of INCO smelter emissions on wet and dry deposition in the Sudbury area. *Atmospheric Environment*; **18** (5); 1001–1008

Chaney, S, Blomquist, W, Muller, K and Dewitt, P (1980) Biochemical effects of sulfuric acid mist inhalation by human subjects while at rest. *Archives of Environmental Health*; **35** (5); 270–275

Chang, SG, Toossi, R and Novakov, T (1981) The importance of soot particles and nitrous acid in oxidising SO_3 in atmospheric aqueous droplets. *Atmospheric Environment*; **15** (7); 1287–1292

Chang, TY (1979) Estimate of the conversion rate of SO_2 to SO_4 from the Da Vinci flight data. *Atmospheric Environment*; **13** (12); 1663–1664

Charlson, RJ and Rodhe, H (1982) Factors controlling the acidity of natural rainwater. *Nature*; **295** (5851); 683–685

Charlson, RJ, Covert, DS, Larson, TV and Waggoner, AP (1978) Chemical properties of tropospheric sulfur aerosols. *Atmospheric Environment*; **12** (1–3); 39–54

Chen, NCJ, Saylor, RE and Lindberg, SE (1982) Plume washout around a major coal-fired power plant: results of a single storm event. In: *Energy and environmental chemistry. Volume 2: acid rain*. Keith, LH (ed.) Ann Arbor, MI, Ann Arbor Science Publishers, pp. 11–22

Chu, K-S and Morrison, R (1980) The vapor pressures and latent heat of aqueous sulfuric acid at phase equilibrium. In: *Environmental and climatic impact of coal utilization. Proceedings of the symposium on environmental and climatic impact of coal utilization. Williamsburg, VA, USA, 17–19 Apr 1979*. New York, NY, USA, Academic Press, pp. 293–306

Clark, PA, Fisher, BEA and Marsh, ARW (1982) A case study of sulphur oxides and other acid rain precursors out of 650 km from industrial areas of the United Kingdom. In: *Conference on remote sensing and the atmosphere. Liverpool UK, 15–17 December 1982.* Leatherhead, UK, Central Electricity Research Laboratories, 8 pp.

Clark, PA, Fletcher, IS, Kallend, AS, McElroy, WJ, Marsh, ARW and Webb, AH (1984) Observations of cloud chemistry during long-range transport of power plant plumes. *Atmospheric Environment*; **18** (9); 1849–1858

Clarke, AG (1981) Electrolyte solution theory and the oxidation rate of sulphur dioxide in water. *Atmospheric Environment*; **15** (9); 1591–1595

Clarke, JF, Clark, TL, Ching, JKS, Haagenson, PL, Husar, RB and Patterson, DE (1983) Assessment of model simulation of long-distance transport. *Atmospheric Environment*; **17** (12); 2449–2462

Clarke, AG, Willison, MJ and Zeki, EM (1984) A comparison of urban and rural aerosol composition using dichotomous samplers. *Atmospheric Environment*; **18** (9); 1767–1775

Cobourn, WG and Husar, RB (1982) Diurnal and seasonal patterns of particulate sulfur and sulfuric acid in St Louis, July 1977–June 1978. *Atmospheric Environment*; **16** (6); 1441–1450

Cocks, AT and Fernando, RP (1982) The growth of sulphate aerosols in the human airways. *Journal of Aerosol Science*; **13** (1); 9–19

Cocks, AT and Fletcher, IS (1982) Possible effects of dispersion on the gas phase chemistry of power plant effluents. *Atmospheric Environment*; **16** (4); 667–678

Cocks, AT, Kallend, AS and Marsh, ARW (1983) Dispersion limitations of oxidation in power plant plumes during long-range transport. *Nature*; **305** (5930); 122–123

Cofer, WR, Schryer, DR and Rogowski, RS (1981) The oxidation of SO_2 on carbon particles in the presence of O_3, NO_2 and N_2O. *Atmospheric Environment*; **15** (7); 1281–1286

Cogbill, CV and Likens, GE (1974) Acid precipitation in the Northeastern United States. *Water Resources Research*; **10** (6); 1133–1137

Colome, SD and Spengler, JD (1982) Residential indoor and matched outdoor pollutant measurements with special consideration of wood-burning homes. In: *Proceedings of the 1981 international conference on residential solid fuels: environmental impacts and solutions, Portland, OR, USA, 1–4 Jun 1981.* Beaverton, OR, USA, Oregon Graduate Center, pp. 435–455

Cooke, MJ and Wadden, RA (1981) Atmospheric factors influencing daily sulfate concentrations in Chicago air. *Journal of the Air Pollution Control Association*; **31** (11); 1197–1199

Cope, DM and Spedding, DJ (1982) Hydrogen sulphide uptake by vegetation. *Atmospheric Environment*; **16** (2); 349–353

Countess, RJ, Cadle, SH, Groblicki, PJ and Wolff, GT (1980) Chemical analysis of size-segregated samples of Denver's ambient particulate. *Journal of the Air Pollution Control Association*; **31** (3); 247–252

Covert, DS, Waggoner, AP, Weiss, RE, Ahlquist, NC and Charlson, RJ (1980) Atmospheric aerosols: humidity and visibility. In: *The character and origin of smog aerosols: a digest of results from the California Aerosol Characterization Experiment (ACHEX).* New York, NY, USA, John Wiley, pp. 559–581

Cowling, EB (1982) Acid precipitation in historical perspective. *Environmental Science and Technology*; **16** (2); 110A–123A

Cox, RA (1980) Rates, reactivity and mechanism of homogeneous atmospheric oxidation reactions. In: *Proceedings of the first European symposium on physico-chemical behaviour of atmospheric pollutants, Ispra (Italy) 16–18 Oct 1979.* Brussels, Belgium, Commission of the European Communities, pp. 91–109

Cox, RA and Penkett, SA (1983) Formation of atmospheric acidity. In: *Acid deposition. Proceedings of the CEC workshop held in Berlin, FRG, 9 Sep 1982. EUR 8307.* Dordrecht, Netherlands, D Reidel, pp. 56–81

Crocker, TD, Schulze, W, Ben-David, S and Kneese, AV (1979) *Methods development for assessing air pollution control benefits. Volume 1. Experiments in the economics of air pollution epidemiology.* EPA-600/5-79-001a, PB-293 615, Laramie, WY, USA, Wyoming University, 177 pp.

Cullis, CF and Hirschler, MM (1980) Atmospheric sulphur: natural and man-made sources. *Atmospheric Environment*; **14**(11); 1263–1278

Dana, MT (1980) Overview of wet deposition and scavenging. In: *Atmospheric sulfur deposition: environmental impact and health effects. Proceedings of the second life sciences symposium, Potential environmental and health consequences of atmospheric sulfur deposition, Gatlinberg, TN, 14–18 Oct 1979.* Ann Arbor, MI, Ann Arbor Science Publishers, pp. 263–274

Dasch, JM and Cadle, SH (1984) The effect of local emissions on wet and dry deposition in southeastern Michigan. *Atmospheric Environment*; **18** (5); 1009–1015

Dasch, JM and Cadle, SH (1985) Wet and dry deposition monitoring in southeastern Michigan. *Atmospheric Environment*; **19** (5); 789–796

Daum, PH, Kelly, TJ, Schwartz, SE and Newman, L (1984) Measurements of the chemical composition of stratiform clouds. *Atmospheric Environment*; **18** (12); 2671–2684

Davidson, CI, Miller, JM and Pleskow, MA (1982) The influence of surface structure on predicted particle dry deposition to natural grass canopies. *Water, Air, and Soil Pollution*; **18** (1/2/3); 25–43

Davis, DD (1977) *OH radical measurements: impact on power plant plume chemistry, EPRI-EA-465*, Palo Alto, CA, USA, Electric Power Research Institute, 51 pp.

Davies, TD (1979a) Dissolved sulphur dioxide and sulphate in urban and rural precipitation (Norfolk, UK). *Atmospheric Environment*; **13** (9); 1275–1285

Davies, TD (1979b) The contribution of rainborne sulphur dioxide to total wet sulphur deposition. In: *Proceedings of the International symposium on sulphur emissions and the environment, London, UK, 8–10 May 1979*. London, UK, Society of the Chemical Industry, pp. 212–214

Davies, TD, Abrahams, PW, Tranter, M, Blackwood, I, Brimblecombe, P and Vincent, CE (1984) Black acidic snow in the remote Scottish Highlands. *Nature*; **312** (5989); 58–61

Davis, DD, Heaps, W, Philen, D and McGee, T (1979) Boundary layer measurements of the OH radical in the vicinity of an isolated power plant plume: SO_2 and NO_2 chemical conversion times. *Atmospheric Environment*; **13** (8); 1197–1203

Davis, DD, Chameides, WL and Kiang, CS (1982) Measuring atmospheric gases and aerosols. *Nature*; **295** (5846); 186

Delmas, RJ (1982) Antarctic sulphate budget. *Nature*; **299** (5885); 677–678

Delmas, R, Baudet, J, Servant, J and Baziard, Y (1980) Emissions and concentrations of hydrogen sulfide in the air of the tropical forest of the Ivory Coast and of temperate regions in France. *Journal of Geophysical Research*; **85** (C8); 4468–4474

Delmas, R and Boutron, C (1978) Sulfate in antarctic snow spatio-temporal distribution. *Atmospheric Environment*; **12** (1–3); 723–728

Delmas, R and Gravenhorst, G (1983) Background precipitation acidity. In: *Acid deposition: Proceedings of the CEC workshop held in Berlin, FRG, 9 Sep 1982. EUR 8307*. Dordrecht, Netherlands, D Reidel Publishing Co.

Derwent, RG and Hov, O (1980) The contribution from natural hydrocarbons to photochemical air pollution formation in the United Kingdom. In: *Proceedings of the first European symposium on physico-chemical behaviour of atmospheric pollutants, Ispra (Italy) 16–18 Oct 1979. EUR 6621 DE, EN, FR*. Brussels, Belgium, Commission of the European Communities, pp. 367–382

Diederen, HSMA, Guicherit, R and Hollander, JCT (1985) Visibility reduction by air pollution in the Netherlands. *Atmospheric Environment*; **19** (2); 377–383

Dittenhoefer, AC and de Pena, RG (1978) A study of production and growth of sulfate particles in plumes from a coal-fired power plant. *Atmospheric Environment*; **12** (1–3); 297–306

Dittenhoefer, AC and de Pena, RG (1980) Sulfate aerosol production and growth in coal-operated power plant plumes. *Journal of Geophysical Research*; **85** (C8); 4499–4506

Dlugi, R and Jordan, S (1980) Versuchen zur Bildung von Sulfaten auf Partikeln. In: *Proceedings of the first European symposium on physico-chemical behaviour of atmospheric pollutants, Ispra (Italy) 16–18 Oct 1979. EUR 6621 DE, EN, FR*. Brussels, Belgium, Commission of the European Communities, pp. 293–297

Dockery, DW and Spengler, JD (1981a) Indoor–outdoor relationships of respirable sulfates and particles. *Atmospheric Environment*; **15** (3); 335–343

Dockery, DW and Spengler, JD (1981b) Personal exposure to respirable particulates and sulfates. *Journal of the Air Pollution Control Association*; **31** (2); 153–159

Dollard, GJ and Unsworth, MH (1983) Field measurements of turbulent fluxes of wind-driven fog drops to a grass surface. *Atmospheric Environment*; **17** (4); 775–780

Dollard, GJ and Vitols, V (1980) Wind tunnel studies of dry deposition of SO_2 and H_2SO_4 aerosols. In: *Ecological impact of acid precipitation. Proceedings of an international conference, Sandefjord, Norway, 11–14 March 1980*. Oslo, Norway, SNSF Project, pp. 108–109

Dollard, GJ, Unsworth, MH and Harve, MJ (1983) Pollutant transfer in upland regions by occult precipitation. *Nature*; **302** (5905); 241–243

Dolske, DA and Gatz, DF (1985) A field intercomparison of methods for the measurement of particle and gas dry deposition. *Journal of Geophysical Research*; **90** (D1); 2076–2084

Domaracki, AJ, Sistla, G and Putta, SN (1983) A receptor climatology of trajectories originating in New York State. *Journal of the Air Pollution Control Association*; **33** (11); 1068–1072

Dovland, H and Semb, A (1980) Atmospheric transport of pollutants. In: *Ecological impact of acid precipitation. Proceedings of an international conference, Sandefjord, Norway, 11–14 March 1980.* Oslo, Norway, SNSF Project, pp. 14–21

Downs, JL (1981) *Measurement of sulfate aerosols from Western low-rank coal-fired boilers with FGD: results of tests at Clay Boswell Unit 4. DOE/FC/10182-T12.* Altadena, CA, Meteorology Research Inc., 50pp.

Dzubay, TG (1980) Chemical element balance method applied to dichotomous sampler data. *Annals of the New York Academy of Sciences*; **338**; 126–144

Dzubay, TG., Stevens, RK, Lewis, CW, Hern, DH, Courtney, WJ, Tesch, JW and Mason, MA (1982) Visibility and aerosol composition in Houston, Texas. *Environmental Science and Technology*; **16** (8); 514–525

Easter, RC, Busness, KM, Hales, JM, Lee, RN, Arbuthnot, DA, Miller, DF, Sverdrup, GM, Spicer, CW and Howes, JE (1980) *Plume conversion rates in the SURE region.* 2 Volumes. EPRI-EA-1498, Palo Alto, CA, USA, Electric Power Research Institute, 177, 246 pp.

Eatough, DJ, Major, T, Ryder, J, Hill, M, Mangelson, NF, Eatough, NL and Hansen, LD (1978) The formation and stability of sulfite species in aerosols. *Atmospheric Environment*; **12** (1–3); 263–271

Eatough, DJ, Lee, ML, Later, DW, Richter, BE, Eatough, NL and Hansen, LD (1981a) Dimethyl sulfate in particulate matter from coal- and oil-fired power plants. *Environmental Science and Technology*; **15** (12); 1502–1506

Eatough, DJ, Richter, BE, Eatough, NL and Hansen, LD (1981b) Sulfur chemistry in smelter and power plant plumes in the Western US. *Atmospheric Environment*; **15** (10/11); 2241–2253

Eatough, DJ, Christensen, JJ, Eatough, NL, Hill, MW, Major, TD, Mangelson, NF, Post, ME, Ryder, JF, Hansen, LD, Meisenheimer, RG and Fischer, JW (1982) Sulfur chemistry in a copper smelter plume. *Atmospheric Environment*; **16** (5); 1001–1015

Eatough, DJ, Arthur, RJ, Eatough, NL, Hill, M, Mangelson, NF, Richter, BE, Hansen, LD and Cooper, JA (1984) Rapid conversion of $SO_2(g)$ to sulfate in a fog bank. *Environmental Science and Technology*; **18** (11); 855–859

Economist (1984) Raining acid on trees. *Economist*; **290** (7334); 78–79

Eggleton, AEJ and Cox, RA (1978) Homogeneous oxidation of sulphur compounds in the atmosphere. *Atmospheric Environment*; **12** (1–3); 227–230

Ehrlich, R (1980) Interaction between environmental pollutants and respiratory infections. *Environmental Health Perspectives*; **35**; 89–100

Eldred, RA, Asbaugh, LL, Cahill, TA, Flocchini, RG and Pitchford, ML (1983) Sulfate levels in the southwest during the 1980 copper smelter strike. *Journal of the Air Pollution Control Association*; **33** (2); 110–113

Eliassen, A and Saltbones, J (1983) Modelling of long-range transport of sulphur over Europe: a two-year model run and some model experiments. *Atmospheric Environment*; **17** (8); 1457–1473

Ellestad, TG (1980) Aerosol composition of urban plumes passing over a rural monitoring site. *Annals of the New York Academy of Science*; **338**; 202–218

Elshout, AJ, Viljeer, JW and Van Duren, H (1978) Sulphates and sulphuric acid in the atmosphere in the years 1971–1976 in the Netherlands. *Atmospheric Environment*; **12** (1–3); 785–790

EMEP (1982) EMEP: the cooperative programme for monitoring and evaluation of long-range transmission of air pollutants in Europe. *Economic Bulletin for Europe*; **34** (1); 29–40

Enger, L and Hoegstroem, U (1979) Dispersion and wet deposition of sulfur from a power plant plume. *Atmospheric Environment*; **13** (6); 797–810

Environmental Protection Agency, USA (1982) *Air quality criteria for particulate matter and sulfur oxides. 3 vols. EPA-600/8-82-029A-C, PB84-156785, PB84-156793, PB84-156801.* Research Triangle Park, NC, USA, Environmental Protection Agency, Environmental Criteria and Assessment Office, 209, 624 and 686 pp.

Eriksson, E (1960) The yearly circulation of chloride and sulfur in nature: meteorological, geochemical and pedological implications. Part II *Tellus*; **12** (1); 63–109

Everett, RG, Hicks, BB, Berg, WW and Winchester, JW (1979) An analysis of particulate sulfur and lead gradient data collected at Argonne National Laboratory. *Atmospheric Environment*; **13** (7); 931–934

Falconer, RE and Falconer, PD (1980) Determination of cloud water acidity at a mountain observatory in the Adirondack Mountains of New York State. *Journal of Geophysical Research*; **85** (C12); 7465–7470

Falk, HL and Jurgelski, W (1979) Health effects of coal mining and combustion: carcinogens and cofactors. *Environmental Health Perspectives*; **33**; 203–226

Fassina, V and Lazzarini, L (1981) Sulphur dioxide atmospheric oxidation in the presence of ammonia. *Water, Air and Soil Pollution*; **15** (3); 343–352

Feeley, JA and Liljestrand, HM (1983) Source contributions to acid precipitation in Texas. *Atmospheric Environment*; **17** (4); 808–814

Ferguson, P and Lee, JA (1983) Past and present sulphur pollution in the southern Pennines. *Atmospheric Environment*; **17**)6); 1131–1137

Ferman, MA, Wolff, GT and Kelly, NA (1981) The nature and sources of haze in the Shenandoah Valley/Blue Ridge Mountains area. *Journal of the Air Pollution Control Association*; **31**; (10); 1074–1082

Ferris, BG, Speizer, FE, Spengler, JD, Dockery, D, Bishop, YMM, Wolfson, M and Humble, C (1979) Effects of sulfur oxides and respirable particles on human health: methodology and demography of populations in study. *American Review of Respiratory Diseases*; **120**; 767–779

Ferris, BG, Speizer, FE, Bishop, YMM, Spengler, JD and Ware, JH (1980) The six-city study: a progress report. In: *Atmospheric sulfur deposition: environmental impact and health effects. Proceedings of the second life sciences symposium, Potential environmental and health consequences of atmospheric sulfur deposition, Gatlinberg, TN, 14-18 Oct 1979*. Ann Arbor, MI, Ann Arbor Science Publishers, pp. 99–107

Ferris, BG, Dockery, D, Ware, JH, Speizer, FE and Spiro, R (1983) The six-city study: examples of problems in analysis of the data. *Environmental Health Perspectives*; **52**; 115–123

Fisher, BEA (1978) The calculation of long term sulphur deposition in Europe. *Atmospheric Environment*; **12** (1–3); 489–501

Fisher, BEA (1982) Deposition of sulphur and the acidity of precipitation over Ireland. *Atmospheric Environment*; **16** (11); 2725–2734

Fisher, BEA and Clark, PA (1983) Testing a statistical long-range transport model on European and North American observations. In: *Proceedings of the 14th international technical meeting on air pollution modelling and its application, Copenhagen, Denmark, Sep 1983*, pp. 471–485

Fisher, BEA and Callander, BA (1984) Mass balances of sulphur and nitrogen oxides over Great Britain. *Atmospheric Environment*; **18** (9); 1751–1757

Flack, WW and Matteson, MJ (1980) Mass transfer of gases to growing water droplets. In: *Polluted rain: proceedings of the 12th Rochester International conference on environmental toxicity, Rochester, NY, USA, 21–23 May 1979*. New York, NY, USA, Plenum Press, pp. 61–85

Forrest, J, Schwartz, SE and Newman, L (1979a) Conversion of sulfur dioxide to sulfate during the Da Vinci flights. *Atmospheric Environment*; **13** (1); 157–167

Forrest, J, Garber, R and Newman, L (1979b) Formation of sulfate, ammonium and nitrate in an oil-fired power plant plume. *Atmospheric Environment*; **13** (9); 1287–1297

Forrest, J, Garber, R and Newman, L (1981) Conversion rates in power plant plumes based on filter pack data: the coal-fired Cumberland plume. *Atmospheric Environment*; **15**(10/11); 2273–2282

Fowler, D (1980a) Wet and dry deposition of sulphur and nitrogen compounds from the atmosphere. In: *Effects of acid precipitation on terrestrial ecosystems. Proceedings of the NATO conference on effects of acid precipitation on vegetation and soils, Toronto, Canada, 21–27 May 1978*. New York, NY, USA, Plenum Press, pp. 9–27

Fowler, D (1980b) Removal of sulphur and nitrogen compounds from the atmosphere in rain and by dry deposition. In: *Ecological impact of acid precipitation. Proceedings of an international conference, Sandefjord, Norway, 11–14 March 1980*. Oslo, Norway, SNSF Project, pp. 22–32

Fowler, D (1984) Transfer to terrestrial surfaces. *Philosophical Transactions of the Royal Society of London; B305 (1124);* 281–297

Fowler, D and Cape, JN (1984) On the episodic nature of wet deposited sulphate and acidity. *Atmospheric Environment*; **18** (9); 1859–1866

Fowler, D, Cape, JN, Leith, ID, Paterson, IS, Kinnaird, JW and Nicholson, IA (1982) Rainfall acidity in northern Britain. *Nature*; **297** (5865); 383–386

Freiberg, JE (1983) *The effects of humidity and temperature on the conversion of SO_2 to particulate sulfate and sulfite. EPRI-EA-3310*. Palo Alto, CA, USA, Electric Power Research Institute, 160 pp.

Freiberg, JE and Schwartz, SE (1981) Oxidation of SO_2 in aqueous droplets: mass-transport limitation in laboratory studies and the ambient atmosphere. *Atmospheric Environment*; **15** (7); 1145–1154

Friend, JP (1973) The global sulphur cycle. In: *Chemistry of the lower atmosphere*. New York, NY, USA, Plenum Press, pp. 177–201

Friend, JP and Barnes, RA (1979) Sulfur dioxide photo-oxidation in air and the creation of nuclei by free radicals. *American Chemical Society, Division of Environmental Chemistry, Preprints*; **19** (1); 587–591

Galloway, JN and Likens, GE (1981) Acid precipitation: the importance of nitric acid. *Atmospheric Environment*; **15** (6); 1081–1085

Galloway, JN and Whelpdale, DM (1980) An atmospheric sulfur budget for eastern North America. *Atmospheric Environment*; **14** (4); 409–417

Galloway, JN, Cowling, EB, Gorham, E and McFee, WW (1978). *National program for assessing the problem of atmospheric deposition (acid rain). Final report. PB80-178767*, Raleigh, NC, USA, North Carolina Agricultural Experiment Station, 103 pp.

Galloway, JN, Whelpdale, DM and Wolff, GT (1984a) The flux of S and N eastward from North America. *Atmospheric Environment*; **18** (12); 2595–2607

Galloway, JN, Likens, GE and Hawley, ME (1984b) Acid precipitation: natural versus anthropogenic components. *Science*; **226** (4676); 829–831

Garber, R, Forrest, J and Newman, L (1981) Conversion rates in power plant plumes based on filter pack data: the oil-fired Northport plume. *Atmospheric Environment*; **15** (10/11); 2283–2292

Gardner, DE, Miller, FJ, Illing, JW and Kirtz, JM (1977) Increased infectivity with exposure to ozone and sulfuric acid. *Toxicology Letters*; **1** (2); 59–64

Garland, JA (1978) Dry and wet removal of sulphur from the atmosphere. *Atmospheric Environment*; **12** (1–3); 349–362

Garland, JA (1981) Enrichment of sulphate in maritime aerosols. *Atmospheric Environment*; **15** (5); 787–791

Garland, JA and Cox, LC (1982) Deposition of small particles to grass. *Atmospheric Environment*; **16** (11); 2699–2702

Garrels, RM, MacKenzie, FT and Hunt, C (1975) *Chemical cycles and the global environment: assessing human influences*. Los Altos, CA, USA, William Kaufmann Inc., 205 pp.

Gatz, DF (1980a) An urban influence on deposition of sulfate and soluble metals in summer rains. In: *Atmospheric sulfur deposition: environmental impact and health effects. Proceedings of the second life sciences symposium, Potential environmental and health consequences of atmospheric sulfur deposition, Gatlinberg, TN, 14–18 Oct 1979*. Ann Arbor, MI, Ann Arbor Science Publishers, pp. 245–260

Gatz, DF (1980b) Associations and mesoscale spatial relationships among rainwater constituents. *Journal of Geophysical Research*; **85** (C10); 5588–5598

Gauri, KL (1980) Deterioration of architectural structures and monuments. In: *Polluted rain: Proceedings of the 12th Rochester International conference on environmental toxicity, Rochester, NY, USA, 21–23 May 1979*. New York, NY, USA, Plenum Press, pp. 125–145

Gauri, KL and Holdren, GO (1981) Pollutant effects on stone monuments. *Environmental Science and Technology*; **15** (4); 386–390

GCA Corporation (1981) *Acid rain information book. Final report. DOE/EP-0018, DE81-024267*, Bedford, MA, USA, GCA Corporation, 235 pp.

Georgii, H-W (1981) Review of the acidity of precipitation according to the WMO-network. *Idojaras*; **85** (1); 1–9

Georgii, HW, Perseke, C and Rohbock, E (1984) Deposition of acidic components and heavy metals in the Federal Republic of Germany for the period 1979–1981. *Atmospheric Environment*; **18** (3); 581–589

Gillani, NV (1978) Project MISTT: mesoscale plume modeling of the dispersion, transformation and ground removal of SO_2. *Atmospheric Environment*; **12** (1–3); 569–588

Gillani, NV and Wilson, WE (1980) Formation and transport of ozone and aerosols in power plant plumes. *Annals of the New York Academy of Science*; **338**; 276–296

Gillani, NV, Husar, RB, Husar, JD, Patterson, DE and Wilson, WE (1978) Project MISTT: kinetics of particulate sulfur formation in a power plant plume out to 300 kM. *Atmospheric Environment*; **12** (1–3); 589–598

Gillani, NV, Kohli, S and Wilson, WE (1981) Gas-to-particle conversion of sulfur in power plant plumes - I Parametrization of the conversion rate for dry, moderately polluted ambient conditions. *Atmospheric Environment*; **15** (10/11); 2293–2313

Gislason, KB and Prahm, LP (1983) Sensitivity study of air trajectory long-range transport modelling. *Atmospheric Environment*; **17** (12); 2463–2472

Gorham, E (1980) *Chemical composition of atmospheric precipitation in Minnesota and North Dakota. Final Report July 1, 1977–Jun 30, 1980. DOE/EV/04327-1*, Minneapolis, MN, USA, Minnesota University, 31 pp.

Graedel, TE (1980) Atmospheric photochemistry. In: *Handbook of Environmental Chemistry, Volume 2, Part A, Reactions and processes*. Berlin, FRG, Springer-Verlag, pp. 107–143

Granat, L (1972) *Deposition of sulfate and acid with precipitation over Northern Europe. AC-20*, Stockholm, Sweden, University of Stockholm, Institute of Meteorology; International Meteorological Institute in Stockholm, 49 pp.

Granat, L, Rodhe, H and Hallberg, RO (1976) The global sulphur cycle. In: Nitrogen, phosphorus and sulphur—global cycles. SCOPE 7. *Ecological Bulletin*; **22**; 89–134

Granat, L (1978) Sulfate in precipitation as observed by the European atmospheric chemistry network. *Atmospheric Environment*; **12** (1–3); 707–713

Gravenhorst, G, Beilke, S, Betz, M and Georgii, H-W (1980) Sulfur dioxide absorbed in rainwater. In: *Effects of acid precipitation on terrestrial ecosystems. Proceedings of the NATO conference on effects of acid precipitation on vegetation and soils, Toronto, Canada, 21–27 May 1978*. New York, NY, USA, Plenum Press, pp. 41–55

Groblicki, PJ, Wolff, GT and Countess, RJ (1981) Visibility-reducing species in the Denver 'brown cloud' – I Relationships between extinction and chemical composition. *Atmospheric Environment*; **15** (12); 2473–2484

Hackney, JD, Linn, WS, Jones, MP, Bailey, RM, Julin, DR and Kleinman, MT (1980) Short-term respiratory effects of sulfur-containing pollutant mixtures: some recent findings from controlled clinical studies. In: *Atmospheric sulfur deposition: environmental impact and health effects. Proceedings of the second life sciences symposium, Potential environmental and health consequences of atmospheric sulfur deposition, Galinberg, TN, 14–18 Oct 1979*. Ann Arbor, MI, Ann Arbor Science Publishers, pp. 77–83

Hackney, JD, Linn, WS and Avol, EL (1984) Assessing health effects of air pollution. *Environmental Science and Technology*; **18** (4); 115A–122A

Haines, BL, Nabholz, JV and Dubois, S (1980) *Rainfall element content and acidity from Apr 30, 1976 to February 17, 1978, Athens, Georgia. DOE/EV/00641–42*, Athens GA, USA, Georgia University, Institute of Ecology, 23 pp.

Hales, JM (1978) Wet removal of sulfur compounds from the atmosphere. *Atmospheric Environment*; **12** (1–3); 389–399

Hales, JM (1979) How the air cleans itself. In: *Proceedings of the International symposium on sulphur emissions and the environment, London, UK, 8–10 May 1979*. London, UK, Society of the Chemical Industry, pp. 193–207

Hales, JM (1982a) Precipitation scavenging pathways and mechanisms: a qualitative overview. In: *Acid forming emissions in Alberta and their ecological effects. Proceedings of the symposium/workshop, Edmonton, Alberta, Canada, 9–12 Mar 1982*. Edmonton, Canada, Alberta Department of Environment, pp. 3–40

Hales, JM (1982b) Mechanistic analysis of precipitation scavenging using a one-dimensional, time-variant model. *Atmospheric Environment*; **16**(7); 1775–1783

Hales, JM and Dana, MT (1979) Regional-scale deposition of sulfur dioxide by precipitation scavenging. *Atmospheric Environment*; **13** (8); 1121–1132

Hamill, P and Yue, GK (1980) A simplified model for the production of sulfate aerosols. In: *Environmental and climatic impact of coal utilization. Proceedings of the symposium on environmental and climatic impact of coal utilization, Williamsburg, VA, USA, 17–19 Apr 1979*. New York, NY, USA, Academic Press, pp. 255–272

Hamilton, LD (1984) Health and environmental risks of energy systems. In: *International symposium on the risks and benefits of energy systems, Julich, FRG, 9–13 Apr 1984. IAEA-SM-273/51, BNL-35459, DE85-004540,* Upton, NY, USA, Brookhaven National Laboratory, 40 pp.

Hamilton, WC and Cooper, DE (1979) Atmospheric sulfates and mortality: the phantom connection. *Coal Mining and Processing*; **16** (3); 88–94

Hansen, DA and Hidy, GM (1982) Review of questions regarding rain acidity data. *Atmospheric Environment*; **16** (9); 2107–2126

Harker, AB, Haynie, F, Mansfeld, F, Strauss, DR and Landis, DA (1982) Measurement of the sulfur dioxide and sulfuric acid aerosol induced corrosion of zinc in a dynamic flow system. *Atmospheric Environment*; **16** (11); 2691–2698

Harrison, RM and Sturges, WT (1984) Physico-chemical speciation and transformation reactions of particulate atmospheric nitrogen and sulphur compounds. *Atmospheric Environment*; **18** (9); 1829–1833

Harrison, H, Larson, TV and Monkman, CS (1982) Aqueous phase oxidation of sulfites by ozone in the presence of iron and manganese. *Atmospheric Environment*; **16** (5); 1039–1941

Harte, J (1983) An investigation of acid precipitation in Qinghai Province, China. *Atmospheric Environment*; **17** (2); 403–408

Haury, G, Jordan, S and Hofmann, C (1978) Experimental investigation of the aerosol-catalyzed oxidation of SO_2 under atmospheric conditions. *Atmospheric Environment*; **12** (1–3); 281–287

Hazrati, AM and Peters, LK (1981) Particulate sulfate in Lexington, Kentucky and its relation to major surrounding SO_2 sources. *Atmospheric Environment*; **15** (9); 1623–1631

Hegg, DA (1985) The importance of liquid-phase oxidation of SO_2 in the troposphere. *Journal of Geophysical Research*; **90** (D2); 3773–3779

Hegg, DA and Hobbs, PV (1978) Oxidation of sulfur dioxide in aqueous systems with particular reference to the atmosphere. *Atmospheric Environment*; **12** (1–3); 241–253

Hegg, DA and Hobbs, PV (1979) The homogeneous oxidation of sulfur dioxide in cloud droplets. *Atmospheric Environment*; **13** (7); 981–987

Hegg, DA and Hobbs, PV (1980) Measurements of gas-to-particle conversion in the plumes from five coal-fired electric power plants. *Atmospheric Environment*; **14**(1); 99–116

Hegg, DA and Hobbs, PV (1981) Cloud water chemistry and the production of sulphates in clouds. *Atmospheric Environment*; **15** (9); 1597–1604

Hegg, DA and Hobbs, PV (1982) Measurements of sulfate production in natural clouds. *Atmospheric Environment*; **16** (11); 2663–2668

Hegg, DA and Hobbs, PV (1983) Particles and trace gases in the plume from a modern coal-fired power plant in the western United States and their effects on light extinction. *Atmospheric Environment*; **17** (2); 357–368

Hegg, DA, Hobbs, PV and Lyons, JH (1985) Field studies of a power plant plume in the arid southwestern United States. *Atmospheric Environment*; **19** (7); 1147–1167

Heintzenberg, J, Hansson, H.-C and Lannefors, H (1981) The chemical composition of arctic haze at Ny-Alesund, Spitsbergen. *Tellus*; **33** (2); 162–171

Henderson, RG and Weingartner, K (1982) Trajectory analysis of MAP3S precipitation chemistry data at Ithaca, NY *Atmospheric Environment*; **16** (7); 1657–1665

Hendry, CD, Edgerton, ES and Brezonik, PL (1981) Acidity and related minerals in precipitation across Florida: spatial and temporal variations. *American Chemical Society, Division of Environmental Chemistry, Preprints*; **21** (1); 207–209

Henry, RC and Hidy, GM (1979) Multivariate analysis of particulate sulfate and other air quality variables by principal components - Part I Annual data from Los Angeles and New York. *Atmospheric Environment*; **13** (11); 1581–1596

Henry, RC and Hidy, GM (1980) Potential for atmospheric sulfur from microbiological sulfate reduction. *Atmospheric Environment*; **14**(9); 1095–1103

Henry, RC and Hidy, GM (1982) Multivariate analysis of particulate sulfate and other air quality variables by principal components - II Salt Lake City, Utah and St Louis, Missouri. *Atmospheric Environment*; **16** (5); 929–943

Hering, SV and Friedlander, SK (1982) Origins of aerosol sulfur size distributions in the Los Angeles basin. *Atmospheric Environment*; **16** (11); 2647–2656

Hering, SV, Bowen, JL, Wengert, JG and Richards, WL (1981) Characterization of the regional haze in the southwestern United States. *Atmospheric Environment*; **15** (10/11); 1999–2009

Herron, MM (1982) Impurity sources of F^-, Cl^-, NO_3^-, and SO_4^{2-} in Greenland and Antarctic precipitation. *Journal of Geophysical Research*; **87** (C4); 3052–3060

Hicks, BB and Wesely, ML (1980) Turbulent transfer processes to a surface and interaction with vegetation. In: *Atmospheric-sulfur deposition: environmental impact and health effects. Proceedings of the second life sciences symposium, Potential environmental and health consequences of atmospheric sulfur deposition, Gatlinberg, TN, 14–18 Oct 1979.* Ann Arbor, MI, Ann Arbor Science Publishers, pp. 199–206

Hicks, BB and Williams, RM (1980) Transfer and deposition of particles to water surfaces. In: *Atmospheric sulfur deposition: environmental impact and health effects. Proceedings of the second life sciences symposium, Potential environmental and health consequences of atmospheric sulfur deposition, Gatlinberg, TN, 14–18 Oct 1979.* Ann Arbor, MI, Ann Arbor Science Publishers, pp. 237–244

Hicks, BB, Wesely, ML, Durham, JL and Brown, MA (1982) Some direct measurements of atmospheric sulfur fluxes over a pine plantation. *Atmospheric Environment*; **16** (12); 2899–2903

Hidy, GM (1984) Source-receptor relationships for acid deposition: pure and simple? *Journal of the Air Pollution Control Association*; **34** (5); 518–531

Hidy, GM, Mueller, PK and Tong, EY (1978) Spatial and temporal distributions of airborne sulfate in parts of the United States. *Atmospheric Environment*; **12** (1–3); 735–752

Hidy, GM, Mueller, PK, Lavery, TF and Warren, KK (1979) Assessment of regional air pollution over the eastern United States: results from the sulfate regional experiment (SURE). In: *Papers presented at the WMO symposium on the long-range transport of pollutants and its relation to general circulation including stratospheric/tropospheric exchange processes, Sofia, Bulgaria, 1–5 October 1979. WMO-538.* Geneva, Switzerland, World Meteorological Organization, 65–76

Hitchcock, DR, Spiller, LL and Wilson, WE (1979) Chemical evidence of acid biogenic sulfate aerosols in coastal environments. *American Chemical Society, Division of Environmental Chemistry, Preprints*; **19** (1); 748–751

Hobbs, PV and Hegg, DA (1982) Sulfate and nitrate mass distributions in the near fields of some coal-fired power plants. *Atmospheric Environment*; **16** (11); 2657–2662

Hobbs, PV, Hegg, DA, Eltgroth, MW and Radke, LF (1979) Evolution of particles in the plumes of coal-fired power plants - I Deductions from field measurements. *Atmospheric Environment*; **13** (7); 935–951

Hobbs, PV, Stith, JL and Radke, LF (1980) Cloud-active nuclei from coal-fired electric power plants and their interactions with clouds. *Journal of Applied Meteorology*; **19** (4); 439–451

Hoffer, T, Kliwer, J and Moyer, J (1979) Sulfate concentrations in the Southwestern Desert of the United States. *Atmospheric Environment*; **13** (5); 619–627

Hoffman, MR and Jacob, DJ (1984) Kinetics and mechanisms of the catalytic oxidation of dissolved sulfur dioxide in aqueous solution: an application to nighttime fog water chemistry. In: *SO_2, NO and NO_2 oxidation mechanisms: atmospheric considerations.* Boston, MA, USA, Butterworth, pp. 101–172

Hoegstroem, U, Smedman, A.-S and Ellijeff, M (1981) *Beraekning av svavelnedfall i nationell skala (Sulphur deposition in Sweden). KHM-TR16*, Vaellinby, Sweden, Statens Vattenfallswerk, Projekt Kol - Haelsa - Miljoe, 40 pp.

Holt, BD, Cunningham, PT and Kumar, R (1979) Use of oxygen isotopy in the study of transformations of SO_2 to sulfates in the atmosphere. In: *Papers presented at the WMO symposium on the long-range transport of pollutants and its relation to general circulation including stratospheric/tropospheric exchange processes, Sofia, Bulgaria, 1–5 October 1979.* WMO-538, Geneva, Switzerland, World Meteorological Organization, 207–212

Holt, BD, Kumar, R and Cunningham, PT (1981) Oxygen-18 study of the aqueous phase oxidation of sulfur dioxide. *Atmospheric Environment*; **15** (4); 557–566

Holt, BD, Kumar, R and Cunningham, PT (1982) Primary sulfates in atmospheric sulfates: estimation by oxygen isotope ratio measurements. *Science*; **217** (4554); 51–53

Holt, BD, Nielson, E and Kumar, R (1983) Oxygen-18 estimation of primary sulfate in total sulfate scavenged by rain from a power plant plume. In: *Precipitation scavenging, dry deposition, and resuspension. Vol. 1 Precipitation scavenging. Proceedings of the fourth international conference, Santa Monica, CA, USA, 29 Nov–3 Dec 1982.* New York, NY, USA, Elsevier, pp. 357–368

Homolya, JB, Barnes, HM and Fortune, CR (1976) Characterization of the gaseous sulfur emissions from coal and oil-fired boilers. In: *Energy and the environment. Fourth national conference on energy and the environment, Cincinnati, Ohio, USA*, 490–494

Homolya, JB and Cheney, JL (1979) A study of primary sulfate emissions from a coal-fired boiler with FGD. *Journal of the Air Pollution Control Association*; **29** (9); 1000–1004

Homolya, JB and Lambert, S (1981) Characterization of sulfate emissions from nonutility boilers firing low-S residual oils in New York City. *Journal of the Air Pollution Control Association*; **31** (2); 139–143

Horvath, L and Bonis, K (1980) An attempt to estimate the rate constant of sulfur dioxide-sulfate conversion in the urban plume of Budapest. *Idojaras*; **8** (4); 190–195

Horvath, H and Pirich, R (1983) Remarks on the nomenclature in atmospheric optics. *Atmospheric Environment*; **17** (12); 2625–2627

Hov, O and Isaksen, ISA (1981) Generation of secondary pollutants in a power plant plume: a model study. *Atmospheric Environment*; **15** (10/11); 2367–2376

Hung, RJ and Liaw, GS (1980) Advection fog formation associated with atmospheric aerosols due to combustion-related pollutants. *Water, Air and Soil Pollution*; **14**; 267–285

Husain, L and Samson, PJ (1979) Long-range transport of trace elements. *Journal of Geophysical Research*; **84** (C3); 1237–1240

Husain, L, Webber, JS, Canelli, E, Dutkiewicz, VA and Halstead, JA (1984) Mn/V ratio as a tracer of aerosol sulfate transport. *Atmospheric Environment*; **18** (6); 1059–1071

Husar, RB, Patterson, DE, Husar, JD, Gillani, NV and Wilson, WE (1978) Sulfur budget of a power plant plume. *Atmospheric Environment*; **12** (1–3); 549–568

Husar, RB and Patterson, DE (1980) Regional scale air pollution: sources and effects. *Annals of the New York Academy of Science*; **338**; 399–417

Husar, RB, Holloway, JM and Patterson, DE (1981) Spatial and temporal pattern of Eastern US haziness: a summary. *Atmospheric Environment*; **15** (10/11); 1919–1928

Hutcheson, MR and Hall, FP (1974) Sulphate washout from a coal fired power plant plume. *Atmospheric Environment*; **8** (1); 23–28

Ibrahim, M, Barrie, LA and Fanaki, F (1983) An experimental and theoretical investigation of the dry deposition of particles to snow, pine trees and artificial collectors. *Atmospheric Environment*; **17** (4); 781–788

Inn, ECY, Farlow, NH, Russell, PB, McCormick, MP and Chu, WP (1982) Observations. In: *The stratospheric aerosol layer*. Berlin, FRG, Springer-Verlag, pp. 15–68

International Electric Research Exchange (1981) *Effects of SO_2 and its derivatives on health and ecology. Volume 1. Human health. Report of a Working Group sponsored by the International Electric Research Institute, The Canadian Electrical Association, The Japan IERE Council, The International Union of Producers and Distributors of Electrical Energy*. 246 pp.

Isaksen, ISA, Hesstvedt, E and Hov, O (1978) A chemical model for urban plumes: test for ozone and particulate sulfur formation in St Louis urban plume. *Atmospheric Environment*; **12** (1–3); 599–604

Ivanov, MV (1983) Major fluxes of the global biogeochemical cycle of sulphur. In: *The global biogeochemical sulphur cycle. SCOPE 19.* Chichester, John Wiley, pp. 449–463

Jaenicke, R (1978) Physical properties of atmospheric particulate sulfur compounds. *Atmospheric Environment*; **12** (1–3); 161–170

Jaenicke, R (1980a) Natural aerosols. *Annals of the New York Academy of Science*; **338**; 317–329

Jaenicke, R (1980b) Atmospheric aerosols and global climate. *Journal of Aerosol Science*; **11** (5/6); 577–588

Jaroslav, S and Dusan, Z (1979) Sulphur compounds in background air pollution in Czechoslovakia. In: *Papers presented at the WMO symposium on the long range transport of pollutants and its relation to general circulation including stratospheric/tropospheric exchange processes, Sofia, Bulgaria, 1–5 October 1979. WMO-538*, Geneva, Switzerland, World Meteorological Organization, 45–51

Jickells, T, Knap, A, Church, T, Galloway, J and Miller, J (1982) Acid rain on Bermuda. *Nature*; **297** (5861); 55–57

Johannes, AH, Altwicker, ER and Clesceri, NL (1981) *Characterization of acidic precipitation in the Adirondack region. EPRI-EA-1826, DE81-903318*, Palo Alto, CA, USA, Electric Power Research Institute, 200 pp.

Johnson, WB (1983) Interregional exchanges of air pollution: model types and applications. *Journal of the Air Pollution Control Association*; **33** (6); 563–574

Joranger, E and Ottar, B (1984) Air pollution studies in the Norwegian Arctic. *Geophysical Research Letters*; **11** (5); 365–368

Josephson, J (1983) Exposure pathways of workplace contaminants. *Environmental Science and Technology*; **17** (4); 168A-172A

Junge, CE (1963) *Air chemistry and radioactivity*. New York, NY, USA, Academic Press, 382 pp.

Kallend, AS, Marsh, ARW, Pickles, JH and Proctor, MV (1983) Acidity of rain in Europe. *Atmospheric Environment*; **17** (1); 127–137

Kaplan, DJ, Himmelblau, DM and Kanaoka, C (1981) Oxidation of sulphur dioxide in aqueous ammonium sulfate aerosols containing manganese as a catalyst. *Atmospheric Environment*; **15** (5); 763–773

Kasina, S (1980) On precipitation acidity in southeastern Poland. *Atmospheric Environment*; **14** (11); 1217–1221

Keesee, RG and Castleman, AW (1982) The chemical kinetics of aerosol formation. In: *The stratospheric aerosol layer*. Berlin, FRG, Springer-Verlag, pp. 69–92

Kelkar, DN and Ashawa, SC (1979) Sulphate in summer monsoonal precipitation over India. In: *Proceedings of the International symposium on sulphur emissions and the environment, London, UK, 8–10 May 1979* London, UK, Society of the Chemical Industry, pp. 67–71

Kellogg, WW, Cadle, RD, Allen, ER, Lazrus, AL and Martell, EA (1972) The sulfur cycle. *Science*; **175** (4022); 587–596

Kellogg, WW (1980) Aerosols and Climate. In: *Interactions of energy and climate. Proceedings of an international conference, Munster, FRG, 3–6 Mar 1980*. Dordrecht, Holland, D Reidel, pp. 281–296

Kelly, NA, Wolff, GT and Ferman, MA (1982) Background pollutant measurements in air masses affecting the eastern half of the United States - I Air masses arriving from the northwest. *Atmospheric Environment*; **16** (5); 1077–1088

Kerr, HD, Kulle, TJ, Farrell, BP, Sauder, LR, Young, JL, Swift, DL and Borushok, RM (1981) Effects of sulfuric acid aerosol on pulmonary function in human subjects: an environmental chamber study. *Environmental Research*; **26** (1); 42–50

Khalil, MAK and Rasmussen, RA (1984) Global sources, lifetimes and mass balances of carbonyl sulfide (OCS) and carbon disulfide (CS_2) in the Earth's atmosphere. *Atmospheric Environment*; **18** (9); 1805–1813

Khemani, LT, Momin, GA, Naik, MS, Rao, PSP, Kumar, R and Murty, BVR (1985) Impact of alkaline particulates on pH of rain water in India. *Water, Air, and Soil Pollution*; **25** (4); 365–376

Kitagawa, T (1984) Cause analysis of the Yokkaichi asthma episode in Japan. *Journal of the Air Pollution Control Association*; **34** (7); 743–746

Kleinman, LI (1983) A regional scale modeling study of the sulfur oxides with a comparison to ambient and wet deposition monitoring data. *Atmospheric Environment*; **17** (6); 1107–112

Kleinman, MT, Linn, WS, Bailey, RM, Anderson, KR, Whynot, JD, Medway, DA and Hackney, JD (1981) Human exposure to ferric sulfate aerosol: effects on pulmonary function and respiratory symptoms. *American Industrial Hygiene Association Journal*; **42** (4); 298–304

Klemm, RF and Gray, JML (1982) Acidity and chemical composition of precipitation in central Alberta, 1977–78. In: *Acid forming emissions in Alberta and their ecological effects. Proceedings of the symposium/workshop, Edmonton, Alberta, Canada, 9–12 Mar 1982*. Edmonton, Canada, Alberta Department of Environment, pp. 153–180

Ko, MKW and Sze, ND (1980) The CS_2 and COS budget. In: *Environmental and climatic impact of coal utilization. Proceedings of the symposium on environmental and climatic impact of coal utilization, Williamsburg, VA, USA, 17–19 Apr 1979*. New York, NY, USA, Academic Press, pp. 323–328

Koerner, RM and Fisher, D (1982) Acid snow in the Canadian high Arctic. *Nature*; **295** (5845); 137–140

Kowalczyk, GS, Gordon, GE and Rheingrover, SW (1982) Identification of atmospheric particulate sources in Washington, DC, using chemical element balances. *Environmental Science and Technology*; **16**(2); 79–90

Kramer, JR (1978) Acid precipitation. In: *Sulfur in the environment. Part I: the atmospheric cycle*. New York, NY, USA, John Wiley, pp. 325–370

Kucera, V (1983) The effect of acidification of the environment on the corrosion in the atmosphere, water and soil. In: *Proceedings of the ninth Scandinavian corrosion congress, Copenhagen, Denmark, Sep 1983*, pp. 153–170

Kurtz, J and Scheider, WA (1981) An analysis of acidic precipitation in South-Central Ontario using air parcel trajectories. *Atmospheric Environment*; **15** (7); 1111–1116

Kurtz, J, Tang, AJS, Kirk, RW, Chan, WH (1984) Analysis of an acidic deposition episode at Dorset, Ontario. *Atmospheric Environment*; **18** (2); 387–394

Lau, N.-C and Charlson, RJ (1977) On the discrepancy between background atmospheric ammonia gas measurements and the existence of acid sulfates as a dominant atmospheric aerosol. *Atmospheric Environment*; **11** (5); 475–478

Lave, LB and Seskin, EP (1977) *Air pollution and the human health*. Baltimore, Johns Hopkins, 388 pp.

Lavery, TF, Baskett, RL, Thrasher, JW, Lordi, NJ, Lloyd, AC and Hidy, GM (1981) Development and validation of a regional model to simulate atmospheric concentrations of sulfur dioxide and sulfate. In: *Second joint conference on applications of air pollution meteorology*, New Orleans, LA, USA, 24–27 March 1980 Boston, MA, USA, American Meteorological Society, pp. 236–247

Lawson, DR and Winchester, JW (1979) Background fine particle mode sulfur aerosol concentrations on the South American continent. *American Chemical Society, Division of Environmental Chemistry, Preprints*; **19** (1); 400–403

Leaderer, BP, Tanner, RL and Holford, TR (1982) Diurnal variations, chemical composition and relation to meteorological variables of the summer aerosol in the New York subregion. *Atmospheric Environment*; **16** (9); 2075–2087

Leaderer, BP, Tanner, RL, Lioy, PJ and Stolwijk, JA (1981) Seasonal variations in light scattering in the New York region and their relation to sources. *Atmospheric Environment*; **15** (12); 2407–2420

Lee, DO (1983) Trends in summer visibility in London and southern England 1962–1979. *Atmospheric Environment*; **17** (1); 151–159

Lee, ML, Later, DW, Rollins, DK, Eatough, DJ and Hansen, LD (1980) Particulate matter. *Science*; **207** (4427); 186–188

Legrand, MR and Delmas, RJ (1984) The ionic balance of Antarctic snow: a 10-year detailed record. *Atmospheric Environment*; **18** (9); 1867–1874

Leonard, RL, Goldman, CR and Likens, GE (1981) Some measurements of the pH and chemistry of precipitation at Davis and Lake Tahoe, California. *Water, Air and Soil Pollution*; **15**(2); 153–167

Leone, I and Brennan, E (1985) An update on forest decline in Germany. *Journal of the Air Pollution Control Association*; **35** (3); 189

Leslie, ACD (1980) Comments on an aerosol sulfur/water relationship in the eastern United States. In: *Long range aerosol transport of metals and sulphur. Proceedings of a workshop in Lund, Sweden, 13 June 1980. Report SNV PM-1337*, Solna, Sweden, National Swedish Environment Protection Board, pp. 46–53

Leslie, ACD, Ahlberg, MS, Winchester, JW and Nelson, JW (1978) Aerosol characterization for sulfur oxide health effects assessment. *Atmospheric Environment*; **12** (1–3); 729–733

Lewin, EE and Torp, U (1982) Influence of contamination on the analysis of precipitation samples. *Atmospheric Environment*; **16** (4); 795–800

Lewis, JE, Moore, TR and Enright, NJ (1983) Spatial-temporal variations in snowfall chemistry in the Montreal region. *Water, Air and Soil Pollution*; **20** (1); 7–22

Lewis, WM and Grant, MC (1980) Acid precipitation in the western United States. *Science*; **207** (4427); 176–177

Liberti, A, Brocco, D and Possanzini, M (1978) Adsorption and oxidation of sulfur dioxide on particles. *Atmospheric Environment*; **12** (1–3); 255–261

Liebsch, EJ and de Pena, RG (1982) Sulfate aerosol production in coal-fired power plant plumes. *Atmospheric Environment*; **16** (6); 1323–1331

Likens, GE and Butler, TJ (1981) Recent acidification of precipitation in North America. *Atmospheric Environment*; **15** (7); 1103–1109

Likens, GE, Bormann, FH, Waton, JS, Pierce, RS and Johnson, NM (1976) Hydrogen ion input to the Hubbard Brook experimental forest, New Hampshire, during the last decade. *Water, Air and Soil Pollution*; **6**; 435–445

Likens, GE, Bormann, FH and Eaton, JS (1980) Variations in precipitation and streamwater chemistry at the Hubbard Brook exprimental forest during 1964 to 1977. In: *Effects of acid precipitation on terrestrial ecosystems. Proceedings of the NATO conference on effects of acid precipitation on vegetation and soils, Toronto, Canada, 21–27 May 1978*. New York, NY, USA, Plenum Press, pp. 443–464

Likens, GE, Bormann, FH, Pierce, RS, Eaton, JS and Munn, RE (1984) Long-term trends in precipitation chemistry at Hubbard Brook, New Hampshire. *Atmospheric Environment*; **18** (12); 2641–2647

Liljestrand, HM and Morgan, JJ (1981) Spatial variations of acid precipitation in Southern California. *Environmental Science and Technology*; **15** (3); 333–338

Lindberg, SE (1981) The relationship between manganese and sulfate ions in rain. *Atmospheric Environment*; **15** (9); 1749–1753

Lindberg, SE (1982) Factors influencing trace metal, sulfate and hydrogen ion concentrations in rain. *Atmospheric Environment*; **16** (7); 1701–1709

Linn, WS, Kleinman, MT, Bailey, RM, Medway, DA, Spier, CE, Whynot, JD, Anderson, KR and Hackney, JD (1981) Human respiratory responses to an aerosol containing zinc ammonium sulphate. *Environmental Research*; **25**; 404–414

Lioy, PJ and Morandi, MT (1982) Source related winter and summer variations in SO_2, SO_4^{2-} and vanadium in New York City from 1972–1979. *Atmospheric Environment*; **16** (6); 1543–1549

Lioy, PJ, Wolff, GT, Rahn, KA, Bernstein, DM and Kleinman, MT (1979a) Characterization of aerosols upwind of New York City: II. Aerosol composition. *Annals of the New York Academy of Sciences*; **322**; 73–85

Lioy, PJ, Wolff, GT and Leaderer, BP (1979b) A discussion of the New York summer aerosol study, 1976. *Annals of the New York Academy of Sciences*; **322**; 153–165

Lioy, PJ, Samson, PJ, Tanner, RL, Leaderer, BP, Minnich, T and Lyons, W (1980) The distribution and transport of sulfate 'species' in the New York Metropolitan area during the 1977 summer aerosol study. *Atmospheric Environment*; **14** (2); 1391–1407

Lioy, PJ, Mallon, RP, Lippmann, M, Kneip, TJ and Samson, PJ (1982) Factors affecting the variability of summertime sulfate in a rural area using principal component analysis. *Journal of the Air Pollution Control Association*; **32** (10); 1043–1047

Lipfert, FW (1980) Sulfur oxides, particulates, and human mortality: Synopsis of statistical correlations. *Journal of the Air Pollution Control Association*; **30** (4); 366–371

Lipfert, FW and Dupuis, LR (1985) Comment on local source impacts at Hubbard Brook, New Hampshire. *Journal of the Air Pollution Control Association*; **35** (2); 127–130

Lippmann, M (1980a) Health significance of exposures to sulfur oxide air pollutants. In: *Atmospheric sulfur deposition: environmental impact and health effects. Proceedings of the second life sciences symposium, Potential environmental and health consequences of atmospheric sulfur deposition, Gatlinberg, TN, 14–18 Oct 1979*. Ann Arbor, MI, Ann Arbor Science Publishers, pp. 85–95

Lippmann, M (1980b) *Effects of sulfur oxide pollutants on respiratory deposition and bronchial clearance. Final report. EPA-600/1-80-035, PB81-168288*, New York, NY, USA New York University, Institute of Environmental Medicine, 52 pp.

Lippmann, M, Kleinmann, MT, Bernstein, DM, Wolff, GT and Leaderer, BP (1979) Size-mass distributions of the New York summer aerosol. *Annals of the New York Academy of Sciences*; **322**; 29–44

Lippmann, M, Schlesinger, RB, Laikauf, G, Spektor, D and Albert, RE (1982) Effects of sulphuric acid aerosols on respiratory tract airways. *Annals of Occupational Hygiene*; **26** (1–4); 677–690

Livingston, RA and Baer, NS (1983) Mechanisms of air pollution-induced damage to stone: In: *Proceedings of the sixth world conference on air quality. Paris, France, 16–20 May 1983*. Vol. 3. pp. 33–40

Lodge, JP, Waggoner, AP, Klodt, DT and Crain, CN (1981) Non-health effects of airborne particulate matter. *Atmospheric Environment*; **15**(4); 431–482

Logan, RM, Derby, JC and Duncan, LC (1982) Acid precipitation and lake susceptibility in the central Washington Cascades. *Environmental Science and Technology*; **16** (11); 771–775

Lovett, GM, Reiners, WA and Olson, RK (1982) Cloud droplet deposition in subalpine balsam fir forests: hydrological and chemical inputs. *Science*; **218** (4579); 1303–1304

Lukow, TE and Cooper, WA (1980) Some properties of particulate plumes from coal-fired power plants. In: *Environmental and climatic impact of coal utilization. Proceedings of the symposium on environmental and climatic impact of coal utilization, Williamsburg, VA, USA, 17–19 Apr 1979*. New York, NY, USA, Academic Press, pp. 21–32

Luria, M, Stockburger, L, Olszyna, KJ and Meagher, JF (1982) Dynamics of sulfate particle production and growth in smog chamber experiments. *Atmospheric Environment*; **16** (4); 697–708

Lyons, WA (1980) Evidence of transport of hazy air masses from satellite imagery. *Annals of the New York Academy of Science*; **338**; 418–433

Lyons, WA, Dooley, JC and Whitby, KT (1978) Satellite detection of long-range pollution transport and sulfate aerosol hazes. *Atmospheric Environment*; **12** (1–3); 621–631

Maahs, HG (1983a) Kinetics and mechanism of the oxidation of S(IV) by ozone in aqueous solution with particular reference to SO_2 conversion in nonurban tropospheric clouds. *Journal of Geophysical Research*; **88** (C15); 10 721–10 732

Maahs, HG (1983b) Measurements of the oxidation rate of sulfur (IV) by ozone in aqueous solution and their relevance to SO_2 conversion in nonurban tropospheric clouds. *Atmospheric Environment*; **17** (2); 341–345

McCormick, MP, Chu, WP, Grams, GW, Hamill, P, Herman, BM, McMaster, LR, Pepin, TJ, Russell, PB, Steele, HM and Swissler, TJ (1981) High-latitude stratospheric aerosols measured by the SAM II satellite system in 1978 and 1979. *Science*; **214** (4518); 328–331

Macias, ES, Blumenthal, DL, Anderson, JA and Cantrell, BK (1980) Size and composition of visibility-reducing aerosols in southwestern plumes. *Annals of the New York Academy of Science*; **338**; 233–257

Macias, ES, Zwicker, JO, Ouimette, JR, Hering, SV, Friedlander, SK, Cahill, TA, Kuhlmey, GA and Richards, LW (1981a) Regional haze case studies in the Southwestern US - I Aerosol chemical composition. *Atmospheric Environment*; **15** (10/11); 1971–1986

Macias, ES, Zwicker, JO and White, WH (1981b) Regional haze case studies in the Southwestern US - II Source contributions. *Atmospheric Environment*; **15** (10/11); 1987–1997

McInnes, G (1980) *Sulphate in particulate survey: analysis of first two year's results. LR 348 (AP)*, Stevenage, UK, Warren Spring Laboratory, 71 pp.

McInnes, G (1982) *Multi-element and sulphate in particulate surveys: summary and analysis of five years' results (1976–81). LR 435 (AP)*, Stevenage, UK, Warren Spring Laboratory, 51 pp

McMurry, PH (1980) The dynamics of secondary sulfur aerosols. In: *Atmospheric sulfur deposition: environmental impact and health effects. Proceedings of the second life sciences symposium, Potential environmental and health consequences of atmospheric sulfur deposition, Gatlinberg, TN, 14–18 Oct 1979*. Ann Arbor, MI, Ann Arbor Science Publishers, pp. 153–160

McMurry, PH and Friedlander, SK (1979) New particle formation in the presence of an aerosol. *Atmospheric Environment*; **13** (12); 1635–1651

McMurry, PH and Wilson, JC (1982) Growth laws for the formation of secondary ambient aerosols: implications for chemical conversion mechanisms. *Atmospheric Environment*; **16** (1); 121–134

Malm, WC and Johnson, CE (1984) Optical characteristics of fine and coarse particulates at Grand Canyon, Arizona. *Atmospheric Environment*; **18** (6); 1231–1237

Mamane, Y and Pueschel, RF (1979) A study of individual sulfate particles generated in a coal-fired power plant plume. *American Chemical Society, Division of Environmental Chemistry, Preprints*; **19** (1); 604–607

Mamane, Y and Pueschel, RF (1980) Formation of sulfate particles in the plume of the Four Corners power plant. *Journal of Applied Meteorology*; **19** (7); 779–790

Mamane, Y, Ganor, E and Donagi, AE (1980) Aerosol composition of urban and desert origin in the eastern Mediterranean. I Individual particle analysis. *Water, Air and Soil Pollution*; **14**; 29–43

The MAP3S/RAINE Research Community (1982) The MAP3S/RAINE precipitation chemistry network: statistical overview for the period 1976–1980. *Atmospheric Environment*; **16** (7); 1603–1631

Marsh, ARW (1978) Sulphur and nitrogen contributions to the acidity of rain. *Atmospheric Environment*; **12** (1–3); 402–406

Martin, A (1979) Distribution of sulphur in rain. In: *Proceedings of the international symposium on sulphur emissions and the environment, London, UK, 8–10 May (1979)* London, UK, Society of the Chemical Industry, pp. 49–66

Martin, A (1982) A short study of the influence of a valley on the composition of rainwater. *Atmospheric Environment*; **16**(4); 785–793

Martin, A (1984) Sulphate and nitrate related to acidity in rainwater. *Water, Air and Soil Pollution*; **21** (1–4); 271–277

Martin, LR (1983) Comment on measurements of sulfate production in natural clouds. *Atmospheric Environment*; **17** (8); 1603–1604

Martin, LR (1984) Kinetic studies of sulfite oxidation in aqueous solution. In: *SO_2, NO and NO_2 oxidation mechanisms: atmospheric considerations*. Boston, MA, USA, Butterworth, pp. 63–100

Martin, A and Barber, FR (1984) Acid gases and acid in rain monitored for over 5 years in rural east-central England. *Atmospheric Environment*; **18** (9); 1715–1724

Martin, A and Barber, FR (1985) Particulate sulphate and ozone in rural air: preliminary results from three sites in central England. *Atmospheric Environment*; **19** (7); 1091–1102

Martin, LR, Judeikis, HS, Hwang, WC and Wren, AG (1979) Aqueous oxidation rates of SO_2 in mixed catalyst systems. *American Chemical Society, Division of Environmental Chemistry, Preprints*; **19** (1); 611–613

Martin, LR and Damschen, DE (1981) Aqueous oxidation of sulfur dioxide by hydrogen peroxide at low pH. *Atmospheric Environment*; **15** (9); 1615–1621

Martin, LR, Damschen, DE and Judeikis, HS (1981) The reactions of nitrogen oxides with SO_2 in aqueous aerosols. *Atmospheric Environment*; **15**; 191–195

Maul, PR (1982) A time dependent model for the atmospheric transport of gaseous pollutants: Part 2 - application to the long-range transport of sulphur compounds. *Environmental Pollution (Series B)*; **4** (1); 1–25

Mazumdar, S, Schimmel, H and Higgins, ITT (1981) Daily mortality, smoke and SO_2 in London, England 1959 to 1972. In: *Proposed SOx and particulate standard. Proceedings of a conference held at Atlanta, GA, USA, 16–18 Sep 1980*. Pittsburgh GA, USA, Air Pollution Control Association, pp. 219–239

Mazumdar, S, Schimmel, H and Higgins, ITT (1982) Relation of daily mortality to air pollution: an analysis of 14 London winters, 1958/59-1971/72. *Archives of Environmental Health*; **37** (4); 213–220

Meagher, JF and Luria, M (1982) Model calculations of the chemical processes occurring in the plume of a coal-fired power plant. *Atmospheric Environment*; **16** (2); 183–195

Meagher, JF, Stockburger, L, Bonanno, RJ, Bailey, EM and Luria, M (1981) Atmospheric oxidation of flue gases from coal fired power plants—a comparison between conventional and scrubbed plumes. *Atmospheric Environment*; **15**(5); 749–762

Meagher, JF, Bailey, EM and Luria, M (1982) The impact of mixing cooling tower and power plant plumes on sulfate aerosol formation. *Journal of the Air Pollution Control Association*; **32** (4); 389–391

Melo, OT (1977) *Nanticoke GS brown plume study*. Report 77-250-K, Ontario, Canada, Ontario Hydro Research, 77 pp.

Melo, OT (1979) *Airborne plume measurements Nanticoke GS - 1978*. 2 Volumes. Report 79-50-K, Ontario, Canada, Ontario Hydro Research, 44, 377 pp.

Melo, OT (1981) Ontario Hydro's acid rain monitoring network. *Ontario Hydro Research Review*; **2**; 29–38

Memorandum of Intent Work Group 2: Atmospheric Sciences and Analysis (1982a) *Monitoring and Interpretation Subgroup Report: Final report ˇ technical basis, United States–Canada Memorandum of Intent on Transboundary Air Pollution. 2F - I*, Toronto, Canada, Atmospheric Environment Service, 250 pp.

Memorandum of Intent Work Group 2: Atmospheric Sciences and Analysis (1982b) *Local and Mesoscale Analysis Subgroup Report: Final report - technical basis. United States–Canada Memorandum of Intent on Transboundary Air Pollution. 2F - L*, Toronto, Canada, Atmospheric Environment Servie, 50 pp.

Memorandum of Intent Work Groups (1983) Executive summaries: work group reports. United States–Canada Memorandum of Intent on Transboundary Air Pollution. Toronto, Canada, Atmospheric Environment Service, 65 pp.

Messer, JJ (1983) Geochemically alkaline snowpack in northern Utah mountains in spring of 1982. *Atmospheric Environment*; **17** (5); 1051–1054

Meszaros, E (1978) Concentration of sulfur compounds in remote continental and oceanic areas. *Atmospheric Environment*; **12** (1–3); 699–705

Meyer, B (1977) *Sulfur, energy, and environment*. Amsterdam, Netherlands, Elsevier, 459 pp.

Middleton, P (1980) A re-examination of atmospheric sulfuric acid aerosol formation and growth. *Journal of Aerosol Science*; **11** (4); 411–414

Middleton, P and Kiane, CS (1978) Experimental and theoretical examination of the formation of sulfuric acid particles. *Atmospheric Environment*; **12** (1–3); 179–186

Miles, LJ and Yost, KJ (1982) Quality analysis of USGS precipitation chemistry data for New York. *Atmospheric Environment*; **16** (12); 2889–2898

Millan, MM, Barton, SC, Johnson, ND, Weisman, B, Lusis, M, Chan, W and Vet, R (1982) Rain scavenging from tall stack plumes: a new experimental approach. *Atmospheric Environment*; **16** (11); 2709–2714

Miller, DF (1978) Precursor effects on SO_2 oxidation. *Atmospheric Environment*; **12** (1–3); 273–280

Miller, DF and Alkezweeny, AJ (1980) Aerosol formation in urban plumes over Lake Michigan. *Annals of the New York Academy of Sciences*; **338**; 219–232

Miller, HG and Miller, JD (1979) Sulphur content and the acidity of rainwater at six rural sites across Scotland. In: *Proceedings of the International symposium on sulphur emissions and the environment, London, UK, 8–10 May 1979* London, UK, Society of the Chemical Industry, pp. 77–80

Miller, JM (1979) The acidity of Hawaiian precipitation as evidence of long-range transport of pollutants. In: *Papers presented at the WMO symposium on the long-range transport of pollutants and its relation to general circulation including stratospheric/tropospheric exchange processes, Sofia, Bulgaria, 1–5 October 1979. WMO-538*, Geneva, Switzerland, World Meteorological Organization, pp. 231–237

Miller, JM and Yoshinaga, AM (1981) The pH of Hawaiian precipitation—a preliminary report. *Geophysical Research Letters*; **8** (7); 779–782

Moeller, D (1984a) Estimation of the global man-made sulphur emission. *Atmospheric Environment*; **18** (1); 19–27

Moeller, D (1984b) On the global natural sulphur emission. *Atmospheric Environment*; **18** (1); 29–39

Morandi, MT, Kneip, TJ, Cobourn, WG, Husar, RB and Lioy, PJ (1983) The measurement of H_2SO_4 and other sulfate species at Tuxedo, New York with a thermal analysis flame photometric detector and simultaneously collected quartz filter samples. *Atmospheric Environment*; **17** (4); 843–848

Morgan, MG, Henrion, M, Morris, SC and Amaral, DAL (1985) Uncertainty in risk assessment. *Environmental Science and Technology*; **19** (8); 662–667

Moss, MR (1978) Sources of sulfur in the environment: the global sulfur cycle. In: *Sulfur in the environment. Part I: the atmospheric cycle*. New York, NY, USA, John Wiley, pp. 23–50

Mueller, PK and Hidy, GM (1983) *The Sulfate Regional Experiment: report of findings. 3 Vols. EPRI-EA-1901*, Palo Alto, CA, USA, Electric Power Research Institute, 352, 231 and 395 pp.

Mueller, PK, Hidy, GM, Warren, K, Lavery, TF and Baskett, RL (1980) The occurrence of atmospheric aerosols in the northeastern United States. *Annals of the New York Academy of Sciences*; **338**; 463–482

Munger, JW (1982) Chemistry of atmospheric precipitation in the north-central United States: influence of sulfate, nitrate, ammonia and calcareous soil particulates. *Atmospheric Environment*; **16** (7); 1633–1645

Munger, JW and Eisenreich, SJ (1983) Continental-scale variations in precipitation chemistry. *Environmental Science and Technology*; **17** (1); 32A–42A

Munn, RE, Likens, GE, Weisman, B, Hornbeck, JW, Martin, CW and Bormann, FH (1983) A meteorological analysis of the precipitation chemistry event samples at Hubbard Brook (NH). *Atmospheric Environment*; **18** (12); 2775–2779

Murray, LC and Farber, RJ (1982) Time series analysis of an historical visibility data base. *Atmospheric Environment*; **16**(10); 2299–2308

Nader, JS (1980) Primary sulfate emissions from stationary industrial sources. In: *Atmospheric sulfur deposition: environmental impact and health effects. Proceedings of the second life sciences symposium, Potential environmental and health consequences of atmospheric sulfur deposition, Gatlinberg, TN, 14–18 Oct 1979*. Ann Arbor, MI, Ann Arbor Science Publishers, pp. 123–130

Nagamoto, CT, Parungo, F, Reinking, R, Pueschel, R and Gerish, T (1983) Acid clouds and precipitation in eastern Colorado. *Atmospheric Environment*; **17** (6); 1073–1982

National Research Council, Subcommittee on Airborne Particles, USA (1979) *Airborne particles*. Baltimore, MA, University Park Press, 355 pp.

National Research Council, Committee on the Atmosphere and Biosphere, USA (1981) *Atmosphere–biosphere interactions: toward a better understanding of the ecological consequences of fossil fuel combustion*. Washington, DC, USA, National Academy Press, 237 pp.

National Research Council, Committee on Atmospheric Transport and Chemical Transformation in Acid Precipitation (1983) *Acid deposition: atmospheric processes in eastern North America: a review of current scientific understanding*. Washington, DC, USA, National Academy Press, 391 pp.

Neftel, A, Beer, J, Oeschger, H, Zuercher, F and Finkel, RC (1985) Sulphate and nitrate concentrations in snow from South Greenland 1895–1978. *Nature*; **314** (6012); 611–613

Newman, L (1981) Atmospheric oxidation of sulfur dioxide: a review as viewed from power plant and smelter plume studies. *Atmospheric Environment*; **15** (10/11); 2231–2239

Nguyen, BC, Bonsang, B and Gaudry, A (1983) The role of the ocean in the global atmospheric sulfur cycle. *Journal of Geophysical Research*; **88** (C15); 10903–10914

Nielson, T, Ramdahl, T and Bjørseth, A (1983) The fate of airborne polycyclic matter. *Environmental Health Perspectives*; **47**; 103–114

Novakov, T (1979) Role of carbon soot in sulfate formation. *American Chemical Society, Division of Environmental Chemistry, Preprints*; **19** (1); 584–586

Nriagu, JO (1978) Deteriorative effects of sulfur pollution on materials. In: *Sulfur in the environment. Part II: ecological impacts*. New York, NY, USA, John Wiley, pp. 1–59

Oblath, S (1980) Reduction of NObx by SO_2 in aqueous solution. In: *Atmospheric aerosol research FY-1979. LBL-10735*, Berkeley, CA, USA, California University, Lawrence Berkeley Laboratory, pp. 8.4–8.6

O'Connor, BH, Chang, WJ and Martin, DJ (1981) Chemical characterisation of atmospheric aerosol in Perth, Western Australia. In: *Proceedings of the seventh international clean air conference, Adelaide, Australia, 24–28 Aug 1981*. Ann Arbor, MI, USA, Ann Arbor Science Publishers, pp. 639–652

Oden, S (1976) The acidity problem—an outline of concepts. *Water, Air, and Soil Pollution*; **6** (2–4); 137–166

Oden, S and Ahl, T (1980) The sulfur budget of Sweden. In: *Effects of acid precipitation on terrestrial ecosystems. Proceedings of the NATO conference on effects of acid precipitation on vegetation and soils, Toronto, Canada, 21–27 May 1978*. New York, NY, USA, Plenum Press, pp. 111–122

Office of Technology Assessment, USA (1984) *Acid rain and transported air pollutants: implications for public policy. OTA-O-204*, Washington, DC, USA, Office of Technology Assessment, 323 pp.

Oke, TR (1980) Climatic impacts of urbanization. In: *Interactions of energy and climate: Proceedings of an international workshop, Munster, FRG, 3–6 Mar 1980*. Dordrecht, Netherlands, D Reidel Publishing Co, pp. 339-356

Okita, T (1977) *Moist air pollution—acid rain. CE Translation 7483*, London, UK, Central Electricity Generating Board, 50 pp, *Kogai to Taisaku*; **13** (7); 732–750

Okita, T and Ohta, S (1979) Measurements of nitrogenous and other compounds in the atmosphere and in cloudwater: a study of the mechanism of formation of acid precipitation. In: *Nitrogenous air pollutants – chemical and biological implications*. Ann Arbor, MI, USA, Ann Arbor Science, pp. 289–305

Ono, A and Ohtani, T (1980) On the capability of atmospheric sulfate particles as cloud condensation nuclei. *Journal de Recherches Atmospheriques*; **14** (3–4); 235–240

Oppenheimer, M, Epstein, CB and Yuhnke, RE (1985) Acid deposition, smelter emissions, and the linearity issue in the Western United States. *Science*; **229** (4716); 859–862

Organisation for Economic Co-operation and Development (1981) *The costs and benefits of sulphur oxide control: a methodological study*. Paris, France, Organisation for Economic Co-operation and Development, 164 pp.

Ottar, B (1978) An assessment of the OECD study on long range transport of air pollutants (LRTAP). *Atmospheric Environment*; **12** (1–3); 445–465

Ottar, B (1981) The transfer of airborne pollutants to the arctic region. *Atmospheric Environment*; **15** (8); 1439–1445

Ouimette, JR, Flagan, RC and Kelso, AR (1981) Chemical species contributions to light scattering by aerosols at a remote arid site: comparison of statistical and theoretical results. In: *Atmospheric aerosol: source/air quality relationships. Proceedings of a symposium held at the 180th national meeting of the American Chemical Society, Las Vegas, Nevada, USA, 27–29 Aug 1980*. Washington, DC, USA, American Chemical Society, pp. 125–156

Overton, JH and Durham, JL (1982) Acidification of rain in the presence of SO_2, H_2O_2, O_3 and HNO_3. In: *Energy and environmental chemistry. Volume 2: acid rain*. Ann Arbor, MI, Ann Arbor Science Publishers, pp. 245–262

Overton, JH, Aneja, VP and Durham, JL (1979) Production of sulfate in rain and raindrops in polluted atmospheres. *Atmospheric Environment*; **13** (3); 355–367

Pack, DH, Ferber, GJ, Heffter, JL, Telegadas, K, Angell, JK, Hoecker, WH and Machta, L (1978) Meteorology of long-range transport. *Atmospheric Environment*; **12** (1–3); 425–444

Pacyna, JM, Semb, A and Hanssen, JE (1984) Emission and long-range transport of trace elements in Europe. *Tellus*; **36B** (3); 163–178

Panter, R and Penzhorn, R-D (1980) Alkyl sulfonic acids in the atmosphere. *Atmospheric Environment*; **14** (1); 149–151

Pena, JA, de Pena, RG, Bowersox, VC and Takacs, JF (1982) SO_2 content in precipitation and its relationship with surface concentrations of SO_2 in air. *Atmospheric Environment*; **16** (7); 1711–1715

Penkett, SA (1979) Chemical changes in the air. In: *Proceedings of the International symposium on sulphur emissions and the environment, London, UK, 8–10 May 1979*. London, UK, Society of the Chemical Industry, pp. 109–122

Penkett, SA, Jones, BMR, Brice, KA and Eggleton, AEJ (1979a) The importance of atmospheric ozone and hydrogen peroxide in oxidising sulphur dioxide in cloud and rainwater. *Atmospheric Environment*; **13** (1); 123–137

Penkett, SA, Jones, BMR and Eggleton, AEJ (1979b) A study of SO_2 oxidation in stored rainwater samples. *Atmospheric Environment*; **13** (1); 139–142

Penzhorn, RD and Panter, R (1980) On the determination of sulfur containing acids in the atmospheric aerosol. In: *Proceedings of the first European symposium on physico-chemical behaviour of atmospheric pollutants, Ispra (Italy) 16–18 Oct 1979. EUR 6621 DE, EN, FR*. Brussels, Belgium, Commission of the European communities, pp. 80–87

Peters, NE (1984) Quality analysis of US Geological Survey precipitation chemistry data for New York. *Atmospheric Environment*; **18** (5); 1041–1042

Peters, NE, Schroeder, RA and Troutman, DE (1982) *Temporal trends in the acidity of precipitation and surface waters of New York*. Geological Survey Water-Supply Paper 2188, Alexandria, VA, USA, US Geological Survey, 39 pp.

Petrenchuk, OP (1980) On the budget of sea salts and sulfur in the atmosphere. *Journal of Geophysical Research*; **85** (C12); 7439–7444

Pierson, WR, Brachaczek, WW, Truex, TJ, Butler, JW and Korniski, TJ (1980) Ambient sulfate measurements on Allegheny Mountain and the question of atmospheric sulfate in the northeastern United States. *Annals of the New York Academy of Sciences*; **338**; 145–173

Pitchford, A, Pitchford, M, Malm, W, Flocchini, R, Cahill, T and Walther, E (1981) Regional analysis of factors affecting visual air quality. *Atmospheric Environment*; **15** (10/11); 2043–2054

Pratt, GC, Coscio, M, Gardner, DW, Chevone, BI and Krupa, SV (1983) An analysis of the chemical properties of rain in Minnesota. *Atmospheric Environment*; **17** (2); 347–355

Prokop, M (1979) Some results of airborne air pollution monitoring from cross-section flights over Czechoslovakia. In: *Papers presented at the WMO symposium on the long-range transport of pollutants and its relation to general circulation including stratospheric/tropospheric exchange processes, Sofia, Bulgaria, 1–5 October 1979*. WMO-538, Geneva, Switzerland, World Meteorological Organization, pp. 11–15

Pruppacher, HR and Klett, JD (1978) *Microphysics of clouds and precipitation*. Dordrecht, Netherlands, D Reidel, 721 pp.

Pueschel, RF (1976) Aerosol formation during coal combustion: condensation of sulfates and chlorides on fly ash. *Geophysical Research Letters*; **3** (11); 651–653

Pueschel, RF and Mamane, Y (1979) Mechanisms and rates of formation of sulfur aerosols in power plant plumes. In: *Papers presented at the WMO symposium on the long-range transport of pollutants and its relation to general circulation including stratospheric/tropospheric exchange processes, Sofia, Bulgaria, 1–5 October 1979. WMO-538*, Geneva, Switzerland, World Meteorological Organization, pp. 141–148

Pueschel, RF and van Valin, CC (1978) Cloud nucleus formation in a power plant plume. *Atmospheric Environment*; **12** (1–3); 307–312

Radke, LF (1982) Sulphur and sulphate from Mt Erebus. *Nature*; **299** (5885); 710–712

Radke, LF, Lyons, JH, Hegg, DA and Hobbs, PV (1984) Airborne observations of arctic aerosols. I: Characteristics of Arctic haze. *Geophysical Research Letters*; **11** (5); 393–396

Rahn, KA (1981) Relative importances of North America and Eurasia as sources of Arctic aerosol. *Atmospheric Environment*; **15** (8); 1447–1455

Rahn, KA and Lowenthal, DH (1984) Elemental tracers of distant regional pollution aerosols. *Science*; **223** (4632); 132–139

Rahn, KA and Lowenthal, DH (1985) Pollution aerosol in the Northeast: Northeastern-Midwestern contributions. *Science*; **228** (4697); 275–284

Rahn, KA and McCaffrey, RJ (1980) On the origin and transport of the winter Arctic aerosol. *Annals of the New York Academy of Sciences*; **338**; 486–503

Rahn, KA, Joranger, E, Semb, A and Conway, TJ (1980a) High winter concentrations of SO_2 in the Norwegian Arctic and transport from Eurasia. *Nature*; **287** (5785); 824–826

Rahn, KA, Brosset, C, Ottar, B and Patterson, EM (1980b) *Black and white episodes, chemical evolution of Eurasian air masses, and long-range transport of carbon to the Arctic. IVL-B-586*, Gothenburg, Sweden, Institutet foer Vatten- och Luftvaardsforskning, 15 pp.

Rao, ST and Sistla, G (1982) Relationship between urban and rural sulfate levels. *Journal of the Air Pollution Control Association*; **32** (6); 645–648

Raynor, GS and Hayes, JV (1981) Acidity and conductivity of precipitation on central Long Island, New York in relation to meteorological variables. *Water, Air and Soil Pollution*; **15** (2); 229–245

Raynor, GS and Hayes, JV (1982a) Concentrations of some ionic species in Central Long Island, New York precipitation in relation to meteorological variables. *Water, Air and Soil Pollution*; **17** (3); 309–335

Raynor, GS and Hayes, JV (1982b) Variation in chemical wet deposition with meteorological conditions. *Atmospheric Environment*; **16** (7); 1647–1656

Raynor, GS and Hayes, JV (1982c) Effects of varying air trajectories on spatial and temporal precipitation chemistry patterns. *Water, Air, and Soil Pollution*; **18** (1/2/3); 173–189

Raynor, GS and Hayes, JV (1982d) Relationships of chemical wet deposition to precipitation amount and meteorological conditions. In: *Energy and environmental chemistry, Volume 2: Acid rain*. Ann Arbor, MI, Ann Arbor Science Publishers, pp. 189–204

Reddy, MM and Claassen, HC (1985) Estimates of average major ion concentrations in bulk precipitation at two high-altitude sites near the Continental Divide in Southwestern Colorado. *Atmospheric Environment*; **19** (7); 1199–1203

Reisinger, LM and Crawford, TL (1980) Sulphate flux through the Tennessee Valley region. *Journal of the Air Pollution Control Association*; **30** (11); 1230–1231

Reisinger, ML and Crawford, TL (1982) Interregional transport: case studies of measurements versus model predictions. *Journal of the Air Pollution Control Association*; **32** (6); 629–633

Rice, H, Nochumson, DH and Hidy, GM (1981) Contribution of anthropogenic and natural sources to atmospheric sulfur in parts of the United States. *Atmospheric Environment*; **15** (1); 1–9

Richards, LW, Anderson, JA, Blumenthal, DL, Brandt, AA, McDonald, JA, Waters, N, Macias, ES and Bhardwaja, PS (1981) The chemistry, aerosol physics and optical properties of a Western coal-fired power plant plume. *Atmospheric Environment*; **15** (10/11); 2111–2134

Richards, LW, Anderson, JA, Blumenthal, DL, McDonald, JA, Kok, GL and Lazrus, AL (1983) Hydrogen peroxide and sulfur (IV) in Los Angeles cloud water. *Atmospheric Environment*; **17** (4); 911–914

Richter, A and Granat, L (1978) *Some examples of total aerosol weight accounted for by major inorganic soluble compounds. AC-46*, Stockholm, Sweden, University of Stockholm, Department of Meteorology; International Meteorological Institute in Stockholm, 9 pp.

Roberts, DB and Williams, DJ (1979) The kinetics of oxidation of sulphur dioxide within the plume from a sulphide smelter in a remote region. *Atmospheric Environment*; **13** (11); 1485–1499

Robinson, E and Robbins, RC (1972) Emissions, concentrations, and fate of gaseous atmospheric pollutants. In: *Air pollution control. Part II*. New York, USA, Wiley-Interscience, pp. 1–93

Rodhe, H (1977) A study of the sulfur budget for the atmosphere over Northern Europe. *Tellus*; **24** (2); 128–138

Rodhe, H (1978) Budgets and turn-over times of atmospheric sulfur compounds. *Atmospheric Environment*; **12** (1–3); 671–680

Rodhe, H and Isaksen, I (1980) Global distribution of sulfur compounds in the troposphere estimated in a height/latitude transport model. *Journal of Geophysical Research*; **85** (C12); 7401–7409

Rodhe, H and Granat, L (1983) Summer and winter budgets for sulfur over Europe: an indication of large seasonal variations of residence time. *Idojaras*; **87** (1); 1–6

Rodhe, H and Granat, L (1984) An evaluation of sulfate in European precipitation 1955–1982. *Atmospheric Environment*; **18** (12); 2627–2639

Rodhe, H, Persson, C and Aakesson, O (1972) An investigation into regional transport of soot and sulfate aerosols. *Atmospheric Environment*; **6**; 675–693

Rodhe, H, Crutzen, P and Vanderpol, A (1981) Formation of sulfuric and nitric acid in the atmosphere during long-range transport. *Tellus*; **33** (2); 132–141

Rodhe, H, Granat, L and Soederlund, R (1984) *Sulfate in precipitation: a presentation of data from the European Air Chemistry Network, CM-64*, Stockholm, Sweden, University of Stockholm, Institute of Meteorology; International Meteorological Institute in Stockholm, 71 pp.

Rose, WI, Chuan, RL and Kyle, PR (1985) Rate of sulphur dioxide emission from Erebus volcano, Antarctica, December 1983. *Nature*; **316** (6030); 710–712

Ryaboshapko, AG (1983) The atmospheric sulphur cycle. In: *The global biogeochemical sulphur cycle. SCOPE 19*. Chichester, John Wiley, pp. 203–296

Sadasivan, S (1980) Trace constituents in cloud water, rainwater and aerosol samples collected near the west coast of India during the southwest monsoon. *Atmospheric Environment*; **14** (1); 33–38

Salmon, L, Atkins, DHF, Fisher, EMR, Healy, C and Law, DV (1978) Retrospective trend analysis of the content of UK air particulate material 1957–1974. *Science of the Total Environment*; **9** (2); 161–200

Saltzman, ES, Brass, GW and Price, DA (1983) The mechanism of sulfate aerosol formation: chemical and sulfur isotopic evidence. *Geophysical Research Letters*; **10** (7); 513–516

Scheider, WA, Jeffries, DS and Dillon, PJ (1981) Bulk deposition in the Sudbury and Muskoka-Haliburton areas of Ontario during the shutdown of Inco Ltd. in Sudbury. *Atmospheric Environment*; **15** (6); 945–956

Schlesinger, RB, Halpern, M, Albert, RE and Lippmann, M (1979) Effect of chronic inhalation of sulfuric acid mist upon mucociliary clearance from the lungs of donkeys. *Journal of Environmental Pathology and Toxicology*; **2**; 1351–1367

Schroeder, WH and Urone, P (1978) Isolation and identification of nitrosonium hydrogen sulfate ($NOHSO_4$) as a photochemical reaction product in air containing sulfur dioxide and nitrogen dioxide. *Environmental Science and Technology*; **12** (5); 545–550

Schyrer, DR, Rogowski, RS and Cofer, WR (1980) A study of the influence of airborne particulates on sulfate formation. In: *Environmental and climatic impact of coal utilization*. New York, NY, USA, Academic Press, pp. 275–292

Schwartz, SE and Freiberg, JE (1981) Mass-transport limitation to the rate of reaction of gases in liquid droplets: application to oxidation of SO_2 in aqueous solutions. *Atmospheric Environment*; **15** (7); 1129–1144

Scott, WD and Cattell, FCR (1979) Vapor pressure of ammonium sulfates. *Atmospheric Environment*; **13** (8); 307–317

Scriven, RA and Howells, G (1977) Stack emissions and the environment. *CEGB (Central Electricity Generating Board) Research*; **5**; 28–40

Sehmel, GA (1979) Particle and gas dry deposition: a review. *American Chemical Society, Division of Environmental Chemistry, Preprints*; **19** (1); 465–468

Sehmel, GA (1980) Model predictions and a summary of dry deposition velocity data. In: *Atmospheric sulfur deposition: environmental impact and health effects. Proceedings of the second life sciences symposium, Potential environmental and health consequences of atmospheric sulfur deposition, Gatlinberg, TN, 14–18 Oct 1979*. Ann Arbor, MI, Ann Arbor Science Publishers, pp. 223–234

Seigneur, C (1982) A model of sulfate aerosol dynamics in atmospheric plumes. *Atmospheric Environment*; **16** (9); 2207–2228

Semb, A (1979) Emission of gaseous and particulate matter in relation to long-range transport of air pollutants. In: *Papers presented at the WMO symposium on the long-range transport of pollutants and its relation to general circulation including stratospheric/tropospheric exchange processes, Sofia, Bulgaria, 1–5 October 1979. WMO-538*, Geneva, Switzerland, World Meteorological Organization, pp. 1a–1m

Semonin, RG, Bowersox, VC, Gatz, DF, Peden, ME and Stensland, GJ (1981) *Study of atmospheric-pollution scavenging. Nineteenth progress report. DOE/EV/01199-T1, COO-1199-63, DE81-023290*, Champaign, IL, USA, Illinois Institute of Natural Resources, State Water Survey Division, 138 pp.

Sequeira, R (1982) Acid rain: an assessment based on acid-base considerations. *Journal of the Air Pollution Control Association*; **32** (3); 241–245

Shah, JJ, Watson, JG, Cooper, JA and Huntzicker, JJ (1984) Aerosol chemical composition and light scattering in Portland, Oregon: the role of carbon. *Atmospheric Environment*; **18** (1); 235–240

Shannon, JD (1981a) Examination of surface removal and horizontal transport of atmospheric sulfur on a regional scale. In: *Second joint conference on applications of air pollution meteorology, New Orleans, LA, USA, 24–27 March 1980*. Boston, MA, USA, American Meteorological Society, 232–235

Shannon, JD (1981b) A model of regional long-term average sulfur atmospheric pollution, surface removal, and net horizontal flux. *Atmospheric Environment*; **15** (5); 689–701

Shannon, JD (1984) Simulation modeling of atmospheric deposition of sulfur from biogenic and anthropogenic emissions. In: *Environmental impact of natural emissions: transactions of the Air Pollution Control Association specialty conference*, Research Triangle Park, NC, USA, March 1984, pp. 318–325

Shannon, JD (1985) Argonne, IL, USA, Argonne National Laboratory, Environmental Research Division, Private communication (Apr 1985)

Shannon, JD and Voldner, EC (1982) Estimation of wet and dry deposition of pollutant sulfur in eastern Canada as a function of major source regions. *Water, Air, and Soil Pollution*; **18** (1/2/3); 101–104

Shaw, GE (1980) Optical, chemical and physical properties of aerosols over the Antarctic ice sheet. *Atmospheric Environment*; **14**; 911–921

Shaw, GE (1981) Eddy diffusion transport of arctic pollution from the mid-latitudes: a preliminary model. *Atmospheric Environment*; **15** (8); 1483–1490

Shaw, RW (1982) Deposition of atmospheric acid from local and distant sources at a rural site in Nova Scotia. *Atmospheric Environment*; **16** (2); 337–348

Shaw, RW and Rodhe, H (1981) *Non-photochemical oxidation of SO_2 in regionally polluted air during winter. CM-53*, Stockholm, Sweden, University of Stockholm, Department of Meteorology; International Meteorological Institute in Stockholm, 26 pp.

Shaw, RW and Rodhe, H (1982) Non-photochemical oxidation of SO_2 in regionally polluted air during winter. *Atmospheric Environment*; **16** (12); 2879–2888

Shaw, RW and Young, JWS (1983) An investigation of the assumptions of linear chemistry and superposition in LRTAP models. *Atmospheric Environment*; **17** (11); 2221–2229

Sheih, CM, Wesely, ML and Hicks, BB (1979) Estimated dry deposition velocities of sulfur over the eastern United States and surrounding regions. *Atmospheric Environment*; **13** (10); 1361–1368

Sidebottom, H (1980) Photo oxidation of sulphur dioxide. In: *Proceedings of the first European symposium on physico-chemical behaviour of atmospheric pollutants, Ispra (Italy) 16–18 Oct 1979. EUR 6621 DE, EN, FR*. Brussels, Belgium, Commission of the European Communities, 247–253

Sievering, H (1982) Profile measurements of particle dry deposition velocity at an air-land interface. *Atmospheric Environment*; **16** (2); 301–306

Sievering, H, Dave, M, Dolske, DA, Hughes, RL and McCoy, P (1979) *An experimental study of lake loading by aerosol transport and dry deposition in the Southern Lake Michigan BAsin. Final report 1 Jun 1976–31 Jul 1979. EPA-905/4-79-016, PB81-101974*, Park Forest South, IL, USA, Governors State University, 197 pp.

Sievering, H, Cooke, J and Pueschel, R (1981) Importance of deposition velocity for sulfur gas to sulfate particle transformation rates at the Four Corners power plant. *Atmospheric Environment*; **15** (12); 2593–2596

Sisterson, DL, Shannon, JD and Hales, JM (1979) An examination of regional pollutant structure in the lower troposphere – some results of the diagnostic atmospheric cross section experiment (DACSE-I). *Journal of Applied Meteorology*; **18** (11); 1421–1428

Sistla, G, Domaracki, AJ and Putta, SN (1982) Impact of New York State emission sources on Class 1 areas. *Water, Air, and Soil Pollution*; **18** (1/2/3); 123–128

Skoulikidis, Th.N (1983) Effect of primary and secondary air pollutants and acid depositions on (ancient and modern) buildings and monuments. In: *Acid deposition: a challenge for Europe, Proceedings of the symposium held at Karlsruhe, FRG, 19–21 Sep 1983*, pp. 193–226

Slinn, WGN (1982) Predictions for particle deposition to vegetative canopies. *Atmospheric Environment*; **16** (7); 1785–1794

Sloane, CS (1982) Visibility trends – II Mideastern United States 1948–1978. *Atmospheric Environment*; **16** (10); 2309–2321

Sloane, CS (1983) Summertime visibility declines: meteorological influences. *Atmospheric Environment*; **17** (4); 763–774

Sloane, CS (1984) Meteorologically adjusted air quality trends: visibility. *Atmospheric Environment*; **18** (6); 1217–1229

Smith, FB (1979) The role of atmosphere in pollution dispersion. In: *Proceedings of the International symposium on sulphur emissions and the environment*, London, UK, 8–10 May 1979 London, UK, Society of the Chemical Industry, pp. 27–35

Smith, FB and Hunt, RD (1978) Meteorological aspects of the transport of pollution over long distances. *Atmospheric Environment*; **12** (1–3); 461–477

Smith, IM (1980) *Nitrogen oxides from coal combustion – environmental effects. ICTIS/TR 10*, London, UK, IEA Coal Research, 97 pp.

Smith, RA (1872) *Air and rain: the beginnings of a chemical climatology*. London, Longmans Green, 671 pp.

Smith, TB (1981) Some observations of pollutant transport associated with elevated plumes. *Atmospheric Environment*; **15** (10/11); 2197–2203

Smith, TB, Blumenthal, DL, Anderson, JA and Vanderpol, AH (1978) Transport of SO_2 in power plant plumes: day and night. *Atmospheric Environment*; **12** (1–3); 605–611

Soederlund, R (1982) *On the difference in chemical composition of precipitation collected in bulk and wet-only collectors. CM-57*, Stockholm, Sweden, University of Stockholm, Department of Meteorology; International Meteorological Institute in Stockholm, 18 pp.

Soskolne, CL, Zeighami, EA, Hanis, NM, Kupper, LI, Herrmann, N, Amsel, J, Mausner, JS and Stellman, JM (1984) Laryngeal cancer and occupational exposure to sulfuric acid. *American Journal of Epidemiology*; **120** (3); 358–369

Spengler, JD and Thurston, GD (1983) Mass and elemental composition of fine and coarse particles in six US cities. *Journal of the Air Pollution Control Association*; **33** (12); 1162–1171

Spengler, JD, Dockery, DW, Turner, WA, Wolfson, JM and Ferris, BG (1981) Long term measurements of respirable sulfates and particles inside and outside houses. *Atmospheric Environment*; **15** (1); 23–30

Sramek, J and Buzek, F (1983) Sulphur isotope composition within surface crusts on stone monuments. In: *Proceedings of the sixth world conference on air quality, Paris, France, 16–20 May 1983. Vol. 3.* pp. 25–31

Stacy, RW, Seal, Jr, E, House, DE, Green, J, Roger, LJ and Reggio, L (1983) A survey of effects of gaseous and aerosol pollutants on pulmonary function of normal males. *Archives of Environmental Health*; **38** (2); pp. 104–115

Steele, RL, Gertler, AW, Katz, U, Lamb, D and Miller, DF (1981) Cloud chamber studies of dark transformations of sulfur dioxide in cloud droplets. *Atmospheric Environment*; **15** (10/11); 2341–2352

Stensland, GJ (1980) Precipitation chemistry trends in the northeastern United States. In: *Polluted rain: Proceedings of the 12th Rochester international conference on environmental toxicity, Rochester, NY, USA, 21–23 May 1979*. New York, NY, USA, Plenum Press, pp. 87–123

Steudler, PA and Peterson, BJ (1984) Contribution of gaseous sulphur from salt marshes to the global sulphur cycle. *Nature*; **311** (5985); 455–457

Stevens, RK, Dzubay, TG, Shaw, RW, McClenny, WA, Lewis, CW and Wilson, WE (1980) Characterization of the aerosol in the Great Smoky Mountains. *Environmental Science and Technology*; **14** (12); 1491–1498

Stevens, RK, Dzubay, TG, Lewis, CW and Shaw, RW (1984) Source apportionment methods applied to the determination of the origin of ambient aerosols that affect visibility in forested areas. *Atmospheric Environment*; **18** (2); 261–272

Stockwell, WR and Calvert, JG (1983) The mechanism of the HO-SO_2 reaction. *Atmospheric Environment*; **17** (11); 2231–2235

Surprenant, NF, Battye, W, Roeck, D and Sandberg, SM (1981) *Emissions assessment of conventional stationary combustion systems: Volume V Industrial combustion sources. EPA-600/7-81-003c, PB81-225 559*, Redondo Beach, CA, USA, TRW Inc, 203 pp.

Sze, ND and Ko, MKW (1980a) CS_2 and COS in atmospheric sulfur budget. In: *Environmental and climatic impact of coal utilization. Proceedings of the symposium on environmental and climatic impact of coal utilization, Williamsburg, VA, USA, 17–19 Apr 1979.* New York, NY, USA, Academic Press, pp. 309–321

Sze, ND and Ko, MKW (1980b) Photochemistry of COS, CS_2, CH_3SCH_3 and H_2S: implications for the atmospheric sulfur cycle. *Atmospheric Environment*; **14** (11); 1223–1239

Tanaka, S, Darzi, M and Winchester, JW (1980) Sulfur and associated elements and acidity in continental and marine rain from North Florida. *Journal of Geophysical Research*; **85** (C8); 4519–4526

Tang, IN (1980) On the equilibrium partial pressures of nitric acid and ammonia in the atmosphere. *Atmospheric Environment*; **14** (7); 819–828

Tanner, RL and Leaderer, BP (1982) Seasonal variations in the composition of ambient sulfur-containing aerosols in the New York area. *Atmospheric Environment*; **16** (3); 569–580

Tanner, RL, Garber, R, Marlow, W, Leaderer, BP and Leyko, MA (1979) Chemical composition of sulfate as a function of particle size in New York summer aerosol. *Annals of the New York Academy of Sciences*; **322**; 99–113

Tartarelli, R, Davini, P, Morrelli, F and Corsi, P (1978) Interactions between SO_2 and carbonaceous particulates. *Atmospheric Environment*; **12** (1–3); 289–293

Thornton, JD and Eisenreich, SJ (1982) Impact of land-use on the acid and trace element composition of precipitation in the north central US *Atmospheric Environment*; **16** (8); 1945–1955

Thurston, GD and Spengler, JD (1985) A quantitative assessment of source contributions to inhalable particulate matter pollution in metropolitan Boston. *Atmospheric Environment*; **19** (1); 9–25

Tong, EY, Mills, MT, Niemann, BL and Smith, L (1979) Characterization of regional episodes of particulate sulfates and ozone over the Eastern United States and Canada. In: *Papers presented at the WMO symposium on the long-range transport of pollutants and its relation to general circulation including stratospheric/tropospheric exchange processes, Sofia, Bulgaria, 1–5 October 1979. WMO-538,* Geneva, Switzerland, World Meteorological Organization, pp. 85–93

Toon, OB and Pollack, JB (1982) Stratospheric aerosols and climate. In: *The stratospheric aerosol layer.* Berlin, FRG, Springer-Verlag, pp. 121–147

Traegaardh, C (1980) *Air chemistry measurements in the lower atmosphere over Sweden - data evaluation. AC-45*, Stockholm, Sweden, University of Stockholm, Department of Meteorology; International Meteorological Institute in Stockholm, 40 pp.

Trijonis, J (1982) Existing and natural background levels of visibility and fine particles in the rural East. *Atmospheric Environment*; **16** (10); 2431–2445

Tuncel, SG, Olmez, I, Parrington, JR and Gordon, GE (1985) Composition of fine particle region sulfate component in Shenandoah Valley. *Environmental Science and Technology*; **19** (6); 529–537

Turco, RP (1982) Models of stratospheric aerosols and dust. In: *The stratospheric aerosol layer.* Berlin, FRG, Springer-Verlag, pp. 93–119

Turco, RP, Whitten, RC, Toon, OB, Pollack, JB and Hamill, P (1980) Carbonyl sulfide, stratospheric aerosols and terrestrial climate. In: *Environmental and climatic impact of coal utilization. Proceedings of the symposium on environmental and climatic impact of coal utilization, Williamsburg, VA, USA, 17–19 Apr 1979.* New York, NY, USA, Academic Press, pp. 331–352

Tyree, SY (1981) Rainwater acidity measurement problems. *Atmospheric Environment*; **15** (1); 57–60

United Nations Economic Commission for Europe (1985) *Report of the second phase of EMEP* Geneva, Switzerland, UN Economic Commission for Europe, 14 pp.

Urone, P and Schroeder, WH (1978) Atmospheric chemistry of sulfur-containing compounds. In: *Sulfur in the environment. Part I: the atmospheric cycle.* New York, NY, USA, John Wiley, pp. 297–324

Utell, MJ, Morrow, PE and Hyde, RW (1982) Comparison of normal and asthmatic subjects' responses to sulphate pollutant aerosols. *Annals of Occupational Hygiene*; **26** (1–4); 691–697

Utell, MJ, Morrow, PE and Hyde, RW (1984) Airway reactivity to sulfate and sulfuric acid aerosols in normal and asthmatic subjects. *Journal of the Air Pollution Control Association*; **34** (9); 931–935

Uthe, EE and Wilson, WE (1979) Lidar observations of the density and behaviour of the Labadie Power Plant plume. *Atmospheric Environment*; **13** (10); 1395–1412

Vanderpol, AH and Humbert, ME (1981) Coloration of power plant plumes – NO_2 or aerosols? *Atmospheric Environment*; **15** (10/11); 2105–2110

Van Valin, CC and Pueschel, RF (1981) Fine particle formation and transport in the Colstrip, Montana, power plant plume. *Atmospheric Environment*; **15** (2); 177–189

Van Valin, CC, Pueschel, RF and Parungo, FP (1980) Sulfate and nitrate in plume aerosols from a power plant near Colstrip, MT. In: *Environmental and climatic impact of coal utilization. Proceedings of the symposium on environmental and climatic impact of coal utilization, Williamsburg, VA, USA, 17–19 Apr 1979*. New York, NY, USA, Academic Press, pp. 35–45

Varhelyi, G (1985) Continental and global sulfur budgets – I Anthropogenic SO_2 emissions. *Atmospheric Environment*; **19** (7); 1029–1040

Venkatram, A and Pleim, J (1985) Analysis of observations relevant to long-range transport and deposition of pollutants. *Atmospheric Environment*; **19** (4); 659–667

Venkatram, A, Ley, BE and Wong, SY (1982) A statistical model to estimate long-term concentrations of pollutants associated with long-range transport. *Atmospheric Environment*; **16** (2); 249–257

Viggiano, AA and Arnold, F (1981) Extended sulfuric acid vapor concentration measurements in the stratosphere. *Geophysical Research Letters*; **8** (6); 583–586

Voldner, EC, Olson, MP, Oikawa, K and Loiselle, M (1981) Comparison between measured and computed concentrations of sulphur compounds in Eastern North America. *Journal of Geophysical Research*; **86** (C6); 5339–5346

Vukovitch, FM (1979) A note on air quality in high pressure systems. *Atmospheric Environment*; **13** (2); 255–265

Waggoner, AP, Weiss, RE, Ahlquist, NC, Covert, DS, Will, S and Charlson, RJ (1981) Optical characteristics of atmospheric aerosols. *Atmospheric Environment*; **15** (10/11); 1891–1909

Waggoner, AP, Weiss, RE and Ahlquist, NC (1983) The color of Denver haze. *Atmospheric Environment*; **17** (10); 2081–2086

Waldman, JM, Munger, JW, Jacob, DJ, Flagan, RC, Morgan, JJ and Hoffmann, MR (1982) Chemical composition of acid fog. *Science*; **218** (4573); 677–679

Waller, RE (1979) The effect of sulphur dioxide and related urban air pollutants on health. In: *Proceedings of the International symposium on sulphur emissions and the environment, London, UK, 8–10 May 1979* London, UK, Society of the Chemical Industry, pp. 171–177

Waller, RE (1980) The assessment of suspended particulates in relation to health. *Atmospheric Environment*; **14** (9); 1115–1118

Waller, RE (1983) The influence of urban air pollution on the development of chronic respiratory disease. In: *Proceedings of the sventh world congress on air quality. Paris, France, 16-20 May 1983*. Paris, France, SEPIC, pp. 51–57

Waller, RE (1985) Field investigations of air. In: *Oxford textbook of public health*. OXford, UK, Oxford University Press, pp. 300–312

Webber, JS, Dutkiewicz, VA and Husain, L (1985) Identification of submicrometer coal fly ash in a high-sulfate episode at Whiteface Mountain, New York. *Atmospheric Environment*; **19** (2); 285–292

Weiss, RE, Waggoner, AP, Charlson, RJ and Ahlquist, NC (1977) Sulfate aerosol: its geographical extent in the Midwestern and Southern United States. *Science*; **195**; 979–981

Weiss, RE, Larson, TV and Waggoner, AP (1982) *In situ* rapid response measurement of H_2SO_4 $(NH_4)_2SO_4$ aerosols in rural Virginia. *Environmental Science and Technology*; **16** (8); 525–532

Wesely, ML, Cook, DR, Hart, RL and Speer, RE (1985) Measurements and parameterization of particulate sulfur dry deposition over grass. *Journal of Geophysical Research*; **90** (D1); 2131–2143

Whelpdale, DM (1978) Large-scale atmospheric sulfur studies in Canada. *Atmospheric Environment;* **12** (1–3); 661–670

Whelpdale, DM (1985) Downsview, Ontario, Canada, Atmospheric Environment Service, Private communication, (Apr 1985)

Whelpdale, DM and Barrie, LA (1982) Atmospheric monitoring network operations and results in Canada. *Water, Air, and Soil Pollution*; **18** (1/2/3); 7–23

Whitby, KT (1978) The physical characteristics of sulfur aerosols. *Atmospheric Environment*; **12** (1–3); 135–160

Whitby, KT (1980) Aerosol formation in urban plumes. *Annals of the New York Academy of Sciences*; **338**; 258–275

Whitby, KT, Cantrell, BK, Husar, RB, Gallani, NV, Anderson, JA, Blumenthal, DL and Wilson, WE (1976) Aerosol formation in a coal fired power plant plume. *American Chemical Society, Division of Environmental Chemistry, Preprints*; **16** (1); 49–52

Williams, DJ, Carras, JN, Milne, JW and Heggie, AC (1981) The oxidation and long-range transport of sulphur dioxide in a remote region. *Atmospheric Environment*; **15** (10/11); 2255–2262

Wilson, WE and Gillani, NV (1979) Transformation during transport: a state of the art survey of the conversion of SO_2 to sulfate. In: *Papers presented at the WMO symposium on the long-range transport of pollutants and its relation to general circulation including stratospheric-tropospheric exchange processes, Sofia, Bulgaria, 1–5 October 1979. WMO-538*, Geneva, Switzerland, World Meteorological Organization, pp. 157–164

Wilson, JW and Mohnen, VA (1982) An analysis of spatial variability of the dominant ions in precipitation in the eastern United States. *Water, Air, and Soil Pollution*; **18** (1/2/3); 199–213

Wilson, JW, Mohnen, V and Kadlecek, J (1980) *Wet deposition in the northeastern United States, DOE/EV/02986-1, ASRC 796*, Albany, NY, USA, State University of New York, Atmospheric Sciences Research Center, 143 pp.

Wilson, JW, Mohnen, VA and Kadlecek, JA (1982) Wet deposition variability as observed by MAP3S. *Atmospheric Environment*; **16** (7); 1667–1676

Wilson, WE (1981) Sulfate formation in point source plumes: a review of recent field studies. *Atmospheric Environment*; **15** (12); 2573–2581

Wilson, WE and McMurry, PH (1981) Studies of aerosol formation in power plant plumes – II Secondary aerosol formation in the Navajo generating station plume. *Atmospheric Environment*; **15** (10/11); 2329–2339

Winchester, JW (1980a) Sulfate formation in urban plumes. *Annals of the New York Academy of Sciences*; **338**; 297–308

Winchester, JW (1980b) A sulfuric acid formation mechanism. In: *Long range aerosol transport of metals and sulphur. Proceedings of a workshop in Lund, Sweden, 13 June 1980*. Report SNV PM-1337, Solna, Sweden, National Swedish Environment Protection Board, pp. 54–58

Winchester, JW (1983) Sulfur, acidic aerosols, and acid rain in the eastern United States. In: *Trace atmospheric constituents: properties, transformations and fates*. New York, NY, USA, John Wiley, pp. 269–301

Winchester, JW (1984) Effects of acid deposition: is the southeast especially vulnerable? *Proceedings of the conference on acid rain impact on Florida and the southeast, University of Central Florida, Orlando, FL, USA, 27–28 Apr 1984*. 20 pp.

Winchester, JW, Leysieffer, FW and Park, YC (1981) *Geographic distributions of lung cancer mortality and estimated sulfuric acid aerosol concentration in southeastern US Atlantic coastal counties*. Tallahassee, FL, USA, Florida State University, Departments of Oceanography and Statistics, 79 pp.

Winner, WE, Smith, CL, Koch, GW, Mooney, HA, Bewley, JD and Krouse, HR (1981) Rates of emission of H_2S from plants and patterns of stable sulphur isotope fractionation. *Nature*; **289** (5799); 672–673

Wisniewski, J (1982) The potential acidity associated with dews, frosts, and fogs. *Water, Air, and Soil Pollution*; **17** (4); 361–377

Wolff, GT (1980) Mesoscale and synoptic scale transport of aerosols. *Annals of the New York Academy of Sciences*; **338**; 379–388

Wolff, GT, Lioy, PJ, Leaderer, BP, Bernstein, DM and Kleinman, MT (1979a) Characterization of aerosols upwind of New York City: I Transport. *Annals of the New York Academy of Sciences*; **332**; 57–71

Wolff, RK, Silbaugh, SA, Brownstein, DG, Carpenter, RL and Mauderley, JL (1979b) Toxicity of 0.4- and 0.8-μm sulfuric acid aerosols in the guinea pig. *Journal of Toxicology and Environmental Health*; **5** (6); 1037–1047

Wolff, GT, Countess, RJ, Groblicki, PJ, Ferman, MA, Cadle, SH and Muhlbaier, JL (1981) Visibility-reducing species in the Denver 'brown cloud' II. Sources and temporal patterns. *Atmospheric Environment*; **15** (12); 2485–2502

Wolff, GT, Kelly, NA and Ferman, MA (1982a) Source regions of summertime ozone and haze episodes in the eastern United States. *Water, Air, and Soil Pollution*; **18** (1–3); 65–81

Wolff, GT, Ferman, MA, Kelly, NA, Stroup, DP and Ruthkosky, MS (1982b) The relationships between the chemical composition of fine particles and visibility in the Detroit Metropolitan Area. *Journal of the Air Pollution Control Association*; **32** (12); 1216–1220

Young, GS and Winchester, JW (1980) Association of non-marine sulfate aerosol with sea breeze circulation in Tampa Bay. *Journal of Applied Meteorology*; **19** (4); 419–425

Yue, GK and Hamill, P (1980) The formation of sulfate aerosols through heteromolecular nucleation process. In: *Environmental and climatic impact of coal utilization. Proceedings of the symposium on environmental and climatic impact of coal utilization, Williamsburg, VA, USA, 17–19 Apr 1979*. New York, NY, USA, Academic Press, pp. 49–79

Zak, BD (1981) Lagrangian measurements of sulfur dioxide to sulfate conversion rates. *Atmospheric Environment*; **15** (12); 2583–2591

Zak, BD, Homann, PS and Holland, RM (1981) *Project Da Vinci: a study of long-range air pollution using a balloon-borne Lagrangian measurement platform. 2 Volumes. SAND-78-0403, DE81-028186*, Albuquerque, NM, USA, Sandia National Laboratories, 314, 144 pp.

Zehnder, AJB and Zinder, SH (1980) The Sulfur Cycle. In: *Handbook of Environmental Chemistry. Volume 1. Part A The natural environment and the biogeochemical cycles*. Berlin, FRG, Springer-Verlag, pp. 105–145

Zinder, SH and Brock, TD (1978) Microbial transformations of sulfur in the environment. In: *Sulfur in the environment. Part II Ecological impacts*. New York, NY, USA, John Wiley, pp. 445–458

Index

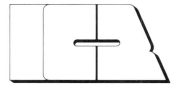